学术引领系列

国家科学思想库

中国学科发展战略

极端地质环境微生物学

国家自然科学基金委员会
中 国 科 学 院

科 学 出 版 社
北 京

内 容 简 介

极端地质环境微生物代表着地球生命的极限和边界，对探索生命起源、促进地球科学与生命科学和行星科学的交叉融合具有重要意义。这些微生物拥有特殊的基因和酶，是重要的种质资源，具有广泛的应用前景，对推动深地、深海、深空等领域关键技术的突破具有重要的战略意义。

本书分别从学科领域的科学意义和战略价值，研究特点、发展规律和发展趋势，关键科学问题、发展思路、发展目标和重要研究方向，国内外发展现状和发展态势，未来发展的有效资助机制及政策建议等方面进行了深入阐述。

本书适合地球科学、生命科学和行星科学领域的高等院校师生、科研机构研究人员，以及高层次战略研究和管理专家参考。

图书在版编目（CIP）数据

极端地质环境微生物学 / 国家自然科学基金委员会，中国科学院编. —北京：科学出版社，2022.3
（中国学科发展战略）
ISBN 978-7-03-071373-5

Ⅰ. ①极… Ⅱ. ①国… ②中… Ⅲ. ①环境地质学—微生物学 Ⅳ. ①Q939.99

中国版本图书馆CIP数据核字（2022）第020002号

丛书策划：侯俊琳 牛 玲
责任编辑：张 莉 姚培培 / 责任校对：韩 杨
责任印制：李 彤 / 封面设计：黄华斌 陈 敬

科学出版社 出版
北京东黄城根北街16号
邮政编码：100717
http://www.sciencep.com

北京虎彩文化传播有限公司 印刷
科学出版社发行 各地新华书店经销
*
2022年3月第 一 版 开本：720×1000 1/16
2023年1月第二次印刷 印张：22 1/4 插页：6
字数：400 000

定价：148.00元

（如有印装质量问题，我社负责调换）

中国学科发展战略
联合领导小组

联合工作组

中国学科发展战略·极端地质环境微生物学
项 目 组

项目专家组

组　长：殷鸿福

成　员（以姓氏拼音为序）：

陈　骏　邓子新　董海良　黄　力　蒋宏忱　焦念志
刘　羽　刘双江　罗根明　潘永信　王成善　王风平
王红梅　向　华　肖　龙　谢树成　朱永官

秘　书：龚剑明　邱　轩

项目工作组

组　长：谢树成

成　员（以姓氏拼音为序）：

董海良　黄　婷　蒋宏忱　李　超　李　平　李江涛
李文均　李一良　林　巍　刘　邓　刘　红　刘常宏
刘翠艳　刘勇勤　陆现彩　罗根明　牟伯中　邵宗泽
佘振兵　束文圣　汤　凯　田　丰　王风平　王红梅
肖　龙　肖　湘　许学伟　张瑞峰

秘　书：邱　轩

总　序

白春礼　杨　卫

17世纪的科学革命使科学从普适的自然哲学走向分科深入，如今已发展成为一幅由众多彼此独立又相互关联的学科汇就的壮丽画卷。在人类不断深化对自然认识的过程中，学科不仅仅是现代社会中科学知识的组成单元，同时也逐渐成为人类认知活动的组织分工，决定了知识生产的社会形态特征，推动和促进了科学技术和各种学术形态的蓬勃发展。从历史上看，学科的发展体现了知识生产及其传播、传承的过程，学科之间的相互交叉、融合与分化成为科学发展的重要特征。只有了解各学科演变的基本规律，完善学科布局，促进学科协调发展，才能推进科学的整体发展，形成促进前沿科学突破的科研布局和创新环境。

我国引入近代科学后几经曲折，及至上世纪初开始逐步同西方科学接轨，建立了以学科教育与学科科研互为支撑的学科体系。新中国成立后，逐步形成完整的学科体系，为国家科学技术进步和经济社会发展提供了大量优秀人才，部分学科已进入世界前列，有的学科取得了令世界瞩目的突出成就。当前，我国正处在从科学大国向科学强国转变的关键时期，经济发展新常态下要求科学技术为国家经济增长提供更强劲的动力，创新成为引领我国经济发展的新引擎。与此同时，改革开放40多年来，特别是21世纪以来，我国迅猛发展的科学事业蓄积了巨大的内能，不仅重大创新成果源源不断产生，而且一些学科正在孕育新的生长点，有可能引领世界学科发展的新方向。因此，开展学科发展战略研究是提高我国自主创新能力、实现我国科学由“跟跑者”向“并行者”和“领跑者”转变的一项基础工程，对于更好把握世界科技创新发展趋势，发挥科技创新在全面创新中的引领作用，具有重要的现实意义。

学科发展战略研究的核心是结合科学技术和经济社会的发展需求，在分析科学前沿发展趋势的基础上，寻找新的学科生长点和方向。在这个过程中，战略科学家的前瞻引领作用十分重要。科学史上这样的例子比比皆是。在 1900 年 8 月巴黎国际数学家代表大会上，德国数学家戴维·希尔伯特发表了题为“数学问题”的著名讲演，他根据过去特别是 19 世纪数学研究的成果和发展趋势，提出了 23 个最重要的数学问题，即“希尔伯特问题”。这些“问题”后来成为许多数学家力图攻克的难关，对现代数学的研究和发展产生了深刻的影响。1959 年 12 月，美国物理学家、诺贝尔奖得主理查德·费曼在加利福尼亚理工学院举行的美国物理学会年会上发表了题为“物质底层大有空间——一张进入物理新领域的请柬”的经典讲话，对后来出现的纳米技术作出了天才的预见。

学科生长点并不完全等同于科学前沿，其产生和形成不仅取决于科学前沿的成果，还决定于社会生产和科学发展的需要。1841 年，佩利戈特用钾还原四氯化铀，成功地获得了金属铀，可在很长一段时间并未能发展成为学科生长点。直到 1939 年，哈恩和斯特拉斯曼发现了铀的核裂变现象后，人们认识到它有可能成为巨大的能源，这才形成了以铀为主要对象的核燃料科学的学科生长点。而基本粒子物理学作为一门理论性很强的学科，它的新生长点之所以能不断形成，不仅在于它有揭示物质的深层结构秘密的作用，而且在于其成果有助于认识宇宙的起源和演化。上述事实说明，科学在从理论到应用又从应用到理论的转化过程中，会有新的学科生长点不断地产生和形成。

不同学科交叉集成，特别是理论研究与实验科学相结合，往往也是新的学科生长点的重要来源。新的实验方法和实验手段的发明，大科学装置的建立，如离子加速器、中子反应堆、核磁共振仪等技术方法，都促进了相对独立的新学科的形成。自 20 世纪 80 年代以来，具有费曼 1959 年所预见的性能、微观表征和操纵技术的仪器——扫描隧道显微镜和原子力显微镜终于相继问世，为纳米结构的测量和操纵提供了“眼睛”和“手指”，使得人类能更进一步认识纳米世界，极大地推动了纳米技术的发展。

作为国家科学思想库，中国科学院（以下简称中科院）学部的基本职责和优势是为国家科学选择和优化布局重大科学技术发展方向提供科学依据、发挥学术引领作用，国家自然科学基金委员会（以下简称基金委）则承担着协调学科发展、夯实学科基础、促进学科交叉、加强学科建设的重大责任。继基金委和中科院于2012年成功地联合发布“未来10年中国学科发展战略研究”报告之后，双方签署了共同开展学科发展战略研究的长期合作协议，通过联合开展学科发展战略研究的长效机制，共建共享国家科学思想库的研究咨询能力，切实担当起服务国家科学领域决策咨询的核心作用。

基金委和中科院共同组织的学科发展战略研究既分析相关学科领域的发展趋势与应用前景，又提出与学科发展相关的人才队伍布局、环境条件建设、资助机制创新等方面的政策建议，还针对某一类学科发展所面临的共性政策问题，开展专题学科战略与政策研究。自2012年开始，平均每年部署10项左右学科发展战略研究项目，其中既有传统学科中的新生长点或交叉学科，如物理学中的软凝聚态物理、化学中的能源化学、生物学中的生命组学等，也有面向具有重大应用背景的新兴战略研究领域，如再生医学，冰冻圈科学，高功率、高光束质量半导体激光发展战略研究等，还有以具体学科为例开展的关于依托重大科学设施与平台发展的学科政策研究。

学科发展战略研究工作沿袭了由中科院院士牵头的方式，并凝聚相关领域专家学者共同开展研究。他们秉承“知行合一”的理念，将深刻的洞察力和严谨的工作作风结合起来，潜心研究，求真唯实，“知之真切笃实处即是行，行之明觉精察处即是知”。他们精益求精，“止于至善”，“皆当至于至善之地而不迁”，力求尽善尽美，以获取最大的集体智慧。他们在中国基础研究从与发达国家“总量并行”到“贡献并行”再到“源头并行”的升级发展过程中，脚踏实地，拾级而上，纵观全局，极目迥望。他们站在巨人肩上，立于科学前沿，为中国乃至世界的学科发展指出可能的生长点和新方向。

各学科发展战略研究组从学科的科学意义与战略价值、发展规律和研究特点、发展现状与发展态势、未来5～10年学科发展的关键科学问题、发展思路、发展目标和重要研究方向、学科发展的有效资助机制与政策建议等方面进行分析阐述。既强调学科生长点的科学意义，也考虑其重要的社会价值；既着眼于学科生长点的前沿性，也兼顾其可能利用的资源和条件；既立足于国内的现状，又注重基础研究的国际化趋势；既肯定已取得的成绩，又不回避发展中面临的困难和问题。主要研究成果以“国家自然科学基金委员会-中国科学院学科发展战略”丛书的形式，纳入“国家科学思想库-学术引领系列”陆续出版。

基金委和中科院在学科发展战略研究方面的合作是一项长期的任务。在报告付梓之际，我们衷心地感谢为学科发展战略研究付出心血的院士、专家，还要感谢在咨询、审读和支撑方面做出贡献的同志，也要感谢科学出版社在编辑出版工作中付出的辛苦劳动，更要感谢基金委和中科院学科发展战略研究联合工作组各位成员的辛勤工作。我们诚挚希望更多的院士、专家能够加入到学科发展战略研究的行列中来，搭建我国科技规划和科技政策咨询平台，为推动促进我国学科均衡、协调、可持续发展发挥更大的积极作用。

前　言

极端地质环境微生物包括现代和地质时期各类极端环境的微生物，涉及深地、深海、深空、深时和当代地表等各类极端地质环境。这些环境中的微生物无论是在微生物本身方面还是在地质环境条件方面，均具有一些共性特征，其研究都需要依托技术方法的突破和创新，也都面临同样的重大基础科学前沿难题。极端地质环境微生物的研究对界定生物圈的边界，理解地球生命的起源，引发重大科学发现，促进地球科学、行星科学与生命科学的交叉融合都具有重要的科学意义。同时，极端地质环境微生物蕴含着极其宝贵的生物种质资源，关系着国家的生物安全、生态健康和资源保障，其研究将推动深地、深海和深空等领域相关技术的突破与创新，因而具有重要的战略价值。

然而，人们对极端地质环境微生物，特别是与深地、深海、深空和深时等极端环境有关的微生物的丰度、组成、多样性、代谢功能、空间分布和时间演化等了解得十分有限，对微生物适应极端地质环境的方式、机制、功能，微生物在这些极端环境中的能量来源和物质循环，以及对环境的影响还知之甚微。为了分析当前极端地质环境微生物学科领域的研究特点、发展特征、发展趋势和发展规律，提炼重大基础科学前沿问题和面向国家重大需求所面临的重大难题，提出符合国际发展趋势和我国发展特点的研究方向、发展思路、发展目标和资助政策建议，提高我国在极端地质环境微生物领域的国际影响力，做好相关战略布局，中国科学院学部和国家自然科学基金委员会于 2018 年 12 月联合批准资助了“极端地质环境微生物学科发展战略研究”项目。殷鸿福院士担任该项目的负责人。

这是继殷鸿福院士领衔组织完成由中国科学院学部资助的“地球生物学”和“深部地下生物圈”的学科发展战略项目之后的第三个学科发展战略研究项目。在前两个战略研究项目积累的经验基础上，为了进一步推进这一战略研究工作的纵深发展，项目专门成立了专家组和工作组，由谢树成担任工作组组长。专家组和工作组成员均来自地球科学和生命科学两大领域。

2018～2020年，项目组在武汉召开了三次专家组会议和工作组会议。2018年11月召开的第一次会议主要讨论了项目的总体框架、极端地质环境的定义以及5个分学科领域（深地、深海、深空、深时和地表极端环境）的初步分工及其相关负责人。2019年9月的第二次会议主要研讨了深地、深海、深空、深时和地表极端环境5个分学科领域的国内外总体发展状况、关键科学难题和战略报告框架。2020年9月的第三次会议主要对已经提交的战略报告初稿进行系统的完善和提升，对一些关键的资助政策和机制进行深入的研讨。殷鸿福、陈旭、金振民、陈骏、王成善、焦念志、邓子新、潘永信、朱永官等院士先后参加了这些讨论会，他们为推进极端地质环境微生物研究，特别是本书涉及的一些关键内容的撰写提出了许多建设性的宝贵意见，在很大程度上提高了本书的前瞻性、战略性和系统性。除这三次全体成员参与的讨论会以外，5个分学科领域也各自组织了多次的分学科领域内部讨论会，在很大程度上促进了各分学科领域的研讨。

在项目执行过程中，在总项目的三次研讨会以及5个分学科领域的讨论会基础上，项目工作组的所有成员都先后参与了本书的撰写。本书由殷鸿福和谢树成统稿，中英文摘要由谢树成撰写。第一章、第五章、第六章由谢树成联合6个版块负责人（肖湘、肖龙、董海良、王红梅、王风平、罗根明、蒋宏忱、李文均）重点撰写，其他成员参与。第二章、第三章、第四章由6个版块负责人组织撰写。具体撰写分工如下：①地表极端环境微生物由蒋宏忱和李文均负责组织，撰写人员包括蒋宏忱和杨渐（盐湖）、李文均和周恩民（热泉）、李文均和董雷（干旱）、刘勇勤和邓永翠（雪冰冻土）、黄立南和陆现彩（酸性矿坑水）。②海洋极端环境微生物由王风平负责组织，撰写人员包括王风平和刘常宏（海洋沉积物）、方家松（深海沟）、邵宗泽（深海热液）、汤凯（低/缺氧区）、许学伟（海山与多金属结合区）、李江涛（洋壳岩石圈）、张瑞峰（寡营养开放大洋）。③深地环境微生物由董海良和王红梅负责组织，撰写人员包括董海良（深地岩石），刘翠艳和张更新（深地钻探），王红梅（洞穴），牟伯中、刘一凡和马挺（油藏），李平（地下水），何环和黄再兴（煤层）。④深时极端地质环境微生物由罗根明负责组织，撰写人员包括罗根明、李超、刘邓、佘振兵。⑤深空环境的生命由肖龙负责组织，撰写人员包括肖龙（行星类比）、李一良（行星保护）、田丰（系外行星宜居性）、刘红（空间舱）、佘振兵（深空环境的生命与生命边界）、林巍（临近空间）、黄婷（火星宜居性）。⑥高压环境微生物由肖湘负责撰写。这三章的总体概述部分由谢树成撰写。参与资料收集、编辑和绘图的其他人员（排名不分先后）还有黄力、向华、束文圣、李丽、侯卫国、吴耿、房保柱、杨欢、黄柳琴、李高远、刘晓磊、曾强、花正双、焦建宇、鲜文东、王北辰、韩明贤、李帅、连文慧、沈亮、刘付业、董良、张晓雯、侯佳林、牛明杨、章陶亮、陈云如、衡曦彤、陶秀祥、程晓钰。

“极端地质环境微生物学科发展战略研究”项目的立项和各项活动都得到了中国科学院学部常务委员会和学部工作局、国家自然科学基金委员会、战略研究专家组的支持和指导。项目执行过程中得到许多院士、专家和同行的大力支持和帮助。从2018年底立项开始，项目组成员兢兢业业，排除万难，尤其是克服新冠肺炎疫情的影响，克服时间紧、任务重等多种困难，有声有色地组织了多项学科战略活动，按时完成了该项目。在此谨向他们的大力支持、指导和辛勤劳动致以诚挚的谢意。极端地质环境微生物是一个多学科交叉、科学与技术融合、科学前沿与国家重大需求结合的学科领域，希望本书对推动我国相关学科的发展、相关核心技术的突破发挥重要作用。本书涉及一个全新领域，难免挂一漏万，不当之处，敬请批评指正。

殷鸿福

2020年11月

摘 要

广义的微生物是指借助显微镜才能看得到或看得清的微小生物的总称，它并不是一个分类学上的名字，而是病毒、细菌、古菌和大部分单细胞真核生物的总称。极端地质环境微生物是指那些能适应并栖息在极端地质环境的微生物，包括栖息在当代地表极端环境、陆地深部地下环境、海洋极端环境、深时极端地质环境和深空环境的微生物。

一、科学意义与战略价值

（一）科学意义

极端地质环境微生物的研究对理解地球生命起源、界定生物圈的边界条件、引发生命科学的重大科学发现与突破、促进地球科学与生命科学以及行星科学之间的交叉融合都做出了不可替代的贡献，因而具有重要的科学意义。

生命起源是全球最具挑战性的科学难题之一，极端地质环境微生物为解决这一难题提供了关键抓手。由于极端地质环境具备与生命起源时相似的环境特征，如高温、高压、黑暗、缺氧、含硫等，其微生物蕴含着生命起源和进化历程的丰富信息，是研究地球生命起源的关键所在。位于生命系统发育树根部的生物大都具有嗜热的生长特性，且分支较短。研究高温等极端环境的微生物有利于人们理解地球早期生命起源与演化以及探讨地外生命等问题。极端地质环境条件还代表着生命对环境的极限适应能力，界定了生物圈的边界，引发了一系列重大科学发现。黑暗生态系统、深部生物圈、嗜极微生物的发现都来自对极端环境微生物的研究。

极端地质环境微生物的研究极大地促进了地球科学、生命科学和行星科学的交叉融合，推动了地球生物学和天体生物学等新型交叉学科体系的形成与发展，在现代科学体系中占据着举足轻重的地位。地球科学与生命科学的交叉融合长期以来一直是自然科学领域的巨大挑战，极端地质环境微生物将微生物的地质过程与现代过程整合到同一个平台上进行研究，促进了这两大基础学科的交叉融合，形成了地球生物学的重要内容。极端地质环境微生物之于地球系统，就像微生物之于人体系统一样关键。极端地质环境微生物的研究还促进了生命科学、地球科学与行星科学的交叉，推动了天体生物学的发展，为寻找地外生命和宜居环境提供了关键的学科支撑，推动了工程技术的进步和探测任务的实施。

（二）战略价值

极端地质环境微生物蕴含着极其宝贵的种质资源，推动了深地、深海和深空等重大工程计划关键技术的突破与创新，极大地影响着生物安全、生态安全和资源能源安全，由此决定了它对实施《国家中长期科学和技术发展规划纲要（2006—2020 年）》以及国家其他科技政策目标具有强大的支撑作用，因而具有重要的战略价值。

极端地质环境微生物是一类极其宝贵的种质资源，有许多非常重要的特殊基因和独特的生物酶，可以很好地应用到环保、冶金、能源、电子、材料、医药、食品等各类工业的生产上。各类嗜盐微生物、耐高温微生物、耐低温微生物、嗜酸微生物等都是工农业生产上广泛应用的关键种质资源。特别是，包括物种资源、基因资源和代谢产物资源等在内的深海微生物资源已成为国家重要的战略资源储备，也是各国海洋战略的核心关注点。

极端地质环境微生物关系到国家的生物安全、生态安全与资源能源安全。洞穴等极端地质环境存在许多种类丰富、功能各异的细菌、古菌和病毒，这些微生物可以通过中间宿主传给人，影响人类社会安全。深地微生物会对地下核废料和二氧化碳（CO_2）的储存等产生影响，还能够调节和影响气候环境演变，影响生态安全。深地、深海微生物还能参与地下资源能源的形成、转化、运移和保存，影响资源能源安全。

极端地质环境微生物又能与深地、深海和深空计划等国家重点工程计划相衔接，推动相关技术的突破。当前，我国特别重视将科学前沿研究与国家目标结合，特别重视“卡脖子”技术的突破以及“从0到1”的科学突破。极端地质环境微生物研究与这些工程计划结合，可以推动我国深部钻探技术、深地原位保真取样技术、深地原位观测技术、极端环境条件下微生物培养与分析技术等关键技术的大发展，从而促进科学与工程的结合，提升学科服务社会的整体能力。

二、研究特点、发展规律和发展趋势

极端地质环境微生物的研究包括微生物的一些基本特征和属性，微生物与环境的相互作用与协同演化，微生物适应机理的生产应用与各类生物安全、生态安全这三个层次的内涵。不同极端地质环境微生物的研究尽管水平很不平衡，却具有一些共性的发展规律，都受到工程技术创新、工农业应用前景、国际重大科学计划以及多学科交叉和共性科学难题的多重推动。

（一）受工程技术创新的推动，体现了前沿科学与工程计划的结合

极端地质环境微生物的科学突破离不开技术的支撑，这充分体现了科学与技术的结合。实际上，人们关注极端地质环境微生物也是从特殊技术的突破开始的，包括保真取样、原位培养、微区分析、大数据构建等核心技术。如果说一系列的分子和同位素技术、微区分析和示踪技术是地球科学的常规技术，那么极端环境微生物的保真取样和原位分析等则需要特色的技术支撑。深部生物圈的提出就得益于大陆超深钻技术以及深海海底的原位观测技术的突破。

（二）受工农业应用前景的推动，体现了基础研究与应用服务的结合

极端地质环境微生物因要适应特殊环境而形成了很多特殊的功能基因、具有重要功能的酶和特殊的蛋白质。微生物的这些特性决定了从这些环境分离出来的微生物是一类极其重要的种质资源，具有很大的工农业应用前景。实际上，从极端地质环境分离的微生物已经在环保、冶金、能源、电子、材料、医药、食品等各类工业上

得到了很广泛的应用。这些微生物的广泛工业应用，促进了从自然界中分离出越来越多具有不同功能的微生物。

（三）受国际重大科学计划的推动，体现了学科领域的国际前沿性

极端地质环境微生物同时受到地球科学和生命科学两大学科领域重大国际科学计划，特别是国际微生物组计划和国际大洋发现计划（International Ocean Discovery Program，IODP）的推动。美国于2007年率先启动了一项微生物组地球计划，2010年启动了地球微生物组计划，2016年又宣布启动国家微生物组计划。当前，受微生物组学技术发展的推动，极端地质环境微生物迎来了新的机遇。与此同时，国际大洋发现计划推动了对海洋深部生物圈的探索和研究。科学家依托国际大洋发现计划（2013～2023年）及其前身大洋钻探计划（Ocean Drilling Program，ODP，1985～2003年）和综合大洋钻探计划（Integrated Ocean Drilling Program，IODP，2003～2013年）组织的科学考察航次获取了大洋沉积物样品，促进了深部生物圈的系统性研究。

（四）具有多学科交叉和共性科学难题等特点而相互促进地发展

极端地质环境微生物的研究明显具有多学科交叉的特点，涉及地球科学、生命科学和化学，以及相关技术科学等多学科的交叉融合。这就要求培养大量具有学科交叉知识背景的学者。一方面，需要建立多学科交叉的科学共同体和交流平台，促进学科之间的交叉和碰撞；另一方面，需要资助机构改变管理模式，以便及时资助那些具有学科交叉特质但又没有很好共识的前瞻性研究。

地表极端环境、大陆深部地下环境、海洋极端环境、深空环境和深时极端地质环境等极端环境的微生物存在密切联系。它们具有一些共性特征，包括生物的生活环境条件比较恶劣或极端、生物代谢缓慢乃至处于休眠状态、代表了比较原始的生物类型等。极端地质环境微生物因而具有类似的重大科学难题，包括微生物在极低能量、极低营养等极端环境中是如何获得足够的能量以维持生命及演化的，古菌、细菌和病毒等生命系统是如何相互关联的，又是如何

改善这些恶劣环境条件的。

三、关键科学问题、发展思路、发展目标和重要研究方向

（一）关键科学问题

极端地质环境微生物的科学问题主要涉及科学前沿问题以及围绕重大需求产生的重大科学难题这两大类。

1. 科学前沿问题

首先，微生物的时空分布、丰度和多样性是所有环境都涉及的有关微生物的最基本问题。有关极端地质环境微生物的丰度和多样性的认识存在很大的争议，主要是由它们存在巨大的空间差异以及研究程度低造成的。在海洋极端地质环境，需要加强海底沉积物及其不同深度的微生物研究。在大陆深部地下环境，需要加强超深钻探微生物的原位监测与研究。对于深时极端地质环境，需要加强各地质时代，特别是前寒武纪微生物的空间分布、丰度和多样性研究，并特别需要相关新技术的突破。

其次，微生物的生态学和生物地球化学功能及其对极端地质环境的改造作用也是一个核心问题。生物与环境之间是存在相互作用的。需要强调的是，微生物不是在被动地适应环境，而是具有主动改造环境的功能。地下油气藏的异常环境和地表的酸性矿坑水环境都是不同微生物作用于极端地质环境而形成的特殊环境。在不同极端地质环境条件下，微生物与微生物之间、微生物与地质环境之间的物质和能量是如何传递与交换的？调控机制又是什么？有哪些生态效应？在全球背景下，极端地质环境微生物如何反作用于气候环境变化？这些都是核心的前沿科学难题。

最后，微生物与地质环境的协同演化是另一个关键科学问题。地球的宜居性与微生物的正面或负面作用关系密切，氧气的产生和含量就是一例。尽管“生命之树”根部的微生物大多起源于极端地质环境，但人们对具体的起源时间和各类微生物功能群不同属种的演化等关键问题都不清楚，这直接影响了某类特定微生物功能群演化树的构建。可以说，目前还没有构建出任何一类微生物功能群可

靠的演化树，因此对这些微生物起源和演化的环境背景以及它们之间的协同演化也很不清楚。

2. 围绕重大需求产生的重大科学难题

极端地质环境的微生物种质资源库有许多特殊基因和独特的生物酶，可以应用到各类工业生产上。它们有待被研究，包括模式生物、各种基因和酶的类型、物理化学特性与功能、提取方法、工农业生产应用中的驯化等科技难题。

微生物对地下油气藏、天然气水合物、页岩气等能源以及金属矿产资源的形成、勘探、开发利用过程中的作用机制还有待深入解剖，包括这些微生物的作用机制、驯化条件、作用效率等问题，特别是在微生物浸矿方面。

微生物作用具有正负两面性，地下 CO_2 封存会受到地下深部微生物的影响，需要研究微生物与微生物之间、微生物与地下岩石矿物之间、微生物与地下流体之间的相互作用及其对地下 CO_2 封存的影响。地质储存也是目前公认的、唯一可行的长期安全处置长半衰期、高放射性核废料的方法，但深地微生物同样会影响核废料储存库的可靠性和安全性，需要详细了解可能发生的微生物过程。

生命是否可以在其他天体上起源、演化和生存？哪些天体适合生命的存在？宇宙中的生命有没有可能突破地球生命的边界？宇宙中是否存在碳基生命之外的其他生命？太空环境下微生物群落物种多样性、适应机理、动力学特征以及维持机制如何？太空环境下有益的植物-微生物如何相互作用，如何建立抑制植物致病菌的生物方法？这些都是在深空探测中面临的重大科学难题。

（二）发展思路和目标

针对我国的实际情况，按照“有限目标、突出重点”的总体要求，未来5～10年需要加强以下三方面的工作。

1. 与深地、深海和深空等重大科学工程计划相结合，突破“卡脖子”技术

我国的深地计划、深海计划和深空计划正在大力推进，这为开

展极端地质环境微生物的研究创造了极其难得的机遇，需要我国科学工作者积极参与国家大科学工程的前期论证、设计和后续的研究。极端地质环境微生物的突破还需要技术上的革新。在钻探和保真取样技术、野外原位观测和模拟技术方面，我国还缺乏自主知识产权的技术体系。

2. 加强不同学科之间的大交叉与大联合，打通学科壁垒并克服零散的研究局面

我国在极端地质环境微生物方面的一个现状是研究力量分散，学术共同体没有很好形成，这需要通过不同学科之间的交叉联合才能实现。其中既需要加强地球科学与生命科学的交叉联合，还需要加强科学研究与工程技术的交叉联合。

3. 实现不同极端地质环境的对比和联合研究，以形成带动整体突破的一体化工作

在我国，各类极端地质环境都开展了点上的微生物工作，但没有形成一体化的系统性研究，导致各类环境的研究难以衔接，没有达到相互借鉴、形成合力的目的。

4. 通过三方面的发展，实现两大目标

在学科体系和人才队伍层面，力争建立和完善地球生物学和天体生物学等新型交叉学科体系和人才培养体系，建成成熟的教学体系、人才培养模式和科学研究平台，培养地球生物学和天体生物学国际领军人才，发展地球生物学与天体生物学新理论。

在科学和技术层面，力争在极端地质环境微生物方面形成国际领先的一体化成果，突破极端地质环境微生物的保真取样技术、野外原位观测和模拟技术等“卡脖子”技术，努力提升我国深地、深海和深空等国家重大工程计划的科学研究效应。

（三）重要研究方向

根据目前我国的发展状况以及未来发展的需要，结合国际发展趋势，未来5～10年着重开展六个方向的研究工作。前三个方向分别与深海计划、深地计划和深空计划相关，服务国家重大工程计划；第四个和第五个方向分别与微生物种质资源、生物安全和生态安全有关，服务国家重大需求；最后一个方向则服务于地球系统科学的发展。这六个方向既符合国家自然科学基金委员会的“三深一系统”的长远规划，又体现了基础研究与国家重大需求结合的国家发展战略。

1. 深海计划的微生物及其资源环境效应

需要继续开展海底沉积物、热液、冷泉、深渊、俯冲带、大洋中脊、缺氧带等核心地区的微生物研究。深渊是连接海洋表层和深部的关键纽带，俯冲带和大洋中脊是连接地球表层与地球深部的重要抓手，这三个地区的研究对地球系统科学尤其重要，但目前有关地质微生物的研究基本处于空白状态。

2. 深地计划的微生物及其资源环境效应

陆地深部地下生物圈除了黑暗，还具有一些特殊的环境特征。需要开展典型油气藏、地下煤层、洞穴、深湖底部及其沉积物、泥炭沼泽深部、地下水、超深钻、深大断裂带、地下 CO_2 储层、地下核废料储层等关键地带的微生物研究。

3. 深空计划的天体生物学

天体生物学是典型的交叉学科，需要多个学科团队、多个实验室的通力合作，多角度共同探索地外宜居性环境，为地外生命探测提出可靠建议。通过国际合作并建立合作共享数据库，加快研究进程，并将其运用于本国的空间探测和地外生命探测任务。

4. 极端地质环境微生物与种质资源

陆地热泉、雪冰和盐碱湖的微生物主要与种质资源及其工业利用有关，可以聚焦关键功能基因的研究。酸性矿坑水、重金属污染区、农药及抗菌药污染区等可以聚焦微生物对各类污染物的降解或分解作用，以便为地质环境的保护和治理服务。

5. 极端地质环境微生物与生物安全

目前人们对深地病毒的认识还十分薄弱，但病毒致病性及对人类健康的危害有目共睹。开展深地病毒的多样性及其在生态系统中的作用研究，进而评估它们潜在的致病性等将极大地提高人们对开发地下空间的生物安全意识，合理科学地规避和控制风险。地下空间环境，尤其是深地环境对人类生理、心理以及病理的影响规律及其机制研究尚属空白，深地医学需要尽快发展。

6. 深时极端地质环境微生物与地球系统科学

深时极端地质环境的微生物很好地记录了地球系统各圈层之间的相互作用，是发展地球系统科学的关键抓手。须开展前寒武纪两次大氧化时期、“雪球地球”时期、显生宙生物大灭绝时期、极热时期等关键地质突变期的地质微生物及其地质演化研究，对比研究揭示地质微生物与极端气候环境之间的相互作用，形成系统理论。

四、发展现状与发展态势

（一）不同极端地质环境微生物存在共性规律

不同极端地质环境微生物的研究都经历了从初期的纯培养模式到随后的基于16S核糖体RNA（16S rRNA）基因克隆文库方法，以及近年来的全基因组测序、宏基因组测序、新一代高通量测序和单细胞测序等方法的不断更新。极端地质环境微生物具有一些共性特征，包括代谢缓慢、多样性低、生物地球化学过程耦合、微生物种间互养共生、微生物的空间差异巨大。

（二）微生物与极端地质环境的相互作用亟待全方位深入研究

极端地质环境微生物对大气圈、水圈、生物圈和岩石圈均具有多方面的重大地质作用，微生物与微生物之间也存在密切的相互作用，但这些地质作用的研究都有待加强。微生物的地质演化研究也是一项长期、艰巨的任务。

（三）微生物资源与生物安全

微生物能以多种方式适应极端地质环境而在工业上有广泛的应用前景，环保、冶金、能源、电子、材料、医药、食品等各类工业生产上都有微生物的应用实例。病毒在深海、盐湖、热泉、极地等极端地质环境都有发现，是自然界微生物的“巨人杀手”，对人类社会也产生了重要影响。

（四）分学科之间的对比与国际平台

各分学科领域的发展很不均衡，研究程度从深到浅依次为地表极端环境、深海极端环境、陆地深部环境、深时极端地质环境、深空环境。国际上对极端地质环境微生物做出了最重要贡献的研究平台是国际大洋发现计划。美国启动的“地下科学研究计划”、“极端环境微生物”（Life in Extreme Environments）计划、深碳观测（Deep Carbon Observatory，DCO）计划、暗能量生物圈调查中心也在很大程度上促进了相关研究。国际重要学术组织包括国际地下微生物学会、国际生命起源研究学会、地球生物学会等。与极端地质环境微生物最密切的国际学术刊物有《极端微生物》（*Extremophiles*）、《地质微生物学期刊》（*Geomicrobiology Journal*）、《地球生物学》（*Geobiology*）、《微生物前沿》（*Frontiers in Microbiology*）等。

（五）国内发展现状与发展态势

我国科学家对海洋极端环境、陆地深部地下环境、当代地表极端环境、深时极端地质环境、深空环境等各类极端地质环境的微生物都开展了研究。从这个层面上来说，我国在国际上处于重要的不可替代地位。在我国，相对有微生物研究优势的极端地质环境包括海洋极端环境中的热泉、深渊和洋底沉积物，陆地深部地下环境的油气藏和洞穴，地表极端环境的酸性矿坑水和冰川，以及若干关键地质时期的极端地质环境等，但海底沉积物微生物、包括超深钻在内的深地微生物、深空环境微生物、各类环境的病毒和微生物的地质演化等方面的研究有待加强。

极端地质环境微生物的突破需要技术革新。我国需要在野外钻探与保真取样技术、野外原位观测与模拟实验技术等“卡脖子”技术上进行自主创新。目前我国还没有专门针对极端地质环境微生物的国家重点实验室等国家级平台，没有专门针对极端地质环境微生物的学术组织，没有形成极端地质环境微生物的学术共同体，没有专门针对极端地质环境微生物的单独学科体系或者资助计划。这些不足导致我国相关研究成果分散，没有形成合力。

五、未来发展的有效资助机制及政策建议

极端地质环境微生物的研究亟须教育部、科技部、生态环境部、中国科学院等多部门从战略高度进行前瞻性布局，实施系统工程，在多学科交叉与跨学科人才培养、重大科学研究计划实施、平台条件建设、科学普及四个关键环节上给予全方位的前瞻性投入，以提升我国在这一关键领域的国际地位。

（一）不遗余力地推进学科交叉和跨学科人才培养

极端地质环境微生物的发展，首先必须建立不同于传统学科的交叉学科体系，这需要从高等院校的课程设置与人才培养入手。地球生物学的课程体系不仅要涉及生命科学和地球科学领域的核心课程，更重要的是还要开设那些真正体现这两大学科交叉的课程体系。后者是决定学科交叉成败的关键所在。天体生物学则需要建立涉及地球科学、行星科学和生命科学真正交叉融合的课程体系，而不是这三大学科核心课程的拼盘。

目前，真正具有跨学科知识的人才极少，许多涉及跨学科研究的项目主要是通过不同学科领域的专家学者聚焦在一起而实现联合研究的，不是真正意义上的交叉研究。因此，当前的学科交叉研究还处于第一层次。要把这类学科交叉研究尽快提升到第二层次，就需要加大人才培养力度。这需要教育部门和各高等院校的重视。

（二）加大基础科学研究的资助力度

极端地质环境微生物能够引发一些重大科学发现，科技部等相关政府部门需要在重大科学研究计划等方面加大经费支持力度。同时，在深地、深海、深空、全球变化等领域的国家重大研究计划中，加大对极端地质环境微生物的研究，以推动相关核心技术的突破，并带动科学前沿的研究。

（三）大力推进条件平台建设

室内实验平台需要建立能开展生命科学与地球科学领域交叉研究的仪器设备，实现分子水平代谢过程与长时间尺度的地质过程的全方位衔接。当前，能够真正实现从微生物代谢过程到地质过程衔接的实验平台还极度缺乏。

大力推进野外研究基地的建设。许多超深钻孔、深海计划长期的现场观测研究点、野外地质剖面提供了相当珍贵的野外研究基地。目前，所有这些研究基地都没有单独纳入对微生物的原位观测系统，需要前瞻布局这些野外研究基地，以实现系统化。

（四）加强微生物的科学普及工作

科学普及工作需要相关企事业单位、各类学会以及科学家群体与个体的联动。2019 年，全球 32 位微生物学家联合撰写论文，提出了“提高社会的微生物素养势在必行”的倡议，认为微生物远比互联网对社会的作用大：微生物是人类的“第二皮肤”“第二分泌系统”，总体上是“人类的朋友而非敌人”，是“生物圈运转的齿轮”，是“生物经济的核心”，是“地球上最大的废物回收者和再生者”，是“这个星球上人类唯一可以依赖的能解决污染的朋友”，是“地球上第一个也是最后一个生物”。上至政府官员和决策者、下至普通百姓都应该了解微生物学知识，人们需要在社会决策和个人行动中权衡微生物的利害关系，需要在幼儿园、小学、中学和大学都开设微生物学教育。同时，需要将科学普及工作贯穿到重大科学研究计划、野外研究基地、学科建设和人才培养等一系列工程建设中，使微生物真正成为人类的朋友，实现人类与微生物的和平共处。

Abstract

Microorganisms include viruses, bacteria, archaea and many types of unicellular eukaryotes that can be seen only with the aid of a microscope. They can survive and thrive for long periods in the harsh physiochemical conditions of extreme geo-environments, including on modern land surfaces (i.e., hot springs, glaciers, desert, acid mine drainage, saline lakes, etc.), in the oceans (i.e., subseafloor sediments, hydrothermal vents, cold seeps, abyssal plains, subduction zones, mid-ocean ridges and anoxic zones, etc.), in the continental subsurface, in deep time, and in deep space. This strategic report, which focuses on six aspects of this rapidly growing field, was written by well-known Chinese scientists led by Professor Hongfu Yin through a program jointly funded by Chinese Academy of Sciences and the National Natural Science Foundation of China.

1. Scientific significance and strategic value

(1) Scientific significance

The study of microorganisms in extreme geo-environments is of great scientific significance; it helps to understand the origin of life on Earth, to define the physiochemical conditions of biosphere development, to catalyze medical discoveries, and to generally inspire cutting-edge breakthroughs in the geosciences, life sciences, and planetary science.

The origin of life on Earth is one of the most challenging and fascinating scientific issues, and microorganisms living in extreme geo-environments may provide key clues to this topic. Modern extreme environments may serve as analogs to the physiochemical conditions that existed on the early Earth, including high temperature, high pressure, weak sunlight, anoxia, high sulfur loading, etc. Microorganisms that are isolated from such extreme environments may preserve genetic and biochemical fingerprints related to the origin and evolution of life, because most species at the root of the tree of life are thermophiles with short phylogenetic branches. The study of microorganisms residing in extreme geo-environments is thus helpful for revealing the origin and evolution of life on the early Earth as well as potentially guiding the study of life in space, if it is found to exist. In addition, the physiochemical conditions present in these extreme geo-environments approach the limits of survival and adaptation for life on Earth, defining the physiochemical boundaries of the biosphere. This has led to a series of recent significant discoveries regarding dark ecosystems, deep biospheres, and extremophiles.

Interdisciplinary research on microorganisms in extreme geo-environments concerns the geosciences, life sciences, and planetary science, and it has been a major contributor to the establishment and development of new interdisciplinary branches including, but not limited to, geobiology and astrobiology. These frontier disciplines represent important interdisciplinary nodes in the modern science network. Effective integration of the geosciences with life sciences has long been a challenge. Microorganisms in extreme geo-environments provide a great opportunity for study of both modern and deep-time systems on the same research platform that led to the birth of the field of geobiology. Microorganisms in extreme geo-environments play critical roles in the Earth system, similar to that of viruses in the human body. Interdisciplinary research on microorganisms in extreme geo-environments also facilitated the integration of life sciences, geosciences, and planetary science that led to the birth and development of the field of astrobiology. This synthesis has provided a key scientific basis for the exploration of extraterrestrial life and inhabitable environments in space, and it has greatly promoted the progress of engineering and technologies related to space exploration.

(2) Strategic value

Microorganisms in extreme geo-environments are a collection of invaluable germplasm resources. Study of these microorganisms can facilitate breakthroughs and innovations in key technologies in the fields of deep Earth, deep sea and deep space, and greatly affect issues of national and international security related to organisms, ecosystems, resources and energy. These features endow extreme geo-environmental microorganisms with strategic value in the context of the National Programs for Long- and Medium-term Science and Technology Development, and to achieve the goals proposed in other national science and technology programs.

Microorganisms from extreme geo-environments are a kind of rare germplasm resource because they contain numerous special genes and unique enzymes that can be applied to industries related to environmental protection, metallurgy, energy, electronics, materials, medical care, food, etc. For example, types of halophiles, thermophiles, psychrophiles and acidophiles are all key germplasm resources widely used in industrial and agricultural production. In particular, deep-sea microbial resources, including species, genes and metabolites, constitute national resource reserves of strategic value, and are also the core concern of the marine development strategies of many nations.

Microorganisms in extreme geo-environments show a close connection with national security issues related to organisms, ecosystems, resources and energy. Bacteria, archaea and viruses with a large biodiversity in species and functions inhabit extreme geo-environments such as caves, and they may be transmitted to humans through intermediate hosts and thus become a human health risk. Moreover, microorganisms in the deep subsurface may represent an ecological threat to our society as their biogeochemical processes can affect underground storage of nuclear waste and CO_2. Microorganisms in the deep subsurface also affect resource and energy security due to their roles in formation, migration and preservation of underground petroleum and natural gas.

The study and exploitation of microorganisms in extreme geo-environments can lead to linkages among national key programs in deep-Earth, deep-sea and deep-space research, and also inspire innovations in related technologies. In China, special emphasis is laid on the social outreach of scientific frontier researches and introduction of advanced techniques (known as 'knocking techniques' in China). Linking investigations of extreme geo-environments with engineering projects can also promote the development of advanced technologies in deep drilling, *in-situ* fidelity sampling (i.e., retained physiochemical conditions), on-site observation and monitoring, on-site microbial cultivation, and analysis under extreme physiochemical conditions. It can also aid in integration of science with technology, thereby enhancing the ability of these disciplines to serve society.

2. Development of scientific study

The study of microorganisms in extreme geo-environments concerns three themes, namely, the basic characteristics and attributes of microorganisms, interactions and co-evolution between microorganisms and environments, and practical applications of microorganisms with regard to biosecurity. Although the extant research on microorganisms in various extreme geo-environments has been uncoordinated, the development of the scientific study of such microorganisms can be described within the general framework below.

(1) Research driven by technological innovation

Breakthroughs in the study of microorganisms in extreme geo-environments show a close connection with technological innovations, vividly demonstrating the essential role of integrating science with technology. In fact, breakthroughs in key technologies, including *in-situ* fidelity sampling and cultivation, microanalysis and database construction, have played a pivotal role in research into microorganisms in extreme geo-environments. Although convention techniques of molecular and isotopic analysis are essential to study of microorganisms in extreme geo-environments, innovations in technology related to *in-situ* fidelity sampling, on-site analysis of extreme physiochemical parameters, and molecular and isotopic measurements under extreme physiochemical conditions have been essential in driving this research forward. As an example, investigations of the deep biosphere have benefitted greatly from breakthroughs in continental ultra-deep drilling technology and on-site observation of the deep seabed.

(2) Research driven by industrial applications

Microbes in extreme geo-environments contain abundant genes, enzymes and proteins with unique functions owing to their adaptation to extreme physiochemical conditions, thus making them an extremely important germplasm resource that holds great potential for industrial and agricultural applications. In fact, microbial resources have been extensively applied in various industries, including environmental protection, metallurgy, energy, electronics, materials, medical care, and food. Industrial applications have been a leading driver of research into, and extractive use of, microorganisms with diverse biochemical functions.

(3) Research promoted by major international programs targeting scientific frontiers

The study of microorganisms in extreme geo-environments is promoted by major international programs in both the geosciences and life sciences, including the International Microbiome Program and the International Ocean Drilling Program (IODP). The United States was the first to launch a Microbiome Earth Project in 2007 and the Earth Microbiome Project in 2010. At the end of 2015, a group of scientists jointly published articles in the journals of *Nature* and *Science* to appeal for the launch of the Joint Microbiome Project, after which the United States announced the start of the National Microbiome Program in 2016.

Currently, driven by developments in microbiome technology, the exploration of microorganisms in extreme geo-environments has ushered in new opportunities. Simultaneously, the IODP has promoted exploration and research of the ocean's deep biosphere. Relying on the International Ocean Discovery Program (2013–2023) and its predecessors, the Ocean Drilling Program (1985–2003) and Comprehensive Ocean Drilling Program (2003–2013), scientists organized scientific expeditions to obtain samples of ocean sediments, which has greatly inspired systematic investigation of the deep biosphere.

(4) Research linked to multidisciplinary goals and outcomes

The study of microorganisms in extreme geo-environments has a distinctive multidisciplinary feature; it involves not only the integration of the geosciences and life sciences but also the combination of chemistry with other disciplines. Because of this multidisciplinarity, the research community studying microorganisms in extreme geo-environments needs to train scholars with interdisciplinary knowledge. To achieve this goal, multidisciplinary scientific communities must be established to promote interdisciplinary cooperation and collaboration, and funding systems with an interdisciplinary focus must be established.

Microorganisms may exhibit some common features even though they live in different types of extreme geo-environments, such as those in the surface and deep subsurface, in oceans, and in deep space. Microorganisms in these varying environments tend to develop a slow or dormant biological metabolism based on relatively primitive biochemical mechanisms. These features produce similar scientific issues, including how such microorganisms obtain sufficient energy to sustain life and to reproduce in such extreme (i.e., low energy and low nutrition) environmental conditions, how different types of microorganisms such as archaea, bacteria and viruses are interrelated, and how they change and adapt in the longer term to extreme physiochemical conditions.

3. Key scientific issues, developmental strategies and goals, and important research fields

(1) Key scientific issues

There are a diversity of scientific issues concerning the study of microorganisms in extreme geo-environments that can be grouped into two main categories, i.e., those related to scientific frontiers, and those related to industrial applications.

1) Scientific frontier issues

First, the spatiotemporal distribution, abundance and diversity of microorganisms in extreme geo-environments are basic questions. The abundance and diversity of microorganisms in extreme geo-environments remain controversial, mainly because of huge spatial variation in microorganism types and abundances but also to the lack of intensive investigations. In the extreme geo-environments of the oceans, for instance, it is necessary to strengthen the study of microbes in sediments of different depths below the seabed. In the deep subsurface of the continents, it is especially necessary to strengthen *in-situ* monitoring and research of microorganisms recovered from ultra-deep drilling cores. For extreme geo-environments in deep time, it is particularly difficult to study the geographical distribution, abundance and diversity of microorganisms due to the lack of techniques and methodologies, which awaits long-term developments.

Second, the ecological and biogeochemical functions of these microorganisms and their impact on changing extreme geo-environments are also core issues. Microorganisms interact with their life environments, actively altering the physiochemical conditions in which they live, although they can also show passive adaptation to their environments. The physiochemical conditions of underground oil and gas reservoirs and surficial acid mine drainage systems offer examples of microbial influence on extreme geo-environments. A series of scientific issues remains to be investigated, including how material cycles and energy flows among microorganisms, or between microorganisms and geo-environments under different extreme physiochemical conditions, what is the regulatory mechanisms for these processes, what are their ecological effects, and how microorganisms in extreme geo-environments show a response to and feedback with climatic and environmental changes.

Finally, co-evolution of microorganisms with geo-environments is another key scientific issue. The origins of many microorganisms are far from clear. Although most microorganisms at the root of the tree of life show a possible origin in extreme geo-environments, the timing and physiochemical conditions of their origins remain unclear, and the fossil record of microorganism evolution is remarkably limited. This directly affects the reconstruction of evolutionary phylogenetic trees for some key microbial functional groups. At present, a reliable evolutionary phylogenetic tree for any of microbial functional group has yet to be established. The uncertainties in divergence times and phylogenetic relationships of many microorganisms impede reconstruction of the environmental conditions in which they originated.

2) Major scientific issues arising from industrial applications

First, the basic scientific issues related to germplasm resources need to be further investigated. Most microorganisms in extreme geo-environments are potentially an important microbial germplasm resource. A diversity of important genes and unique enzymes have been utilized in various industries, such as environmental protection, metallurgy, energy, electronics, materials, medical care, food, etc. However, the characteristics of many special genes and unique enzymes extracted from various microorganisms in extreme geo-environments remain to be deciphered, and further research is needed concerning model organisms, types of genes and enzymes, physical and chemical properties and functions, extraction methods, domestication in industrial and agricultural production, and so on.

Second, scientific issues of microorganisms involved in resource exploitation and energy extraction remain to be studied. Microorganisms in extreme geo-environments have a significant impact on fossil fuel reserves, but their specific roles in the formation, exploration and development of underground oil and gas reservoirs, natural gas hydrates, shale gas and other sources of energy need to be further investigated, including the mechanisms, domestication conditions, and efficiency of these microorganisms in resource recovery. In particular, in regard to biometallurgy, investigations are needed concerning the physiological and biochemical characteristics, environmental conditions, and growth and reproduction rate of microorganisms in order to optimize their utility in biometallurgical processes.

Third, scientific issues arising from security considerations related to microorganisms need to be explored. Microbial activity can have both positive and negative effects in nature, with the former to be exploited and the latter avoided. For instance, underground CO_2 sequestration and storage can be affected by microorganisms in the deep subsurface, and it will be necessary to study the interactions between microorganisms, reservoir rocks and minerals, and fluids. Underground storage is also currently recognized as the only viable long-term safe alternative for disposal of long-lived high-level nuclear waste. However, microorganisms in the deep subsurface will affect the reliability and safety of nuclear waste repositories, and a detailed understanding of possible microbial processes is needed for the planning and design of nuclear waste storage.

Finally, there are many scientific issues related to the exploration of extraterrestrial life. Could life have originated, evolved and survived on other celestial bodies? Which celestial bodies are suitable for life? Is it possible for life in the universe to break through the physiochemical boundaries of life on Earth? Is there any life in the universe that is not based on carbon? What are the characteristics, adaptation mechanisms, and biological dynamics of microbial communities in the space, if they exist? How do microbes interact in space environments, and how can biological methods be developed to inhibit plant pathogenic bacteria? These are the major scientific issues related to the exploration for life in deep space.

(2) Developmental strategies and goals

In view of China's situation on basic research, and in accordance with the general requirements of "limited goals and highlighted priorities", the following three strategic measures might be taken to develop this important field in the next 5 to 10 years.

1) Advancement of critical technologies through linkage of major engineering projects to key research fields

Projects related to deep-Earth, deep-sea and deep-space discovery are being vigorously promoted in China, which has created a precious opportunity for research on microorganisms in extreme geo-environments. This will require Chinese scientists to actively participate in the preliminary demonstration, design and follow-up research of the national major science programs related to these engineering projects. Furthermore, achievements in the study of microorganisms in extreme geo-environments will require technical innovation. Among advanced technologies, deep drilling and *in-situ* fidelity sampling, *in-situ* monitoring and simulation should be given a high priority for development in China.

2) Advancement of multidisciplinary research

Guided by technological innovation, it is necessary to conduct and strengthen multidisciplinary research. The current situation of studies on microorganisms in extreme geo-environments in China, in which researchers operate independently in various disciplines, needs to be changed. Multidisciplinary research is urgently needed and should be encouraged. It is necessary to strengthen collaboration not only between the geosciences and life sciences but also between science and engineering technology.

3) Transformative breakthroughs through study of extreme geo-environments

In China, studies of microbes, although conducted within limited areas, have examined many types of extreme geo-environments. Integrative systematic research has not yet emerged because of barriers to multidisciplinary collaboration and a knowledge gap between different disciplines. Microorganisms in extreme geo-environments are not only a bridge to deep-time microbial communities but also to those in deep space. However, interdisciplinary work is needed among research communities studying deep-time geo-environmental microbiology, modern extreme environmental microbiology, and astrobiology in order to achieve transformative breakthroughs.

4) Achievement of key goals through these three strategic measures

A second key goal is to gain the world-class achievements in the field of geomicrobiology of extreme geo-environments through innovations of advanced technologies (e.g. *in-situ* sampling techniques, *in-situ* observation and simulation technologies), and to further promote scientific research related to national major projects targeting the deep Earth, deep sea and deep space. Moreover, we seek to identify microbial germplasm resources with wide application prospects, and to establish a resource repository of microbes from extreme geo-environments; this will allow development of applications in industrial and agricultural production, bioengineering, ecological restoration, pollution treatment, and other areas that will fulfill important strategic goals of China.

(3) Key research fields

On the basis of the current domestic situation and international development trends, the following six fields are proposed as foci for research development over the next 5 to 10 years. The first three fields are related to grand projects of deep-sea, deep-Earth and deep-space research, which can be funded by major national programs. The fourth and fifth fields are connected to microbial germplasm resources and biosecurity and ecological security, whose chief benefits will be to society. The last field serves the development of Earth system science. These fields not only match the long-term plan of National Science Foundation of China but also combine fundamental research with major national goals as advocated by the central government.

1) Deep sea microorganisms and their effect on resources and environments

Deep-sea microbial research is needed to enhance knowledge of microbes in subseafloor sediments, hydrothermal vents, cold seeps, abyssal plains, subduction zones, mid-ocean ridges and anoxic zones. The abyssal plains connect the surface and deep ocean, and subduction zones and mid-ocean ridges offer important windows connecting the surface and the deep Earth. It is important to investigate these three regions from the viewpoint of the Earth system, as only a few studies on microorganisms have been conducted in these regions so far.

2) Deep terrestrial subsurface microorganisms and their effect on resources and environments

In addition to darkness, there are some other special environmental features related to the deep terrestrial subsurface biosphere. Microbial studies are needed with the following focuses, including typical oil-gas reservoirs, underground coal beds, cave systems, sub-lake floor sediments, deep peat sediments, groundwater, ultra-deep drilling, major deep fault zones, deep strata for subsurface CO_2 sequestration, and underground nuclear waste repositories.

3) Astrobiology related to deep space projects

Astrobiology is a typical interdisciplinary branch of science; it requires the collaboration of teams from multiple disciplines and laboratories to jointly explore the habitable environment from various perspectives and provide reliable suggestions for the exploration of extraterrestrial life. Through international cooperation and the establishment of a shared database, research processes will be accelerated and applied to national space exploration and extraterrestrial life exploration missions.

4) Microbial germplasm resources in extreme geo-environments

Microorganisms in terrestrial hot springs, ice and alkali-saline lakes have the potential to contribute germplasm resources for wide industry exploitation, so a focus on the study of key functional genes is critical. Microorganisms isolated from acid mine drainage areas, heavy metal pollution areas, and pesticide and antibiotic contamination zones can be used to degrade or decompose various pollutants, contributing to a comprehensive protection and improvement of geo-environments.

5) Securities and risks of microorganisms in extreme geo-environments

Scientific knowledge of deep subsurface viruses is scarce, and the pathogenicity and potential risks that these viruses pose to humans needs further investigation. Research on the diversity of deep subsurface viruses and their roles in Earth's ecosystem will greatly enhance public awareness of biosafety in exploiting underground spaces and help to avoid and control potential risks. Moreover, researchers have yet to examine the mechanism of the influence of underground spaces, especially deep subsurface environments, on human physiology, psychology and pathology. Thus, there is an urgent need for the development of deep subsurface medical science.

6) Microbiology in deep-time extreme geo-environments related to Earth-system science

Microorganisms in deep-time extreme geo-environments can shed light on the long-term development of the Earth system and its component spheres (atmosphere, hydrosphere, lithosphere, biosphere), which will be the key to further development of Earth system science. Studies of geomicrobes and their geological evolution during some critical periods of Earth history, including the two great oxidation periods, the Snowball Earth period, the biotic mass extinctions of the Phanerozoic, and hyperwarming event periods, need to be conducted to better understand interactions between microorganisms and their extreme physiochemical environments. Comparisons based on these studies will allow development of important systemic theories.

4. Overview of research achievements and perspective

(1) Common features shared by microorganisms in extreme geo-environments

Advancements in the study of microorganisms in extreme geo-environments have been closely connected to the development of new methodologies, including microbial culturing followed by 16S rRNA gene-based clone library, whole-genome sequencing, metagenomic sequencing, next-generation high-throughput sequencing, single-cell sequencing, etc. These techniques led to the finding that microorganisms in extreme geo-environments share common features, including a slow metabolism, low biodiversity, coupling of biogeochemical processes, development of microbial consortia, and a large spatial heterogeneity in their abundance, diversity, and functional groups.

(2) Interactions of microbes with extreme geo-environments

It has been found that microorganisms in extreme geo-environments have important impacts on the atmosphere, hydrosphere, biosphere and lithosphere. Close interactions were also found within microbial communities. However, less is known about how microbes and microbial communities in extreme geo-environments have evolved at long (geological) timescales. This is an area of research that awaits detailed investigation.

(3) Rewards and risks of microbial resources

The adaptation of microorganisms to extreme environmental conditions has led to the production of special genes and enzymes. These represent important microbial resources that can be utilized in industry, including for environmental protection, metallurgy, energy industry, electronics, materials, medical care, and food production. A variety of viruses have been found in the deep sea and oceans, saline lakes, hot springs, and polar regions. Some of these viruses are known killers of microbial cells and represent a potential risk to humans. Much research is needed to fully assess the rewards and risks presented by these diverse types of microbes.

(4) Research progress to date and reporting platforms

The amount of research on microorganisms varies greatly among different extreme geo-environments, with relatively more undertaken in modern surface and marine extreme environments, and relatively less in terrestrial deep subsurface, deep-time, and deep-space extreme environments. The International Ocean Drilling Program is one of the most important platforms for research in this field. Some programs in the USA, including Subsurface Science Program, Life in Extreme Environments, Deep Carbon Observatory, Center for Dark Energy Biosphere Investigations, have also contributed greatly to knowledge concerning microorganisms in extreme geo-environments. The International Society for Subsurface Microbiology, International Society for the Study of the Origin of Life, and The Geobiology Society are international organizations supporting this kind of research, and *Extremophiles*, *Geomicrobiology Journal*, *Geobiology*, *Frontiers in Microbiology*, etc. are important journals in this field.

(5) China's contribution to this field

Chinese scientists have conducted investigations on microorganisms in a variety of extreme geo-environments, including those found in modern surface, marine, terrestrial deep subsurface, deep-time, and deep-space settings. This shows a great contribution of China to this field. The most impressive investigations concern the hydrothermal spring, abyssal and sea-bed sediments, oil and gas reservoirs, caves, acid mine drainage regions, glacial deposits, and some critical periods in Earth history. Investigations on microorganisms in deep-sea sediments, terrestrial drill holes, and deep space, as well as on the evolution of viruses and microbes through geologic time are to be strengthened.

Investigations on these microorganisms awaits the development of technical innovations. In China, some advanced techniques related to *in-situ* fidelity sampling during drilling, and on-site monitoring and modelling systems are in particular to be developed. In this particular field, the state key laboratory is to be ratified, the academic society is to be organized, the academic community is to be built and the funding system is to be proposed.

5. Proposal and suggestion for funding and infrastructure policies

Research on microbes in extreme geo-environments is urgently needed to be strengthened, and a roadmap to develop this particular field is to be planned jointly by the Ministry of Education, Ministry of Science and Technology, Ministry of Ecological Environment, Chinese Academy of Sciences and some other ministries. A full range of prospective investments are needed for the following four key aspects, including education of interdisciplinary talents, launching of major scientific research projects, construction of infrastructure platforms, and popularization of science. These strategic inputs will enhance China's influence on international science in this field, ensure national security in organisms, ecology and energy, and further provide valuable contributions to the achievement of major strategic objectives.

(1) Development of interdisciplinary subjects and education system of interdisciplinary talents

To further research on microbes in extreme geo-environments, it will be necessary to establish an interdisciplinary system that is different from the traditional focus on individual disciplines. In particular, we need to promote the integration of geosciences, life sciences, and planetary science, and further the development of new interdisciplinary fields such as geobiology and astrobiology. The construction of these interdisciplinary fields is a long-term goal. Curriculum and talent training should be conducted in colleges and universities. The curriculum of geobiology should not only involve the core subjects of life sciences and geosciences but also truly embody the cutting-edge subjects of these two disciplines. Such embodiment is the key to the success of interdisciplinarity. Similarly, astrobiology needs a curriculum built on an integrated study of geosciences, planetary science and life sciences, not just a list of the core courses of these three individual disciplines.

After establishing an interdisciplinary curriculum, funding for the education of interdisciplinary talent is needed. At present, there are few talents who can conduct truly interdisciplinary investigations. Many current projects of this type are conducted by joint communities from the different disciplines, not by talents who have been trained specifically in interdisciplinary research. Thus, current interdisciplinary research is still at the lowest level. To promote interdisciplinary research, we need to deliver interdisciplinary training for talents. This requires the attention of institutes and universities in China.

(2) Financial support for basic scientific research

While universities and institutes construct an education system for interdisciplinary training, the Ministry of Science and Technology and other relevant government agencies need to increase the financial support for interdisciplinary research. The study of microbes in extreme geo-environments can yield important scientific discoveries, inspire disciplinary research among the geosciences, life sciences, and planetary science, and play an important role in the development of new fields, such as geobiology and astrobiology. It is thus of importance to launch major scientific programs to promote interdisciplinary studies and to accelerate the expansion of these scientific fields.

Simultaneously, a focus on microbes in extreme geo-environments should be added to major national scientific programs that have been launched, such as those concerning deep space exploration, deep Earth, deep sea and global change. This addition may favor breakthroughs in related key technologies, especially those related to *in-situ* sampling and deep drilling.

(3) Infrastructure construction

Research on microorganisms in extreme geo-environments requires both interdisciplinary studies and innovations in key technologies. In addition to the implementation of major scientific projects, it is necessary to build essential infrastructure. Laboratory experimental platforms need facilities for interdisciplinary research concerning both life sciences and geosciences, which will enable the connection of metabolic processes at the molecular level with long-term geological processes at regional or global scales. However, at present, there is a pronounced shortage of these kinds of experimental platforms.

Along with the construction of laboratory facilities, field research bases should be vigorously established. Ultra-deep boreholes drilled through the deep Earth project provide a unique chance for *in-situ* observation of microorganisms in the deep subsurface. Special ecosystems found via deep-sea projects deserve long-term on-site monitoring. Geological sections in the field provide precious research bases for the exploration of microorganisms in the extreme geo-environments of deep time. However, not all of these important research bases make use of *in-situ* observation systems specially designed for microorganism monitoring. Consequently, a prospective layout of such field bases for investigation and monitoring on microorganisms needs to be scheduled in advance.

(4) Popularization of microbial knowledge and impacts

Unlike the aforementioned three proposals that mainly rely on central government departments including the Ministry of Education and the Ministry of Science and Technology, the popularization of microbial knowledge will require active collaboration among enterprises and institutions, various academic societies and associations, scientist communities and individuals to succeed. In 2019, 32 microbiologists in the world published a paper stating the necessity of improving microbial knowledge among the general public. They argued that microbial knowledge potentially will have a greater societal impact than the internet. For example, microorganisms might serve as a "second skin" and "second secretory system" of human bodies, the friends rather than the enemies of the human species as a whole, the engine of the biosphere, the core of the biological economy, the largest recycler and regenerator of Earth materials, the most reliable solution to pollution issues, and the first and last creatures on Earth. Microbial knowledge is thus necessary for every individual, from government officials and decision-making elites to the public. Both the decision-makers and individuals need to be trained to have the ability to evaluate the positive and negative impacts of microorganisms. Indeed, it is desirable to offer microbiology education in kindergartens, primary schools, middle schools as well as universities.

In pursuit of this goal, microbial knowledge should be popularized through a series of scientific or engineering projects, including major scientific research programs, field research bases, discipline construction, and personnel training, so that human beings can make friends and coexist peacefully with microorganisms throughout planet Earth.

目　录

第一章 科学意义与战略价值

第一节 科学意义

极端地质环境微生物的研究不仅促进了生命科学领域的一些重大科学发现，还推动了地球生物学的发展，更全方位推动了天体生物学的发展。

一、推动了地球科学与生命科学领域的一些重大科学发现

极端地质环境由于具备与生命起源时地球环境相似的特征，如高温、高压、高辐射、强还原等，适应或生长于此的微生物携带了生命进化历程的丰富信息，代表着生命对于环境的极限适应能力，界定了生物圈的行星边界参数（Martin et al.，2008）。极端微生物的生命形式和起源过程的研究，对揭示生物圈起源的奥秘和生物进化规律、阐明生物多样性的形成和演变机制、认识生命与环境的相互作用都具有极为重要的意义。

（一）深化了对生命本质和生命起源的认识

1. 生命的本质

生命起源与进化是一个亘古未解之谜。生命的本质是一种“秩序”（低熵）。要维持生命状态，只能通过能量的不断输入来对抗宇宙中自然的熵增趋势。微生物是地球上存在最古老的生命形式，生命的维持所依靠的是一类

有效的能量转换机制。生命依赖能量这一观点自被薛定谔于1944年提出以来已经得到广泛认可。现代生物学研究表明，这一类能量转换机制是广泛存在并高度保守的。通俗地讲，不断完善中的生物进化理论与中心法则可以看作生命科学的两大基本规律，分别代表了19世纪与20世纪生命科学最重要的发现。进入21世纪，合成生物学研究思想在从“分析”趋于“综合”、从“局部”走向“整体”的认识基础上，上升至复杂生命体系“合成、构建”的更高层次。合成生物学的发展使我们对生命本质的认识有了不断“试错”与完善的条件，也必将提升我们对生命起源与演化的认识。

一般认为，超化学的类生命系统表现出下述生命特征：自我维持的化学网络、自我复制的多聚体、自我生成的囊泡。自催化的过程如何整合到生物化学过程中，是确立代谢网络形成的关键问题。现代生物的代谢网络非常复杂，以大肠杆菌（*Escherichia coli*）为例，在其好氧生长中，其中间代谢产物达到1亿个，酶分子达到240万个。迄今，只有一个实验结果证实了自催化的非生物可能性，即甲醛聚糖。生物中的自催化的遗传过程被发现的只有RNA酶。所以自下而上（bottom-up）代谢网络形成的研究才刚刚起步。在自上而下（top-down）方面，迄今，有一条古老的代谢途径基本得到公认，即厌氧乙酰辅酶A（anaerobic acetyl-CoA）途径（Wood-Ljungdahl途径）。Wood-Ljungdahl途径广泛分布于严格厌氧的细菌和古菌中，同时它也是最短和最简单的固碳途径。在氢气（H_2）存在的情况下，合成生物物质只需要较少的能量。此外，它还是唯一能够通过底物水平磷酸化形成六碳糖和三磷酸腺苷（ATP）的途径，通过硫化镍/硫化亚铁（NiS/FeS）催化剂从一氧化碳（CO）、甲硫醇（CH_3SH）合成碳—碳（C—C）键的研究表明，该途径是古老碳固定途径的遗迹。与其他已知的5条碳固定途径比起来，Wood-Ljungdahl途径有下述反应类似生命系统的特点：①这是一条由三个简单循环组成的线性途径替代典型的自催化循环；②有机金属催化剂替代了合成代谢中的有机物中间体；③使用了四氢呋喃（THF）、类咕啉、Ni-Fe簇等辅酶因子作为催化剂。待解决的科学问题：除碳固定途径以外，其他主要代谢途径的起源，类似RNA酶一类自催化生物大分子的发现与机制的研究。

生命的本质是通过能量的输入来对抗熵增趋势，而生命获取能量离不开细胞膜。依靠细胞膜所形成的质子/钠离子（Na^+）梯度，通过三磷酸腺苷合酶形成ATP。目前对原始细胞膜和生物获取能量的认识主要来源于两个方面：一是对热液生态系统，尤其是碱性热液生态系统的认识（图1-1）；二是对最古老的生命形式，如产甲烷菌和产乙酸菌能量代谢的认识。产甲烷

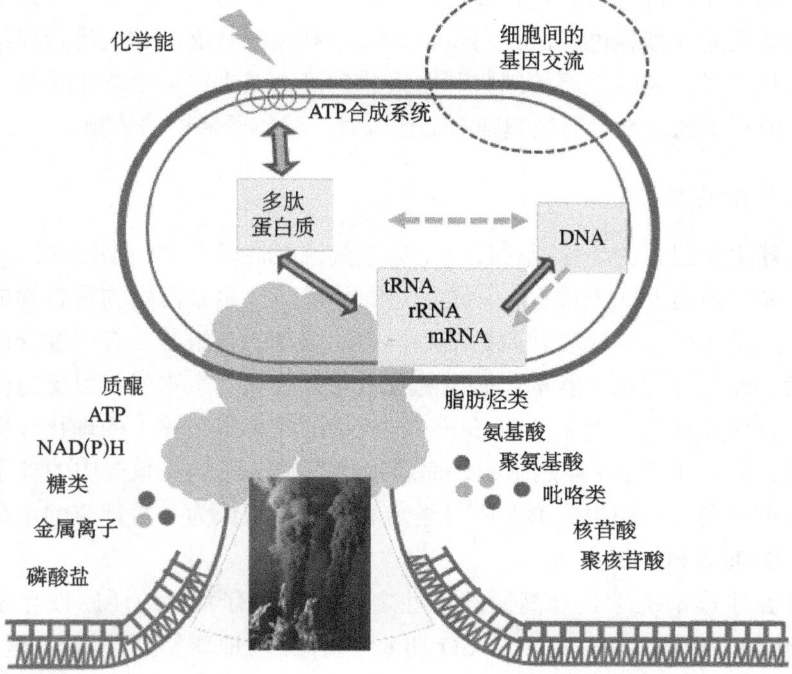

图 1-1　生命可能起源于深海热液环境（文后附彩图）

注：转移 RNA（tRNA）、核糖体 RNA（rRNA）、信使 RNA（mRNA）、三磷酸腺苷（ATP）、还原型辅酶Ⅰ（NADH）、还原型辅酶Ⅱ（NADPH）。

古菌和大肠杆菌每次细胞分裂形成 500 亿～550 亿个 ATP 分子，是形成细胞生物量所需能量的 50～100 倍。如果是在早期的 RNA 世界，那么原始的细胞膜中产生的总能量与形成细胞生物量所需能量的比值将达到 40 000。此外，由于质子 / 钠离子梯度形成后，有机酸会破坏这一梯度，所以在蛇纹石热液系统等极端环境中，微生物倾向于利用钠离子梯度，这可能也是早期细胞膜形态的由来。以失落之城（Lost City）为代表的深海碱性热液系统提供了一个很好的原始细胞膜获得能量的雏形，包括在薄的无机壁两侧的自然质子梯度。烟囱壁内是还原性的气体 H_2，烟囱壁外是氧化性气体 CO_2，烟囱壁内热液 pH 为 9～11，烟囱壁外海水 pH 为 5.5～6.0，形成 3～5 个 pH 单位的质子梯度，相当于质子驱动力 150～300 mV，其 pH 和电势与现代自养生物类似，但这里还存在一个热力学上的谜题，即 CO_2/ 甲酸（HCOOH）的 $E_0'=-0.199$ V 低于 $2H^+/H_2$ 的 $E_0'=0$ V 值，而电子双歧途径的发现使不可能变为可能。电子双歧途径显示，铁氧化还原蛋白中的铁硫簇是比 ATP 更早、更简单的生物能量单元。在最古老的物种，如产乙酸菌和产甲烷菌中，只有一个离子耦合位点而缺乏醌和细胞色素，它们还原 CO_2 是利用低势能的铁氧还蛋白经黄素电子双歧途径实现的。原始细胞利用天然有机或无机壁及地球化学梯度获得能量，但真正的自由生活细胞则需要摆脱对天然质子梯度的依赖。

2. 生命的起源

地球生命起源的环境及过程是全球最具挑战性的科学问题之一。一般认为，生命起源需要两个过程：一是通过非酶反应产生必需生物聚合物的缩合反应，二是将聚合物封装到原始细胞群中的膜室的自组装。有一种普遍接受的假设，即只要有化学能量、简单的有机化合物和液态水就足以支持生命的起源。有研究认为，生命最初形态（有机物的简单组装体）的理化性质都极其脆弱，受离子和 pH 强烈影响，而陆地热泉和海底热液系统中的离子成分比海水更有利于早期生命形态的自组装和聚合，被认为是最理想的生命起源生境（Dodd et al.，2017）。

高温地热系统除具有高温的特性之外，还含有大量的还原性化学成分[如硫化物、甲烷（CH_4）、H_2、CO 和 Fe^{2+} 等]，与地球早期环境相似。部分学者认为，现代陆地热泉系统可能存在与地球早期环境相似的微生物群落，譬如热泉中的蓝细菌菌席可追溯到产氧光合作用（oxygenic photosynthesis）的产生和大氧化事件（great oxidation event，GOE）。生物进化分析显示，在系统进化中大部分位于系统发育树根部的生物都是嗜热菌，且分支较短。因此，研究高温环境及其微生物有利于我们理解地球早期生命起源与演化以及探讨地外生命的问题（Deming and Baross，1993）。

生命的起源离不开地球独特的环境带来的机遇，生命的演化也从来不是独立于地球环境而自发随机产生现在这样的多样性的。无论是突发的地质事件还是漫长的地质时间，都对微生物进行了塑造，深海热液口是生命起源的可能地点。研究深海微生物能够帮助我们理解早期的生命，对研究微生物与环境的协同演化有着重要的意义。深海微生物对高压适应的特性意味着它们可能与陆地微生物有着不同的演化方向，深海环境也保护了生活在其中的微生物，使其免于遭受陨石对地球的冲击。已有多种假说对生命起源的早期环境进行了描述，深海中的热液口与早期的地球环境非常相似，而深渊环境通过蛇纹石化提供能量，也有着与外部水体交流少的特点。其较低的能量输入，使得微生物有较低的活性和较长的代时（200～2000 年，是大肠杆菌生长速率的 10^{-7}），相对封闭的环境使早期的微生物可能得以保存。将深海微生物作为研究材料，必然能够从微生物的基因组上找到信息和线索，从而对环境的变化进行反演。研究演化的方法是通过生物信息学手段进行计算、比较和分析。少数研究表明，超嗜热古菌在应对瞬时温度上升时会改变代谢途径，但是对于它们是否自发突变，以及这种潜力是否源自独特的环境压力，答案还是未知。前期的研究表明，嗜热菌的蛋白质序列在功能上比常温菌更加保守，暗示了突变在嗜热菌株中比在常温菌株中更有害。实验同样验证了较中性的突变在高温下会更加有害。此外，对两株菌株，超嗜热硫化叶菌（*Sulfolobus acidocaldarius*）和嗜热栖热菌（*Thermus thermophilus*）的突变速率估算试验（fluctuation test）表明，超嗜热菌株比常温菌株有较低的突变速率。这支撑了高温倾向于更高的 DNA 复制保真度的假设。

微生物是地球上生物地球化学循环的主要调控者，深地环境的生命活动可以孤立于地表存在，这些生命完全能合成自身的碳氮化合物，根本不用依赖光合作用合成的有机物和氧气进行生命活动。这些“黑暗生命”的发现完全改变了人们以往对“万物生长靠太阳”的认识，也为生命可能起源于地下环境提供了新的证据。早期地球受到陨石的频繁撞击，也因为没有臭氧层的保护而受到紫外线的强烈照射，因此地表环境非常恶劣，不适合生物的起源，即使生物起源后也不适合生存，但是深地环境可以免遭地表的辐射与陨石的碰撞，从而保护了有机分子与原始生命。

（二）对认识生命边界、深部生物圈、黑暗生态系统的科学意义

极端地质环境微生物的持续探索不断地拓展着人们对生命存在边界的认知。研究表明，在海底以下 2.5 km 深处的沉积物中埋藏了数千万年的微生物群落仍然具有活性，是至今发现埋藏最深的微生物群落。深海平原约占全球海底面积的 80%，通常分布于水深大于 4000 m 的寡营养开放大洋，一度被认为是生命的“荒漠”。然而，在全球最为贫瘠的南太平洋环流区沉积物中依然存在微生物群落，其丰度比环流区外同深度沉积物低 3～4 个数量级且呼吸速率较低。微生物群落甚至还存在于年龄超过一亿年的沉积物中。微生物在全球海洋沉积物中的广泛存在暗示了最终决定生命边界的因素很有可能是温度而非能量。与平静的深海平原不同，深海海底热液和冷泉生态系统则是深部生命的“绿洲”。深海海底热液沿大洋中脊分布，流体中携带的大量还原性物质为微生物生长提供了能量来源，也决定了热液口的微生物群落结构。化能自养微生物主要通过还原性三羧酸（rTCA）循环和卡尔文循环将

CO_2固定，生成有机碳，生成的有机碳随着摄食等作用在食物链传递或者变成溶解有机碳回到海水中。冷泉则主要分布在大陆架边缘，沉积物中产生的甲烷和其他有机质流体向海底渗漏或喷发，为此环境中的生命系统提供了物质和能量。

近30年来，人们对海洋沉积物中的生物圈规模有了初步的估算结果。20世纪90年代，多个样品中的微生物丰度首次被用于推测全球海洋沉积物中的微生物总量，并估算出其总量占全球总生物量的1/10，甚至是1/3。随着深海采样技术和细胞计数手段的发展，更加准确且具有代表性的数据得以积累。人们结合模型对沉积物生物圈的规模重新进行了估算，指出海洋沉积物中大约含有2.9×10^{29}个微生物细胞，相当于4.1×10^{15} g的碳，占地球总活体生物量的0.6%（Kallmeyer et al.，2012）。这与近些年重新评估的陆地土壤微生物总量［（2～6）$\times10^{29}$个细胞］相当（Magnabosco et al.，2018），凸显出深部生物圈在全球生物圈中的重要性。

洋壳生物的发现是深部生物圈研究中的一个重要突破。深部生物圈即埋藏在地球深部的黑暗生态系统，一般指海底沉积物1 m以下环境中栖息的生态系统。不同于地表生物圈的光能自养作用，深部微生物则通过化能自养作用提供能量和碳源，形成一个“黑暗食物链”。对于黑暗海底生命的探索可以追溯到20世纪30年代，美国斯克里普斯（Scripps）海洋研究所的科学家发现，在海床下数厘米至数米范围内的海洋沉积物内存在细菌，并开始观测支持这个深部生物圈的产能过程。直到1977年，美国的“阿尔文号”深潜器在太平洋的格拉帕戈斯大洋中脊发现了深海热液喷口及其周围的热液生态系统。后来，科学家通过大洋钻探，发现海床下数百米的沉积物样品中存在具有活性的微生物。这些早期的探索发现驱动了人们对地球深部生命的极大研究兴趣，深部生物圈的发现与研究一方面拓展了人类对地球生命边界的认识，另一方面也为人类认识生命的起源与存在形式，以及开拓利用深海的生物基因资源提供了全新的方向与视角。洋壳中的流体是地球上最大的含水系统，极可能含有规模不可忽视的深部生物圈。在与海水接触的玄武岩洋壳样品中，微生物浓度在每立方厘米10^6～10^9个细胞，比上覆海水高出3～4个数量级，而在下洋壳中，微生物浓度降至每立方厘米小于2000个细胞。据模型估算，全球洋壳微生物的生物量大约为200 Pg，远大于沉积物中的4 Pg。然而由于样品位点和数目非常有限，洋壳生物圈的规模还有待更多的实测数据验证。新生洋壳通过与海水发生水-岩反应而被逐渐氧化、水合，从而持续为洋壳中的化能自养微生物提供能量。除此之外，洋壳生物圈还会参与降解洋壳流体中的溶解有机碳（DOC），占全球深海 DOC 损失的 5% 以上。洋壳生物圈除参与有机碳循环外，细菌群落组成显示其与铁锰（Mn）氧化密切相关，对海底地貌风化也存在潜在的重要作用。

二、促进地球科学与生命科学交叉融合，推动地球生物学的发展

地球科学与生命科学无论是研究对象和研究目标，还是研究的时空尺度和研究方法体系都存在巨大的差异，两者的交叉融合研究长期以来一直是科学领域的巨大挑战。但同时，在学科之间的分界越来越模糊的今天，在全球化的大数据时代，它们之间的交叉融合又孕育着巨大机遇。极端地质环境微生物将微生物的地质过程与现代过程放在同一个平台上进行研究，为地球科学与生命科学的交叉融合提供了难得的机遇和抓手。地球科学从关键地质时期的地质环境条件、地质微生物的演化等一系列地质记录方面探索微生物对极端环境条件的响应与反馈。生命科学则聚焦在现代微生物和现代环境条件，从分子、个体、物种、群落乃至生态系统和生物圈等不同层次阐述微生物对极端环境条件的响应、适应与反馈机制。从这个层面来说，极端地质环境微生物是地球科学与生命科学的交叉融合地带。如果说研究人体系统必须研究微生物，必须研究病毒，那么研究地球系统则必须研究地质微生物，必须研究极端地质环境微生物。极端地质环境微生物对地球系统的重要性，犹如微生物之于人体系统一样。

例如，在距今 2.5 亿年前的二叠纪—三叠纪，全球温度升高导致海洋出现了严重的缺氧和动物大灭绝现象，一些地质微生物开始暴发，导致了碳、氮、硫等元素的生物地球化学循环的异常。在当前全球变暖背景下，现代海洋也出现了类似的变化，比如微生物暴发和海洋缺氧等。微生物暴发的地质过程与现代过程受到了地球科学家和生命科学家的关注，促进了科学共同体的自然构建，使得不同学科领域的科学家能在同一个平台上进行碰撞和对话，这是学科交叉融合并孕育科学突破的关键所在。从这一点来说，极端地质环境微生物的研究蕴含着重大科学突破，极大地促进了这两个学科领域的交叉融合，在现代科学体系中占据着举足轻重的地位。

（一）生物对多种极端地质环境的协同适应

静水压力、温度、pH、水活度、营养、盐度和辐射等都是影响微生物生长的环境因子。当微生物处于非最适生长条件时，会启动胁迫响应机制来帮

助自身适应环境的改变。

最近，越来越多的研究表明环境胁迫可以诱导氧化胁迫产生，在某些环境胁迫因子如高温、干燥、离子辐射下可检测到胞内有活性氧（reactive oxygen species，ROS）产生。同时，微生物还可通过多种抗氧化机制对抗环境胁迫造成的活性氧增加。但是这一过程非常复杂，即使同一株菌对不同的环境胁迫因子的耐受性也有着不同的交叉保护模式。例如，A 胁迫因子可以帮助菌株耐受随后的 B 胁迫因子，但并不意味着 B 胁迫因子的预处理同样可以帮助菌株对抗随后的 A 胁迫因子。人们对大肠杆菌在不同的环境胁迫（高渗透、酸、氧化胁迫、丁醇）下进行了 500 代次的传递，发现酸胁迫下进化的大肠杆菌菌株对高渗透压的耐受性增加，反之亦然。但是，尽管氧化胁迫下进化的大肠杆菌菌株对高渗透压的耐受性增加，但在高渗透压下进化出的菌株对氧化胁迫的耐受性并没有增加。转录组分析发现，高渗透压诱导了铁相关基因的上调，而这会导致胞内游离铁的增加，从而加重了氧化损伤（Dragosits et al.，2013）。总之，细胞抗氧化的过程是生物在逆境中抗胁迫的核心，同时也可以看作是与地球自无氧至氧化演化并行的生命（胞内）演化过程。

对深海细菌耐压希瓦氏菌（*Shewanella piezotolerans*）WP3 的连续传代研究，也证实了抗氧化能力的提升增强了深海微生物压力耐受能力。通过对不同环境胁迫诱导不同抗氧化基因或蛋白质的表达情况进行总结，人们发现几乎所有的环境胁迫因子都可以诱导部分抗氧化基因 / 蛋白质的表达。其中，辐射所诱导的抗氧化基因种类最多（12 种），包括 ROS 清除酶基因 *kat*、*sod*、*bcp*、*ahp*CF、*gpx* 等和 DNA 修复酶基因 *mutSLH*。低温胁迫所诱导的抗氧化酶类数目（11 种）仅次于辐射，主要是 ROS 清除酶基因，其次是酸（8 种）、高压（7 种）、高盐（6 种）、干燥（5 种）、碱（4 种）、低盐（4 种）、低铁（4 种）、高铁（4 种）、热（2 种）胁迫等。虽然环境胁迫所诱导的抗氧化酶种类越多并不完全代表其细胞遭受的氧化胁迫越强，但这些数据间接地反映了细菌的抗氧化系统在应对极端环境胁迫中的重要性。此外，在与地球环境协同演化的过程中，微生物随时面临着各种非生物逆境胁迫，如干旱、高盐、极端温度和辐射等，这些逆境胁迫导致的胞内过氧化是生命演化过程中不可忽视的驱动力，也对早期生物地球化学循环产生了深刻的影响。因此，研究微生物逆境相关基因及逆境信号转导的分子机制，对人们认识理解微生物功能群如何在环境中发挥作用具有重要的理论意义和实际应用价值。

（二）生命起源与演化的地质环境

微生物一方面随着地球环境的变化而持续进行着适应性演化，另一方面通过其自身的代谢活动影响地球元素循环并改造其生存环境。例如，深海微生物可通过改变体内基因表达量、改变代谢通路等多种不同的方式适应高压环境，产生与常压表层生物圈中微生物不同的演化方向。微生物对压力适应方式的多样化，主要来源于演化过程中不断变化的环境因素，以及微生物自身多重驱动力（突变、基因漂变、自然选择和基因重组）的塑造。

高压对生命的起源和演化都有着非常重要的影响。一般认为地球上的生命起源于 3.8 Ga 之前。通过古老的化石证据，推测生命可能在 4.28 Ga 前就已经出现在深海热液口。对于生命起源的方式有多种假说，其中比较受认可的早期生命起源环境是深海环境，可能的地点是深海热液口或深渊。早期地球海洋的面积比现在更加广阔，大于 10 MPa 的深海环境覆盖了早期地球表面超过 2/3 的面积，被淹没的陆地表面平均的静水压力为 7 MPa。压力被认为能够帮助保持细胞膜结构，维持生命的稳态，以及改变微生物体内的化学反应平衡及提高反应速率。高压和低温都能降低生物体的熵值，但与低温不同，高压并不能降低生物化学反应速率，更有利于促进生命的起源。另外，高压环境有利于增强氢键，提高核酸解链温度（Tm）值；提升细胞膜高温耐受性；更高的气体溶解度也有利于自养的生命过程。此外，原始生命可能在深海高压环境中躲过了早期地球表面的高辐射环境与小行星撞击，并免于承受撞击产生的瞬间压力与热胁迫的影响。

据推测，在生命起源的时期，地球表面的水体温度远高于现代海水，早期地质构造活动频繁使得热液口在早期地球广泛存在。现代深海中的热液口环境与早期的地球环境非常相似，从而为探索生命起源提供了良好的窗口。现代深海热液生态系统为微生物的生存提供了多样的环境，包含剧烈波动的温度、化学梯度、氧化还原梯度等，是一种独特的极端环境。此外，尽管深海热液口的理化性质波动剧烈，不同热液口的差异极大，但它们都有一个共同的特征，即均为烟囱体厌氧流体产生的高温区域。对生命的共同祖先（last universal common ancestor，LUCA）的研究结果表明，LUCA 很有可能来源于无氧高温的深海热液口，并且拥有嗜热和厌氧代谢等特点。

另一种可能的生命起源方式类似今天的深渊环境，即通过蛇纹石化提供能量。蛇纹石化产生的偏碱性环境和高氢气浓度为铁氧还蛋白提供了足够的还原力，从而为依赖氢气的化能自养微生物的出现提供了必要的条件。深渊的主要特征有：超高的静水压力、寒冷、无光、有机质匮乏、特殊的海底地形、剧烈的板块构造活动等。其较低的能量输入，使得在此栖息的微生物活性较低且代时较长（200～2000 年）。然而，相对封闭的环境也使得早期的微生物可能得以保存，故对深渊环境微生物的演化史进行研究，能够帮助人们了解海沟的形成过程和其对全球生态环境的意义。

地球大气及海洋的演化历史经历了一个从无氧到氧化的过程。地球出现最早期的生命时，大气及海洋仍是一个还原度较高的环境。27 亿年前演化出了产氧的光合蓝细菌后，其代谢作用推动了大气的氧化。氧气作为最终的电子受体，为微生物的有氧呼吸提供了大量的能量。然而氧分子有两个未成对的电子，是一种活性极强的分子，容易捕获电子生成 ROS。ROS［超氧阴离子（$O_2 \cdot^-$）、过氧化氢（H_2O_2）、羟自由基（·OH）］会氧化破坏细胞中的遗传物质及蛋白质分子，从而对细胞造成伤害，甚至导致其死亡。微生物的基本代谢途径大约起源于 3.8 Ga 之前，因此，尽管有氧呼吸为后期生物发展提供了大量的能量，但起源于厌氧环境中的生物必须抵御氧的胁迫。部分微生物通过生活在厌氧环境中来避免氧的伤害，而大部分微生物则进化出了一套氧化胁迫应答与防御机制来对抗氧的伤害。

三、促进地球科学与行星科学和生命科学的交叉，推动天体生物学的发展

（一）天体生物学的不断发展

寻找地外生命是人类探索宇宙的主要驱动力。20 世纪中叶以来，随着科学技术的进步，人类拥有了探索太空的能力，而寻找地外生命和宜居环境一直是人类进行深空探测的重要科学目标。天体生物学正是在这一背景下产生和发展起来的。早期的天体生物学研究以寻找地外生命为主要目标，随着该学科的不断发展，天体生物学发展成为研究在宇宙演化背景下生命的起源、演化及其在宇宙中的分布和未来的学科（Des Marais et al.，2008）。其研究范畴以地球宜居环境和生命起源为切入点，涵盖生命与环境协同演化、极端环境生命、宜居环境演化及在太阳系内 / 外的分布、生命信号识别、行星保护和开发等内容（林巍等，2020）。

天体生物学有两个主要特征。一是典型的多学科交叉，其涉及的学科领域包括生物学、地球科学、行星科学、天文学、化学、物理学等。二是该学

科与深空探测紧密结合，20世纪50～70年代的美苏太空竞赛直接促进了天体生物学这一学科的正式形成和第一个发展黄金期的到来。自20世纪90年代以来，一系列探测器（如探测火星的“勇气号”“机遇号”“凤凰号”“好奇号”“洞察号”“天问一号”等，探测水星的“信使号”等，探测巨行星及其卫星的“伽利略号”“卡西尼-惠更斯号”“朱诺号”等）的成功发射和大量原位数据的获得，使得天体生物学研究进入了第二个发展黄金期。在此期间，欧美国家成立了以美国国家航空航天局（NASA）天体生物学研究所（Astrobiology Institute）为代表的多个与天体生物学相关的学术和研究机构，相继制定了多个天体生物学的发展路线图和战略规划，加强了学术交流和深空探测合作研究，天体生物学进入了快速发展阶段。

天体生物学的研究拓展了学科之间的联系，促进了不同学科研究方法之间的交流。天体生物学在行星科学中占据着重要的基础地位，尽管行星科学的研究还包括各行星的内部组成、地质、气候、演化等方面的问题，但中心一直是寻找地外生命和宜居环境。作为行星科学的重要分支之一，行星地质学将地质科学应用于太阳系内各行星和卫星，开展地质类比研究，包括确定类地天体的内部结构、考察地质活动（如撞击、流水、热液作用等），为在地外寻找生命宜居环境提供了重要素材。对生物学而言，火星的宜居性探测及生命探测是测试地球微生物能否在地外极端环境中生存的途径之一，探索地外生命需要结合地球微生物的研究结果。同时，极端地质环境微生物是准确认识行星保护的重要基础。天体生物学作为一个交叉学科，能够促进多学科发展，并有望将研究成果应用于人类生活中，促进国民经济社会发展。

天体生物学研究也推动了深空探测工程技术的进步和探测任务的实施。以火星探测为例，早期的“海盗号”任务是为了到火星表面寻找存在生命的证据，后来的“勇气号”“机遇号”“凤凰号”“好奇号”“洞察号”等任务则关注火星早期和现代环境（特别是宜居环境）的演变以及过去和现在是否存在生命等，这些任务都是以天体生物学相关的科学问题为驱动而设立的，同时由这些任务获得的新信息极大地加深了人们对天体生物学的认识，并进一步孕育出了新的科学问题。科学问题和探测任务相辅相成、互相促进，使得天体生物学这门新兴学科的研究深度和广度不断深化，逐渐成熟起来。

此外，天体生物学相较于其他学科的天然优势是对社会公众的巨大吸引力，因此也是开展科普宣传和教育的重要抓手。美国、英国等国家的天体生物学研究机构都把科普宣传、中小学生教育、教师和宇航员培训等作为其重要的工作内容，以提高民众的科学素养、激发青少年的科研热情、培育天体生物学后备人才。

（二）从地球生物学到天体生物学

一些类地行星的地表环境与早期地球环境和地球的地壳深部环境有相似性，如强辐射、高温与频繁的陨石撞击等。因此对地球早期和深部生物圈的生命形式、生存方式、能量代谢等的研究对认识其他行星环境是否有生命也具有重要启示（Joseph et al.，2019）。例如，洞穴、地壳深部、冰川、盐岩等都可能是火星微生物的避难场所。洞穴内部有相对恒定的温度与湿度，能避免地表环境的极端气候、太阳辐射，为微生物的生存提供有利条件。同样地，在地壳深部环境中由各种地球化学反应（辐射分解、蛇纹石化、费-托反应等）产生的还原性与氧化性物质，都可能被微生物利用。需特别指出的是，地球上由蛇纹石化产生的氢气是深地微生物的主要能量来源，而火星的深地环境可能也具有类似的机理来产生氢气与甲烷，为火星深部生物圈提供能量与底物。另外，地壳深部发现的不以太阳为能量的“黑暗生命物质”给传统的生物学和生态学理论带来了挑战，也提供了进一步验证这些理论的机会，将来生态系统的重大发现最有可能来自巨大的、未被探索的深地环境。

不同于深海热液起源说，英国学者菲利普·霍里格尔（Philip Holliger）和詹姆斯·阿特沃特（James Attwater）提出生命也许始于“冰冻的世界”的观点，这个观点是基于沃特·吉尔伯特（Walter Gilbert）提出的 RNA 世界假说（地球上早期的生命分子以 RNA 先出现，之后才是 DNA）所得出的。科学家提出生命起源于寒冷世界是因为在 0℃以下的寒冷环境中，才利于 RNA 的一个关键催化因子进行表达和复制。微生物是生命进化的起始，在长期的进化过程中，低温微生物形成了一系列适应低温的机制。研究低温环境下微生物的生命活动，对揭示地球寒冷环境下生命起源的奥秘和早期的生命形式具有重要意义。在地球以外的宇宙环境中，存在着很多表面温度较低的星球。根据生物进化的历程，在寻找外星生命时，要从存在最广泛且存在可能性最大的微生物开始，因此研究低温微生物对探索外太空某些寒冷外星球生命的存在同样具有重要意义，体现了地球生物学与天体生物学的结合。

实际上，当前天体生物学的诸多研究内容均体现了它与地球生物学的渊源。当前天体生物学关注的科学前沿问题包括生命的起源和演化、行星系统中有机物的来源和迁移、行星系统和行星宜居环境的形成和演化、地球生命生存的极限、地球生命与环境的协同演化、地外生命和宜居环境的识别和探测等。其中的许多内容均涉及地球生物学的工作，为从地球生物学到天体生物学的发展架起了桥梁。

NASA 和欧洲空间局（ESA）在天体生物学研究方面较为成熟，代表了该学科最前沿的研究趋势。NASA 重点开展的六大领域研究包括识别有机化合物的非生物来源、生命起源中大分子的合成与功能、早期生命生存环境和复杂性的变化、生命和环境的协同演化、识别和探究宜居环境的特点、构建太空宜居环境。ESA 当前进行的 2015～2025 年工作规划以“行星形成的条件和生命的存在形式”为核心研究任务。在当前和未来寻找外星环境和生命特征的空间探索中，地球生物学，特别是矿物与微生物的相互作用研究具有重要作用。据此，ESA 成立了专题研究小组“空间探索中的地球生物学”，其未来主要的研究方向包括地面进行的空间环境模拟实验、微生物群落与矿物相互作用、生物膜对空间环境的响应、酶与生化矿物相互作用的特征、微生物学实验仪器的改进技术、现有地基行星模拟设施的整合、实验室和环境生物学的交叉研究、原位样品制备技术、现有分析方法的小型化、分析小体积样品化学特征的新技术、可重复使用的月球及近地天体实验平台的开发、通过地球生物学的研究推测其他天体上可能的生物特征、空间机构和地面研究组织之间的新通信渠道的开发。

第二节　战略价值

极端地质环境微生物既是一类极其宝贵的种质资源，同时又能与深海、深地和深空计划等国家重点工程计划相衔接，这决定了其对实施《国家中长期科学和技术发展规划纲要（2006—2020 年）》以及国家其他科技政策目标具有强大的支撑作用。当前，我国特别重视将科学前沿研究与国家目标结合，特别重视关键“卡脖子”技术的突破以及“从 0 到 1”的科学突破，要将科学研究的论文写在中国大地上，要树立“绿水青山就是金山银山”的理念。极端地质环境微生物的研究及其应用集中地体现了国家的这些科技政策目标和长远发展规划，因而具有重要的战略意义。

一、蕴含丰富的微生物种质资源

极端地质环境微生物是一类极其宝贵的生物种质资源，有许多非常独特的基因和生物酶，可以高效地应用于环保、能源、食品等诸多工业生产上。极端地质环境微生物是微生物与一系列极端环境条件相互作用的体现，充分显示了微生物通过合成各种化学组分来适应各类极端环境条件的机制，可以在各种极端环境条件下得到广泛应用。因此，极端地质环境微生物的这些应用对国民经济和社会发展乃至国防安全都起着重要作用。

例如，低温微生物因能够适应极低温环境而应用到动植物抗冻、医用疫苗等方面。嗜热微生物可以应用于高温快速降解多种天然高分子物质，也可用于湿法冶金以选冶各种低品位矿石。嗜酸微生物因对极低 pH 的适应能力而用来处理酸性矿坑水，包括应用杀菌剂、金属固定化的细菌等消除重金属污染。这些对国民经济的绿色发展至关重要。又如，嗜碱微生物的膜蛋白可以作为光敏材料应用于全息光学、信号存储和处理。耐辐射菌可以应用于处理各种放射性废物。这些又对国家安全起着关键作用。

（一）海洋极端环境微生物资源及其开发应用

海洋的极端环境蕴藏了丰富的物种独特性和多样性，是重要的基因资源地，这些物种往往具有结构功能较为独特的生物大分子（酶等）与小分子（次级代谢产物等）。目前，随着各国对海洋开发利用的重视、采样技术的发展，极端地质环境微生物的研究正从浅海到深海、从边缘海向大洋扩展。深海微生物资源已成为国家重要的战略资源储备，也是各国海洋战略的核心关注点。2018 年调查研究显示，受国际保护的专利涉及 862 个海洋物种和 12 998 条基因序列，其中微生物物种高达634种，微生物相关序列为9468条，包括深海极端环境来源相关的物种，如深海和热液喷口系统等。我国在 2001 年以来的深海微生物资源采探中，已分离鉴定并规范保藏深海极端环境来源的微生物菌株，实现了一定的资源积累。据统计，我国已分离鉴定并规范保藏大洋来源的菌株有 9376 株，分属于 1443 个种和 444 个属（统计至 2018 年）。但总体来看，我国深部微生物资源领域仍然存在采样区域局限、取样手段落后、高压 / 高温环境模拟装备等研制不足等问题，使得目前深部微生物无论是种类还是生物量都很难达到下游资源开发的要求。当前，随着深海微生物资源获取和开发活动的快速发展，海洋环境的保护和保全受到前所未有的关注，联合国的国家管辖海域外生物多样性（BBNJ）养护与可持续利用协定谈判涉及国家管辖范围外海域遗传资源多样性以及惠益分享，会逐步形成相关的法律条文，将对现有的国际海洋秩序产生重要影响。随着谈判进程趋紧，我国深海微生物资源领域依然存在采样区域较为局限、资源量有待扩容、产业化推进机制缺失、知识产权占有量不足等问题。

海洋极端地质环境微生物资源，不仅具有重大的科学研究价值，同时具有潜在的重要经济和战略意义。利用新技术从深海中开发新的生物资源、从技术和资源的源头进行创新研究，既是国际上新资源研究与开发的前沿方向，也是各海洋强国进军深海的重要关切，隐含着各国深海安全和重大战略利益。微生物由于对深海多种极端特殊环境（如低温、高压、高盐、黑暗、寡营养等）的长期适应，具备多种特殊极端条件的耐受性，同时能够产生各类新型可利用的天然产物，在医药、环保、能源、化工生产等领域具有巨大的潜在经济价值。深部微生物由于其环境的特殊性及物种的丰富性，已经成为生物应用研究的重点对象。在医疗方面，已有分离自深海放线菌的治疗多发性骨髓瘤的盐孢酰胺 A（Salinosporamide-A）等药物进入Ⅰ/Ⅱ期临床试验阶段。除此以外，抗真菌、抗病毒药物等抗菌药也从深海微生物活性产物中分离鉴定。在工业用酶方面，已纯化鉴定出具有耐盐、耐低/高温等优势性质的糖苷酶、酯酶、淀粉酶、蛋白酶、几丁质酶等各类具有生产潜力的深海酶类。此外，深海热液口是发现极端酶的资源库，从中分离的嗜热 DNA 聚合酶、嗜热蛋白酶、嗜热淀粉酶、耐酸/碱淀粉酶、氢化酶等具有优质的市场应用性能。在生物修复方面，从深海寻找石油和难降解烃类物质分解菌株和功能酶也是全球研究者的重点工作。

深海微生物资源包括物种资源、基因资源和代谢产物资源等，具有巨大的应用开发潜力。预计 2025 年全球海洋生物科技市场可达到 64 亿美元，涵盖制药、生物燃料、化工等多个商业领域。由于欧盟、美国、日本发达国家和地区在深海极端环境微生物基础研究上的投入，其在产业化方面也较为领先，美国的深海生物基因相关产业的产值达到了数十亿美元，德国化工巨头巴斯夫（BASF）股份公司拥有 5701 条海洋遗传资源专利序列，其中多条与深海极端环境微生物相关。我国深海生物基因资源产业化开发应用还没有形成一定规模，产业化应用的领域也十分有限，主要集中在海洋生物医药方面，工业应用方面还没能产生规模经济效益。预计未来 20 年内，深海微生物资源将在新药开发、工业催化、能源矿产、环境保护等领域形成重要产业。巨大的潜在商业价值将引领深海微生物资源的研究开发活动，并带来巨大的经济效益和难以估量的社会效益。

1. 工业方面的应用

工业生产常常要求一些特殊的反应温度、pH或添加某些有机溶剂，在这种条件下，普通酶无法保持活性，因此，依赖酶的工业必须花费大量资金、采取特殊的工艺，以保持这些酶的活性，从而大大提高了成本。深海微生物体内的各种活性物质具有耐高温、耐高压、耐酸碱和耐盐（包括重金属等）的特性，尤其是一些极端酶，在普通酶失活的条件下仍然能保持较高的活性，它们是对环境友好的高效生物催化剂，在工业上有着广泛的应用前景，可用于特殊要求的化工生产和环保产业，如酶固定化工艺和水处理等。目前，已经有高温聚合酶、糖酶、淀粉酶、蛋白酶等几种极端酶开始工业化生产，并且已经创造了数十亿美元的经济效益。

2. 医药方面的应用

越来越多的病原菌或病毒对目前的药物产生了抗药性，自然界中的病原微生物还是人类潜在的威胁，如新型冠状病毒COVID-19等，因此从海洋中筛选新型生物药物成为药物研究开发的重要方向。深海微生物特殊的化学物质结构为当今科学家筛选新的药物提供了良好的天然化合物，成为抗肿瘤、抗病毒、降压降脂等新型特效药物的重要来源。值得注意的是，病毒在海洋生态系统和全球过程中的真正丰富性和生态重要性，在过去20年才得以认识。病毒通过与各种海洋生物的相互作用，在海洋群落结构、海洋过程和生物地球化学循环中发挥着关键作用。最近的发现使病毒生态学成为海洋科学中最令人兴奋和发展最快的分支之一。据估计，世界海洋含有1030多种病毒，它们是海洋生态系统中最小和最丰富的成员。从细菌到鲸的所有细胞生命形式都可受病毒感染。通过对单个病毒的遗传和基因组分析发现，病毒是一个多样性的储藏库，在通过基因转移的进化中发挥着重要作用。海洋宏基因组分析揭示了巨型病毒和可感染其他病毒的病毒存在。这些发现意味着需要进一步讨论它们在营养循环、特定生物地球化学循环中的活动，重新评估海洋病毒及其致病性和危险性。

3. 能源方面的应用

深海油藏从其环境特征来看也属于极端地质环境。生活在其中的极端地质环境微生物直接或间接地参与了油气资源的形成。例如，产甲烷古菌活动所产生的甲烷气体可直接形成气田和天然气水合物矿藏，资源量巨大的海洋天然气水合物的主要气体来源为微生物成因的甲烷气体。因此，油藏环境微生物结构功能特征可作为寻找油气矿藏的标志之一。对油藏环境微生物的研究，不仅对油气能源勘探具有重要的经济价值，而且对清除石油等污染物、理解和应对气候变化具有重要的战略意义。

4. 矿产方面的应用

深海中存在着如多金属结核、富钴结壳、多金属硫化物等金属富集区，它们具有重要的战略价值和潜在的经济价值。对深海金属富集区微生物金属耐受机制及其在金属循环中的作用的清晰认识，将有助于指导矿产资源勘探和可持续性开发利用，为重金属的清除等生态环境保护和修复提供支撑。

综上，对海洋极端地质环境微生物的研究不仅在微生物多样性、生物地球化学元素循环和海底地貌风化等方面具有重要的科学意义，而且在基因资源、矿产资源的形成与演替中同样具有深远的战略价值。

（二）地表极端环境的微生物资源

地表极端环境具有丰富的特殊微生物种质资源，具有重要的生物工程应用价值与工业价值。

1. 嗜盐微生物资源

嗜盐微生物资源在许多行业具有巨大的应用价值和开发前景。例如，在食品行业应用方面，嗜盐微生物可以产生长链多不饱和脂肪酸，如二十二碳六烯酸（DHA）、二十碳五烯酸（EPA）、花生四烯酸（AA）、γ-亚麻酸（GLA）等人体所必需的营养物质以及对腌制食品的味道起着调节作用的酶类物质。在医药与化妆品应用方面，某些嗜盐或耐盐菌产生的β-胡萝卜素是一种抗氧化剂，具有解毒作用，是维护人体健康不可缺少的维生素，在抗癌、预防心血管疾病、预防白内障及抗氧化上具有显著的功能，并进而能防止老化和衰老引起的多种退化性疾病，其产生的渗透压平衡溶剂四氢嘧啶能使生物酶或核酸免受高盐、热变性、干燥或冷冻等不利影响，也可修复由紫外线照射诱发的皮肤损伤，因此被添加在防衰老或抗干燥的化妆品中。在生物电子行业应用方面，嗜盐杆菌属（*Halobacterium*）产生的一种特有的紫膜蛋白（bacteriorhodopsin，BR）在生物电子领域具有广泛的应用价值：紫膜蛋白作为光驱动的质子泵，可以用于将太阳能转化成电能、人工合成ATP和海水淡化等；可以作为光电转换器，主要应用在生物传感器和生物芯片的研究中；还可以作为光敏材料，应用于全息光学、信号存储和处理以及光学检测领域；能通过构型的改变储存信息，可作为生物计算机芯片的新材料。在生物材料应用方面，某些嗜盐微生物可以产生一种多聚物（poly-β-hydroxyalkanoate，PHA），该物质具有源于石油的塑料和合成橡胶的材料特性，同时具有生物可降解和优良的生物相容性等特征，可以用来制作生物可降解塑料。在生物能源应用方面，杜氏盐藻（*Dunaliella salina*）能产生大量的β-胡萝卜素和甘油，体内含β-胡萝卜素达8%～10%（干重）、36%左右的甘油和36%～40%的蛋白质及脂肪酸，既可以作为β-胡萝卜素的商业来源和甘油生产的潜在来源，也可以作为生物能源的原材料。在上述各种应用中，两类微生物资源的商业应用已经取得了巨大成功：杜氏盐藻生产的β-胡萝卜素被广泛用于医药行业，中度耐盐或嗜盐细菌生产的四氢嘧啶被广泛用于化妆品行业。

2. 高温微生物资源

陆地热泉环境存在着丰富而独特的高温微生物资源，这些高温微生物可以产生功能独特的高温酶类以及结构新颖的生物活性物质，在医药和生物工程方面具有广泛的应用前景。特别是高温酶类能够适应工业生产中较为严苛的环境，提高生产效率。此外，热泉环境孕育着大量的未培养微生物，如基于全球地热系统16S rRNA基因高通量测序结果分析显示，热泉环境在门一级水平达到16.1%，在纲一级水平达34.0%，在目一级水平达42.1%，在科一级水平高达46.9%。新的微生物往往具有新奇的代谢途径，在生产生活和环境治理等方面具有巨大的应用价值。高温纤维素降解菌能将纤维素高效降解为小分子糖类，热解糖热厌氧杆菌（*Thermoanaerobacterium thermosaccharolyticum*）可以高效生产丁醇，激烈火球菌（*Pyrococcus furiosus*）可以高效产生氢气，其氢化酶在85℃时活性最高，这是至今已知的能在NADPH上接受电子产生氢气的两种酶之一。其他常见的高温微生物来源的酶有纤维素酶、木聚糖酶、果胶酶、几丁质酶、淀粉酶、脱支酶、蛋白酶、脂肪酶、葡萄糖异构酶、乙醇脱氢酶和酯酶等。高温微生物还用于生物采矿、生物修复、生物表面活性剂等方面。Plumb等（2007）发现并命名了一株高温古菌食硫嗜酸两面菌（*Acidianus sulfidivorans*），其最适生长温度为45～83℃，最适生长pH为3.0～0.35，可以在各种硫化矿物中生长，包括黄铁矿、砷黄铁矿、黄铜矿，可以在严酷的极端环境中富集金属元素。除此之外，利用高温微生物等嗜极微生物处理重金属污染土壤等污染环境已发展成为一种潜力巨大的生物修复技术。高温微生物还具有毒性低，生物降解能力高，环境兼容性好，起泡能力强，高温、高盐、强 pH 条件下活性高，以及从可再生资源中获取潜力大的特点，在食品、药物、农业、造纸等工业中越来越被青睐。在其他方面，高温微生物在生产耐热、耐极端理化条件等表面活性剂方面有公认的潜力。目前报道的能产生表面活性剂的高温微生物主要是硫胺素芽孢杆菌属（*Aneurinibacillus*）、地芽孢杆菌属（*Geobacillus*）、产碱杆菌属（*Alcaligenes*）、芽孢杆菌属（*Bacillus*）和短芽孢杆菌属（*Brevibacillus*）。

3. 低温微生物资源

低温微生物在环境工程、食品、保健品、洗涤剂、医药和农业的生产中有着较高的应用价值。冰川或冻土等低温环境中存在大量的低温微生物，相比其他中高温环境，分离到低温微生物的概率更大，而且从低温环境中分离到的可培养微生物具有相对较低的最低生长温度和最适温度，可以很好地应用于环境修复中。例如，工业废水中硫酸盐废水的处理一般都需要加温或保温处理，处理成本相对较高，而从丹麦低温环境发现的一种耐冷硫酸盐还原菌（SRB）能在低温环境中还原硫酸盐，处理硫酸盐废水效果很好，不但提升了处理效率，还降低了处理成本。另外，低温微生物产生的低温酶由于催化反应的最适温度较低，可以应用于食品、保健品和药品加工等领域，如从来自青藏高原的一株北见微杆菌（*Microbacterium kitamiense*）中分离出的具有低温活性的纤维素酶 M6A，在 0～5℃仍然可以保持 30%～40% 的活性，这个酶将有可能在纺织、食品饮料和洗涤剂工业生产中发挥重要作用。微生物在应对低温环境压力过程中合成的一些物质则可能具备潜在的保健品功效，如低温微生物合成的抗氧化酶、类胡萝卜素和多不饱和脂肪酸等可以作为一类重要的膳食补充剂或食品添加剂。嗜冷性必需基因可用于为重要的人类病原体开发一类独特的细菌温度敏感疫苗。从青藏高原分离的嗜冷菌芽孢杆菌（*Bacillus* spp.）可以在 10 ℃下促进冬小麦幼苗的生长，这表明从寒冷环境中分离出的细菌可以弥补中温菌在低温条件下促进作物生长效果减弱的缺点。开展冰芯样品中低温微生物的研究，对认识生命的极限及其与环境相互作用的规律等都具有极为重要的科学意义。

4. 嗜酸微生物资源

嗜酸微生物蕴藏着大量的未知功能基因和代谢潜能，在生物资源、生物技术和环境领域具有重要的应用潜力。极端酸性环境的微生物往往具有起源古老、珍稀度高的特点，是宝贵的种质资源，是生物遗传和功能多样性的宝库。利用基因工程技术挖掘存在其中的功能基因对相关行业的发展具有广阔的应用前景。例如，利用嗜酸基因构建出具有高强度耐酸和产酸速率快的工程菌株并将其应用于偏酸饮食工业生产中，将极大地推动新型低能耗偏酸饮食工业的发展。利用嗜酸的铁、硫氧化微生物将贫矿和尾矿中金属元素溶出并回收的生物湿法冶金技术是一门新兴的矿物加工技术。与传统方法相比，生物湿法冶金技术拥有环境友好、耗能低、流程短、资源利用率高等无可比拟的优点，从而具有广阔的应用前景。因此，科学有效地利用嗜酸微生物进行生物冶金将是未来解决矿产资源贫杂化问题的方向之一，也是提升我国矿产资源国家战略地位的重要手段（Liu et al.，2017）。采矿活动导致的矿山酸性废水（acid mine drainage，AMD）是矿区环境治理所面临的重大挑战。这些极端酸性和高重金属毒性的废水，已在全球范围内引发一系列影响深远的生态环境问题。利用嗜酸的硫酸盐还原菌在厌氧条件下通过还原过程产生的负二价硫离子（S^{2-}）与重金属结合形成金属硫化物沉淀，有效去除 AMD 中的重金属离子，同时生成的碱度可以提高废水的 pH。该方法具有处理费用低、无二次污染、有效回收金属的优点，已成为 AMD 治理的研究热点和前沿技术。

二、影响能源与矿产资源

（一）深海资源与能源

深海除蕴含丰富的微生物种质资源外，还蕴藏着丰富的天然气水合物资源和多金属锰结核、黄铁矿等矿产资源。这些资源的形成都与微生物的作用有关。

天然气水合物作为一种新型能源资源存在于大陆坡海底和永久冻土带，是由水和天然气在高压低温下形成的冰态超分子笼状固态物，每立方米的天然气水合物在标准状态下可释放 160～180 m^3 天然气。据最新估算，全球天然气水合物蕴藏的天然气总量约为 1.2×10^{17} m^3，相当于全球已探明传统化石燃料的 10 倍，具有巨大的能源开发前景与潜力，已经成为世界各国尤其是发达国家争相研究开采的战略资源。这些天然气水合物通常分布在海底及海底 500～600 m 的深度范围。

多金属锰结核主要分布在洋底 3500～6000 m 的海底表层，含有锰、铁、镍、钴、铜等几十种元素，据估计储量在 3 万亿 t 以上，已成为重要的矿产资源。多金属锰结核的生长是一种非常缓慢的地质现象，数百万年才增长1 cm左右，以太平洋的多金属锰结核为例，年龄在2 Ma～3 Ma，其形成通常是海水与岩石相互作用的结果，也有热液成因。

海底，尤其是热液系统，作为天然的成矿实验室，在海底热液喷口周围富集的微生物群落为人们提供了研究深部生物圈的窗口，催生了地球科学与生命科学之间新兴的交叉学科：地质微生物学和微生物地质学。一方面，热液周边的流体产生的水岩反应直接促成了矿物（黄铁矿等）的产生；另一方面，在这一过程中产生的物质与能量，促进了以化能合成为初级生产力的热液生物群落的生长。大洋钻探发现，这些矿物不仅分布在表面，还广泛分布在沉积物中。钻探发现，在蚀变的火山岩中出现了以黄铁矿为主的硫化物，而热液脉的主要类型包括硬石膏-二氧化硅脉体、二氧化硅-硬石膏-磁铁矿脉体。其中，黄铜矿、磁黄铁矿和闪锌矿等比较少见，它们可以微量矿物的形式包裹在黄铁矿中。岩心中的热液硫化物也以黄铁矿为主，含有少量的黄铜矿、闪金矿、磁黄铁矿、白铁矿和方铅矿，也有自然金。

海洋极端微生物的酶和代谢产物也已成为重要的战略资源，并已经创造了巨量的财富。目前在国际上已经实现大规模产业化的极端酶有两种，一种是嗜热菌产生的*Taq* DNA聚合酶，它使DNA的体外复制变得异常简便和常规化，大大加快了生物工程、基因组等分子生物学研究的进程，年销售利润已经超过1亿美元；另一种是嗜碱菌产生的一种纤维素酶，其作为洗涤剂的添加剂，也已有数十亿美元的全球市场。此外，嗜盐古菌的质子泵紫膜蛋白也有望用于光学信息处理和光储存、生物芯片和生物计算机、全息照相和存储、仿视觉功能人工视网膜、图像传感器和运动探测等高科技产业领域。在国际上，不可生物降解的化学合成材料也逐渐被极端嗜盐菌产生的可降解材料取代。同时，来自极端微生物的生物催化剂也极大地拓展了催化的环境边界条件，生物催化有望突破高压、高温、低温、高盐碱或有机溶剂等工业过程屏障。

（二）深地资源与能源

除了油、气等可燃矿床，微生物参与了某些砂岩铀（U）矿、微细粒浸染状金矿和红层铜矿以及磷块岩等矿床的形成。微生物成矿作用在矿床学领域已经得到重视，但是，已报道的微生物成矿作用研究大多关注地表或近地表的微生物矿化现象，关于深地微生物成矿作用的专门研究还鲜有报道。

1. 微生物成因铀矿床

放射性元素U的表生地球化学循环主要受控于氧化-还原条件。U（Ⅵ）被还原为U（Ⅳ）时，会形成晶质铀矿、沥青铀矿等矿物沉淀，当规模达到开采水平时即可形成铀矿床。相反，在氧化条件下，U（Ⅳ）被氧化成U（Ⅵ），铀呈离子态而散失或形成铀酰矿物沉淀。自然界中存在的多种厌氧和好氧微生物可以控制铀的氧化还原状态，包括硫杆菌、硫酸盐还原菌、铁还原细菌、硝化细菌、反硝化细菌等。微生物可通过代谢性和非代谢性两种途径富集铀。

在沉积成因铀矿床中的微生物矿化现象多有报道。新疆十红滩和蒙其古尔两个典型砂岩型铀矿床中，微生物类群的分布特征严格受控于氧化还原条件，各带细菌的分布受矿层中有机碳含量、铁的含量和化学形态、溶解氧和硫酸盐含量等的制约，微生物之间也存在着共生、互生、竞争和拮抗的关系，从而影响着矿床的发育。

2. 硫化物次生富集带矿床

富含金属硫化物的岩石和矿石出露地表后会经地表氧化而形成铁帽。与此同时，有用金属和硫酸盐被部分释放至流体中，随着流体的下渗进入还原带，有用金属会以硫化物的形式再次沉淀。硫化物矿床次生富集作用可以成倍提高原生矿石的品位，从而大大提升经济价值。

硫化物次生富集带的形成包括地表矿体风化淋滤和地下深部硫化物的形成。已经证实，微生物在金属硫化物地表氧化过程中发挥着关键作用，使硫化物的风化速率比单纯的化学风化提高1～2个数量级。但是，微生物在地下深部硫化物形成过程中的作用并没有得到充分的认识。事实上，近年来的实验模拟发现，含金属的硫酸盐流体在地下深部经微生物还原作用即可形成硫化物，这类硫化物的硫同位素更轻。但是，相关的研究尚未起步，微生物对金属硫化物次生富集带形成的调控作用和相对贡献还没有得到系统研究。

除金属硫化物富集带之外，还有一些金属的次生富集成矿与微生物有着密切的联系。铁、锰在沉积物中具有相似的地球化学行为，在特定环境中会高度富集甚至形成矿床（如湖沼型铁矿），而微生物矿化被视为湖沼型铁矿的成因机制。

3. 生物成因天然气

生物成因天然气在全世界范围内具有广阔的勘探前景。生物成因天然气主要指烃源岩中以产甲烷菌为主的微生物分解沉积有机质而形成的天然气，主要由甲烷组成，在良好的圈闭和保存条件下可以形成天然气藏。

在沉积盆地里，产甲烷菌主要利用乙酸发酵和氢还原 CO_2 形成甲烷。在深度上，乙酸的微生物降解和 CO_2 还原可在浅部同时出现，但随埋深增加，CO_2 还原作用的贡献逐渐增大。生物成因气的形成是一个持续的过程，形成的甲烷以水溶态或游离态向上运移，运移方向主要受重力和压实体系控制，同时受岩石渗透率、气饱和砂岩的空间展布控制。随着压力下降，游离态甲烷所占比例上升，并聚集于各类构造圈闭中。在生物气藏的保存方面，地层水的流动十分重要，缓慢上涌的地下水可将深部形成的甲烷携带迁移至浅部，构造断层和岩石裂缝是主要的运移通道，所以在构造发育和地下水活跃的地区，生物成因甲烷气难以长期保存。因此，深地生物成因甲烷气从产生到运移再到保存均需要一定的地质条件。

（三）深时极端地质环境的资源与能源

深时极端地质环境微生物在部分重要金属矿产资源和油气资源的形成方面起着重要作用，如前寒武纪铁建造的形成、微生物碳酸盐岩和黑色页岩的沉积以及磷酸盐矿床的形成等。深入开展深时极端地质环境微生物的研究，对这些资源能源的勘探和开发具有重要的指导意义，因而具有重要的战略价值。

铁建造，包括条带状铁建造（banded iron formation，BIF）和颗粒状铁建造（granular iron formation，GIF），是早前寒武纪富 Fe^{2+} 海洋中的一种典型沉积型铁矿产，占国际已探明铁储量的 85% 以上。铁建造主要分布于太古宙和古元古代早期，以及短暂出现于新元古代成冰纪的间冰期，主要的矿物成分是赤铁矿、磁铁矿、黄铁矿和菱铁矿等。前寒武纪铁建造的成因有三种假说：①紫外线（UV）的光氧化作用（非生物成因说）；②产氧光合蓝细菌释放氧气的氧化作用；③非产氧光合铁氧化细菌的氧化作用。除第一种假说外，其他两种假说都与微生物的代谢作用密切相关。目前的认识表明，非产氧光合铁氧化细菌是早前寒武纪铁建造形成的主要原因。除此之外，还有其他类型的铁代谢微生物参与了铁建造形成过程中铁-碳的循环。例如，异化铁还原微生物功能群利用铁氧化物作为电子受体进行厌氧呼吸，将有机质和 Fe^{3+} 转换成 CO_2 和 Fe^{2+}，Fe^{2+} 与环境中的硫化氢（H_2S）、碳酸根离子（CO_3^{2-}）和三氧化二铁（Fe_2O_3）作用进而形成铁建造常见的黄铁矿、菱铁矿和磁铁矿等矿物。因而，查明早前寒武纪富 Fe^{2+} 海洋环境中铁代谢微生物功能群的分布特征对提高铁建造中的铁资源类型、分布和储量的认识具有重要的指导作用。

黑色页岩是指富含有机质的细粒沉积岩，在能源资源领域具有重要的应用价值。一方面，黑色页岩一般都是非常优质的烃源岩，是石油和天然气的重要来源。同时，自 21 世纪以来，页岩气在能源领域所占的比例显著升高，其主要的赋存层位就是黑色页岩。我国四川盆地中商业开采的页岩气主要就是来源于奥陶纪—志留纪之交的黑色页岩。另一方面，黑色页岩还是许多重要金属矿产的富集层位，包括常见的铜（Cu）、铅（Pb）、锌（Zn）、镍（Ni）、钼（Mo）和铀（U）等。例如，我国寒武系底部广泛发育的牛蹄塘组黑色页岩是 Ni 和 Mo 元素的富集层位。虽然目前对黑色页岩的成因机制还存在很多的争议，但已有研究表明，它广泛发育于海洋缺氧程度显著升高的极端地质环境。同时，在一些黑色页岩中检出的大量来自细菌和古菌的生物标志物和同位素记录表明，微生物在黑色页岩的形成中起着重要的作用（Kuypers et al.，2001）。深入理解黑色页岩形成过程中的极端地质环境微生物过程对其所赋存的能源资源的勘探开发亦具有重要意义。

微生物岩主要是由微生物作用所形成的一类岩石类型，在前寒武纪和显生宙重大地质转折期广泛分布。从地球生物学角度看，微生物岩是潜在的优质烃源岩。深入开展微生物岩形成和成岩过程中微生物地质过程的研究，明确微生物岩作为潜在烃源岩的条件对烃源岩的勘探具有重要意义。

磷块岩是指五氧化二磷（P_2O_5）含量 >18% 的含磷自生沉积岩，是重要的农业和化工原料。磷块岩主要产出于显生宙，在前寒武纪相对少见，但是在古元古代晚期（约 20 亿年前）和埃迪卡拉纪（约 6 亿年前）在全球范围内都广泛产出了海相磷块岩。例如，在我国南方埃迪卡拉系陡山沱组地层中，产出了很多超大型-大型磷矿床，其储量超过我国磷矿资源的 51%。前寒武纪两次主要的磷块岩沉积事件分别发生于古元古代大氧化事件以及新元古代氧化事件和“雪球地球”事件之后，指示大气-海洋氧化和深时极端环境变化事件对大规模磷成矿可能起到了控制作用。在磷块岩沉积过程中，除适宜的水体化学、物理条件外，微生物活动可能起到了关键作用。对现代海底成磷环境的研究表明，巨大硫细菌——纳米比亚嗜硫珠菌（*Thiomargarita namibiensis*）可在氧化条件下将聚磷酸盐储存于其细胞内，在还原条件下释放出溶解磷酸盐，从而导致磷的富集和磷酸盐的沉淀，而另外一些微生物也被发现可控制纳米级磷灰石颗粒的沉淀。因而，深入认识极端地质环境微生物磷代谢及其对海洋磷循环的控制作用对理解前寒武纪大规模磷成矿事件的成因机制，以及指导磷块资源开发都具有重要意义。

三、调控与响应气候环境变化

（一）海洋生物地球化学循环与气候变化

海洋约覆盖了地球表面积的 70%，是地球物质和能量循环的重要场所，而深部海洋更是岩石圈和水圈物质交换的关键地带。深部海洋存在大量形态各异的生态环境，包括海洋沉积物、洋壳、热液喷口和冷泉等，这些海洋极端地质环境构成了地球上生物，尤其是微生物最大的栖息地。海洋极端地质环境微生物的研究对人类认识其在地球环境、气候、生物地球化学循环中的作用均具有重要的科学意义。

海洋也是地球表层最大的碳库，其碳储量是大气的 50 倍、陆地的 20 倍；海洋碳库在地质时间尺度上的变动与大气大氧化、“雪球地球”等事件息息相关。由 CO_2 升高导致的全球变暖打破了地球各圈层之间的原有平衡，对包括气候带变化、生态系统变异等与人类社会密切相关的生态环境产生了影响。海洋每年从大气中净吸收大约 22 亿 t 的碳，自工业革命以来，已经吸收了大约 48% 由人类排放的 CO_2，因此，海洋是全球气候变化的“缓冲器”。海洋中的有机碳绝大部分是以 DOC 形式存在的，占总有机碳的 97%，“海洋微型生物碳泵”（microbial carbon pump，MCP）理论框架指出，微生物活动把溶解有机碳从活性有机碳转化为生物难以利用的惰性有机碳（refractory dissolved organic carbon，RDOC），从而使碳长期储存在海洋里。海洋 RDOC 碳库巨大（约 650 Gt），可与大气中 CO_2 总碳量相媲美。气候变化所致的海洋变暖、层化、酸化、脱氧、冰川消融、入海径流量变化、环流改变、干湿沉降等影响到微生物参与的 DOC 循环，从而改变 CO_2 的海气交换，对全球气候变化产生重要的反馈。海洋有机硫循环中，通过微生物转化释放的二甲基硫醚（dimethyl sulfide，DMS）是大气中生物硫的主要来源，其大气氧化产物形成云凝结核，也会对气候产生影响。全球人类活动产生的活性氮，由河流、大气输入海洋，导致海洋氮组分和浓度发生显著变化，微生物参与的海洋氮循环可调控生产力并影响海洋碳循环，造成缺氧、赤潮、温室气体释放等一系列生态与气候问题。海洋氧循环在生物进化中扮演着举足轻重的角色，它与生命大爆发和数次生物大灭绝密切相关。由于温室效应、富营养化、物理作用带来氧交换改变等因素，过去 60 年，全球海洋中的氧气含量下降 2%，贫氧海域面积扩大 4 倍。这将导致海洋生产力、生物多样性以及生物地球化学循环的重大变化，并且对渔业、生态系统和沿海经济产生负面影响。例如，开阔大洋中最小含氧带（oxygen minimum zones，OMZ）已经扩展了几百万平方千米，其中微生物的反硝化和厌氧氨氧化得到加强，导致海洋氮损失增加，这可能会改变海洋的氮储量，并最终改变生物量。因此，微生物不仅是海洋生物量的“巨人”，而且参与调控全球地表生物地球化学过程并对气候变化产生重要影响。

海洋和海洋极端环境孕育的各类生态系统与全球物质循环和气候密切相关。当代海洋的生态系统每年通过光合作用从大气中固定 450 亿 t 碳，其中 160 亿 t 输送至深海，海洋的初级生产者贡献了约一半的全球初级生产力。同时，由于海洋微生物的作用，大洋中存在约 6620 亿 t 的惰性有机碳，这一保留在深海的巨大的碳库近乎接近于大气中二氧化碳的碳库（Zhang et al., 2018）。生命起源于海洋，海洋生态系统中的生命活动也影响着海洋与地球的演化和物质循环。在地质时间尺度上，从海洋中的不产氧光合作用到海洋光合作用，再到陆地植物的演化，导致了地球和海洋的氧化。这一进程导致了铁、氧、硫等元素浓度在海洋中的巨大改变，同时也导致了元素在不同圈层中的分配。据估算，不产氧光合作用将进入生物圈的营养盐和碳量提高到 100 个数量级。除了改变元素的量级，海洋生态系统也作用于元素的比例。当代海洋中氮磷比值遵循 16∶1 的雷德菲尔德（Redfield）化学计量比值，这一比值形成的主要机制是海洋浮游植物按照一定的比例吸收海水中的营养物质，并通过深海的再矿化作用调节海水中氮磷等营养元素的比值。海洋生物圈也与地球生态系统的多圈层在不同的时间和空间维度相互作用。例如，冰期的大气沉降是当代的 10 倍，大气沉降携带的痕量营养元素促进了冰期海洋初级生产力的增加；伴随着洋流输运携带的沉积物，南大洋凯尔盖朗深海高原对海洋存在天然的铁“施肥”效应。并促进浮游植物的生长。工业革命以来，由人类活动导致的海洋层化、海洋酸化、海洋暖化、海洋缺氧、UV 辐射增强等环境变量的变化，都综合地作用于海洋生态系统并接受其反馈。

深部海洋作为岩石圈和海洋物质交换的界面，在地球物质循环中具有至关重要的作用。以碳元素为例，上层沉降和深海底部生物死亡之后掉落到底部沉积物中的有机质，在沉积环境中被进一步降解释放后进入成岩过程，并最终进入岩石圈和地球内部的元素循环中。至今虽然我们对深部生物圈的群落结构、代谢相互作用及其来源知之甚少，但是已有的研究显示，这些微生物的生命活动在近地表的元素循环中起到重要的作用，从而直接影响全球气候和地球表面的氧化还原状态。不同于利用太阳能进行光合作用维持的地表生物圈，深部自养微生物可通过化能合成作用将无机物转化为有机质，并作为初级生产者为消费者提供能量和碳源，形成一个“黑暗食物链”，是深海有机碳的固定和循环的重要组成部分。同时，深部生物圈中的异养微生物也可能以沉积物中的有机质为能量和物质来源，从而改变了沉积物中有机碳的含量和成分。例如，研究发现，深古菌（Bathyarchaeota）可以分解难降解的有机化合物（如木质素）以及利用无机碳自养合成乙酸，是海洋沉积物中碳循环和生态系统的核心驱动者之一（Yu et al.，2018）。根据释放能量的多少，沉积物中的微生物先后利用氧气、硝酸盐、铁锰氧化物、硫酸盐和碳酸盐作为氧化有机质过程中的电子受体。据估计，全球每年有 11.3 Tmol 的硫酸盐被还原，这一过程中被氧化的有机质占海底沉积物从上层水体获得有机质通量的 12%～29%。这一部分活性的有机质被矿化并再次回到活跃的碳循环中，降低了有机质在沉积物中的埋藏效率，从而对全球碳循环产生影响。除此之外，在沉积物更深处的硫酸盐甲烷转换带中，硫酸盐还原与甲烷厌氧氧化（AOM）耦合，这一过程将在深部沉积物中产生并向上溢出的甲烷气体转化为碳酸盐岩沉积下来。全球估算模型的结果显示，全球通过与硫酸盐还原耦合而被氧化的甲烷总量为每年 45～61 Tg，而这一过程 80% 都发生在水深浅于 200 m 的大陆架区域。这个过程作为一道有效的甲烷“屏障”，阻止了超过 80% 的甲烷扩散到水体中，对控制大气中甲烷气体的含量具有重要意义。

（二）地表极端环境微生物对气候和生态环境的响应与反馈

陆地热泉广泛分布于世界各地，彼此之间相似又相对孤立，地球化学特征具有多样性，是研究生态系统响应环境变化规律的绝佳场所。同时，高温环境往往限制了高等生命的存在，其生态系统成员主要由微生物组成，结构相对简单，有利于开展微生物生态学方面的研究工作。高温微生物还参与了 CO_2、CO、CH_4、一氧化二氮（N_2O）等温室气体的循环过程。开展陆地高温热泉环境下微生物多样性和功能对温度响应规律的研究，有利于我们预测全球气候变化背景下生态系统结构和功能变化的趋势。

微生物的膜脂化合物对高温环境变化具有重要的指示作用。随着外界环境的变化，微生物合成的膜脂化合物结构与组成也会发生变化，这种变化记录了环境因子如温度、湿度、pH、大气中 CO_2 浓度等的改变。微生物死亡后，膜脂化合物中的完整极性脂类（intact polar lipids，IPLs）可迅速转化成更为稳定的结构，称为核心脂类（core lipids，CLs），其成分的化学性质比较稳定，能够在漫长的地质时间内（可到上百万年）以“分子化石”的形式保存下来，因此常常被用于古环境的重建，而微生物活体中的膜脂化合物可指示当代的环境变化。因此，无论是重建古环境还是指示当代的环境变化，膜脂化合物均对理解高温环境生态系统如何响应环境变化，具有重要的科学意义。

低温环境微生物群落对全球变化的响应是一个非常复杂的过程。微生物可通过改变生态系统中有机质分解速率和温室气体的排放速率等直接响应全球气候变化，也可通过改变生态系统营养物质有效性及其转化等做出间接响应。北极和南极永久冻土带约有世界土壤有机物总量的25%。全球变暖可能会增加永久冻土中的微生物活性，使有机碳暴露于微生物的分解作用下，从湿地、湖泊、溪流和河流中释放的甲烷和二氧化碳将是永久冻土消融的重要后果，温室气体的产生加速消融的正反馈循环。作为地球第三极的青藏高原高海拔地区，是全球气候变化影响最为显著的地区之一。随着全球变暖，青藏高原冰川正经历着加速退缩和解冻。研究青藏高原冰冻圈中的微生物，不仅是研究全球气候变暖背景下微生物对气候环境变化响应的重要内容，也是极端环境下微生物生态研究的重点。

极端酸性环境越来越成为微生物生态学、生物进化和群落功能定量分析研究的理想目标和模式生态系统。这是因为，极端酸性环境不仅具有相对较低的物种丰富度、有限的资源多样性、明确的生态演替模式和紧密耦合的生物地球化学过程等特点，还具有在全球范围内均有分布、阶段再生能力较强、系统通常长期存在和可供多次采样的独到优势，从而具有较高的生态系统进化、功能和动态的可研究性和可比较性。此外，从低复杂度系统中推断出的生态原理可以预测更复杂系统中的生态过程与模式。研究生态系统如何响应环境变化所面临的最大挑战，是很难从整体代谢过程中梳理出单个有机体的相对贡献的。极端酸性环境作为低复杂度的系统，应用特定组学技术（如基于串联质谱标签技术的宏蛋白质组学方法）可以很好地定量外界环境改变对单个微生物种群的影响。因此，极端酸性环境也成为研究外界环境变化乃至全球气候变化对生态系统机能影响的重要模式生态系统。

干旱区微生物群落不仅是土壤中碳、氮、磷等营养元素生物地球化学循环的主要驱动力，还与成土过程及土壤的保水保肥能力有密切关系。此外，干旱区土壤微生物中存在较丰富的产抗菌药基因。干旱区土壤与岩石表面群落（soil- and rock-surface communities，SRSC）以多种方式对干旱区生态系统的功能做出贡献，可以产生区域和全球影响。这些方式包括增强矿物岩石的风化作用，从而促进当地土壤的形成，防止水和风的侵蚀，以及调节当地的水文和养分循环。因此，加强了解微生物对干旱区生态系统的作用机制、保护其生物多样性势在必行。

干旱区占全球陆地总面积的35%～40%，全球气候变化和人类活动导致其中95%的土地荒漠化。荒漠是干旱区的主要代表性生境，具有植被稀少、常年缺水等特点。研究表明，微生物是荒漠生物量的最大组成部分，且具有非常重要的生态学功能。同时，荒漠环境中还蕴藏着大量的新物种资源，尤其是稀有放线菌类群，可产生多种类型的新结构活性物质，是新型药物先导化合物和抗感染药物研发的新"源泉"。微生物多样性是制约干旱区荒漠化进程的重要因素。生物土壤结皮由蓝藻、地衣和苔藓表层组成，对土壤的稳定性和保水性非常重要，并通过碳氮循环提高土壤肥力。然而，生物土壤结皮很容易受到气候变化，尤其是降水和气候变暖模式变化的影响。这种变化可以改变生命周期、生物土壤结皮微生物的分布和功能，并且扰乱土壤的碳氮平衡。努力保持和恢复有沙漠化风险的干旱区土壤生物结皮十分必要，这不仅有利于已退化土壤的生态修复，同时也能够防止土壤的进一步沙漠化。

（三）深地微生物的地质过程及其环境效应

自从地球上出现微生物以来，地球环境与微生物就发生着交互作用并深刻影响物质循环、生命进化与环境演变。地球环境塑造了自然界丰富多彩的微生物，微生物又通过其新陈代谢活动对一系列地质过程发生作用，深刻影响着矿物岩石风化、土壤形成、元素循环、水质变化、资源开采与污染物治理等重大地质与环境过程。

第一，深地微生物决定了生态系统的演化方向。微生物占据了"生命之树"的绝大部分，在地球历史上占据了30多亿年的统治地位。地球早期产氧光合作用微生物的出现使得地球大气层由厌氧向富氧方向演化，导致了真核生物的出现，形成了现在我们所看到的丰富多彩的地球生态系统。

第二，自38亿年前地球上出现生命以来，生物一直深刻影响着矿物的形成与演化。地质历史时期的众多矿物、岩石、地层和矿床的成因都与生物活动有关。微生物驱动的风化作用甚至还与某些花岗岩和陆地形成有关。

第三，地质微生物改造了地球环境，使地球有别于任何其他星球。深地微生物直接参与岩石矿物的风化、土壤的形成、元素的迁移与转化等地质过程，在维持土壤生态、地下水水质等方面起着重要作用。微生物直接参与了C、N、S、P等重要生源要素的循环，是地球表层生态系统中最活跃的力量，也是地球关键带重要的营力之一。

第四，深地微生物通过温室气体的排放与消耗调控全球气候环境变化。“全球变化与陆地生态系统”（GCTE）、“过去的全球变化”（PAGES）、“陆地生态系统-大气过程集成”（iLEAPS）等，均是“国际地圈生物圈计划”（IGBP）中关注陆地-大气相互作用的国际计划，是当前人类活动与全球变化研究的重要内容，其中气候变化对陆地生态系统的影响及其反馈一直是这些计划研究的核心问题之一。大型沉积盆地深部蕴藏着重要碳库，深部生物圈活动是否会增加或减少碳排放，迫切需要准确评价深部碳的源汇效应，为制定低碳经济和低碳外交政策提供科学基础。

四、与深地、深海、深空计划相衔接而推动相关技术的发展

极端地质环境微生物的研究能够将科学前沿研究与国家重大工程计划相衔接，影响学科发展战略的布局，具有重要的战略价值。习近平在全国科技创新大会等多个重要场合对深地、深海、深空等涉及国家战略部署的重要举措及其重大意义做出了明确指示，我国充分利用中国特色社会主义能够集中力量办大事的优越性，布局各项重大项目与国家科技重大工程计划，这为科学前沿研究提供了难得的发展机遇。极端地质环境微生物研究与深地计划结合，通过探明微生物对深地环境的改造作用，了解一些特殊微生物的生态环境功能，为人类移居地下深部提供服务。极端地质环境微生物研究与深空计划结合，通过探索生命的极端适应能力而为拓展人类的生存空间提供服务。极端地质环境微生物研究与深海计划结合，能够解析关键微生物功能群对海洋环境与海底资源的影响，获得具有战略意义的种质资源。特别重要的是，极端地质环境微生物可以同深地、深海与深空计划进行全方位的衔接，以促进科学与工程的结合，在带动相关技术发展的同时，提升学科服务社会的整体能力。

（一）发展与深海计划相关的技术

在国际上，德国、法国、日本、美国、俄罗斯等海洋强国由于技术的发展而对深海环境的研究起步早，这些国家具备更强的深海勘探能力、装备技术优势。美国、日本等国家早在20世纪后期就获取到深海样品并分离出系列耐/嗜压微生物。截至21世纪20年代，国际上共分离培养和分类鉴定出百余种深海嗜高压微生物。在培养技术方面，国外研究者开创了极限稀释法、

近自然环境模拟法、凝胶微囊法等方法且将其广泛应用于深海微生物分离培养领域，并在营养条件模拟、微生物群体环境模拟、高通量筛选及大规模培养等方面取得了重要进展。我国深海与深部微生物研究起步晚，但自“海洋863计划”启动以来，我国的深海运载装备、深海勘探与取样、实验室装备对接、深海微生物研究等技术获得持续发展。同时，“蛟龙号”“深海勇士号”载人深潜器及多种无人深潜器在多海域的成功下潜作业，标志着我国在深海资源研发中拥有更高自主水平与发展能力，也带动了深海微生物资源开发全链条装备与技术的进步。在深海科学探索方面，我国研究人员对中国南海、马里亚纳海沟、雅浦海沟、印度洋、东太平洋、南太平洋、南大西洋、北极等海域深海环境进行了详细的研究，依托采样技术获得了大量原位样品，样品类型覆盖了深海沉积物、热液口、冷泉、多金属结核区、海沟深渊区等多种深海海域，并对其中的特征微生物完成了分离提纯培养。这些工作持续扩充着我国的深海菌种资源库，为新型优良菌种的发现、深海微生物生理生态功能的研究和新型深海活性物质的发掘奠定了基础。

深海及深部微生物资源研究和开发利用能力是一个国家综合国力的体现之一，这需要一系列技术的突破来带动。深海及深部微生物的特殊性（原位环境条件极端、生长代谢缓慢等），需要专业的研究团队的长期投入，构建特殊的技术方法体系，从而为上游的资源获取和下游的应用开发提供强大的理论支撑和方向指导。深部微生物资源研发具有成本高、周期长、技术链条环节多、高精设备依赖性强、科学专业性要求高等特点，其水平的提升依赖于海洋学、生物地球化学、海洋地质学、海洋工程和生物工程等多学科与技术的创新和发展，需要大量专业团队的支持与投入。“深海进入，深海探测，深海开发”三部曲战略的完成，更有赖于技术先行，从而提升我国深海及深部微生物研究水平、建立更完整的深部资源开发体系，对应对国际竞争具有战略意义。

（二）发展与深地计划相关的技术

深地环境（高温、高压、高放射性、极端pH、寡营养等）的微生物本身是一种重要的自然资源，菌体本身、基因或其代谢产物有重要的生物工程应用价值。深地极端微生物产生的多糖类物质及其物理化学性质非常独特。有些多糖在医药领域已经显示了广阔的应用潜力，可以增加非特异性免疫、促进骨骼修复和再生等。地下极端环境微生物产生的酶在高温和高压下，活性和稳定性增加，具有良好的离体转移性，提高了物质的传输和反应速率，有着良好的应用前景。地下极端环境微生物次生代谢产物的独特性成为抗肿瘤、抗病毒、降压降脂等新型特效药物的重要来源。这些种质资源的开发利用首先需要技术方法的突破。

深地环境微生物还在形成油气、金属矿产、冶金等方面具有广阔的应用前景，这也需要微生物利用技术方法的突破。油气是关系国家社会经济发展的重要战略资源。目前，我国正处在一个工业化加速发展的阶段，油气供需矛盾愈加突显，2010年石油对外依存度达53.7%，且在近期内难以改变。然而，随着我国地质勘探难度不断加大，提升油气储量与产量的难度进一步增大，国际化经营拓展空间又受多因素制约，油气资源保障能力与经济社会发展不协调的问题将长期存在。因此，增强国内油气资源保障能力仍是我国当前重要的战略决策，这就需要继续提高油气田采收率，发现更多、更大的油气田，以保障油气产量的稳步增长。具有清洁、耗能低、效率高等特点的微生物采油（MEOR）技术在西方国家已得到应用，在油田提高采收率方面发挥了重要的作用。由于我国油田具有渗透率低、原油黏稠等特征，MEOR技术一直没有得到推广。因此研究深部生物圈，认识深部微生物与油气、岩石和水的相互作用，对进一步提高微生物采油效率、推广该技术有重要意义。以微生物烃检测技术为核心的油气勘探开发技术，是MEOR技术的延伸，该技术是通过采集地表土壤或海底表层沉积物样品，利用地质微生物学方法检测样品中专属微生物来预测下伏油气藏存在、分布的新一代烃检测（油气预测）技术。该技术可以预测有效的含油气圈闭（烃类富集），是加快勘探节奏、提高勘探成功率的一项经济有效的新关键技术。另外，在已经探明的世界化石能源中，煤炭约占2/3，石油、天然气分别占1/6。在未来相当长时期内，煤炭作为重要化石能源的地位不会削弱。传统的煤炭开发利用存在许多弊端，燃烧取暖、发电等会产生许多对环境和生命有害的污染物。这些迫在眉睫的环境问题催生了一系列煤炭的清洁技术及新型利用技术，其中以微生物将煤转化成甲烷天然气这一技术最为醒目。用高效、经济、干净的微生物方法将褐煤转化成甲烷天然气，也属于深地微生物潜在的应用范畴。

除化石能源外，深地微生物还能直接参与金属元素的迁移、转化、富集与成矿，通过新陈代谢活动将岩石矿物有用的重金属淋滤出来，富集成有经济价值的金属矿床。一些大型的表生矿床，譬如太古代—元古代之交的条带状硅铁建造、砂金矿床、密西西比河谷型铅锌矿、海底沉积喷流型铅锌银矿以及铀、铜、铁矿等许多矿床的成因，可能都与微生物活动密切相关。从深地等极端环境中分离得到的菌种还能在低品位矿石的选冶中发挥重要作用，其中用微生物方法从尾矿和低品位岩石中回收稀土元素与贵金属具有较大的应用潜力，是一种高效、廉价的绿色环保技术。

深地微生物在形成金属矿床、提取贵重金属的同时，还能将岩石矿物中的有害重金属释放到环境中，从而造成环境污染。但某些微生物类群又能在重金属污染治理方面发挥重要作用。深地微生物通过呼吸作用，将地下水中的高价态、高溶解度、高活动性的重金属还原成低价态、低溶解度、低活动性的重金属。如果将这一技术与黏土矿物结合在一起，还能将还原以后的重金属包裹在低渗透性的矿物结构或集合体内，与外界隔离，降低二次污染的发生。因此，微生物在治理有机污染物方面也有着广泛的应用前景。

深地微生物还在近年来发展起来的深地医学中扮演了重要角色。世界人口增长、地球浅部资源竞争激烈，给人类可持续发展带来了前所未有的挑战，“向地下要资源、要空间”已经成为世界各国的基本战略共识，地下空间已被列为继太空空间、海洋空间之后的人类可开发的第三大空间，强化地下空间开发利用是21世纪前沿性的国际战略主张。地下空间的安全健康利用，逐步成为国家城市发展、民生效益的标化体现，必将开启国家主权战略的新时代。充分利用深地空间优势，可使其成为优于太空定居、应对未来地球灾害的最佳人类移居方式。地下空间开发首先要解决好地下空间生态学问题，也就是搞清地下生态系统的能源、空气、水、土、生物等生态要素的结构、功能、演化以及与地上部分的关系，实现可持续化发展。

地下空间的开发对深地微生物的影响以及深地微生物在地下空间生态学中的作用将是重要的研究内容。由此衍生出来的深地医学与深地微生物生态学关系密切。深地环境中低宇宙辐射、增重、恒温、幽闭、岩石、不同深度气压变化等环境因素必然对人体生理及心理产生相应的影响。深地医学是旨在研究地下不同深度赋存环境条件对生命体的生理、病理以及心理的影响及其机制，探索有害因素的应对策略，以及安全高效利用地下有益因素服务于人类社会活动的专门学科。深地医学主要利用天然洞穴、人工矿井等地下空间进行实验研究。在深地环境条件下，观察细胞生长、基因表达、DNA损伤的修复以及酶活性等影响。在这些实验研究中，深地微生物往往是最合适的材料。2010年6月，中国锦屏地下实验室的建成并投入使用标志着我国迈出了向深地科学探索的一步。2018年1月12日，以四川大学华西医院 / 临床医学院为依托、整合多个学院资源优势的四川大学深地医学中心正式成立，标志着我国在深地医学领域迈出了重要的一步。

（三）发展与深空计划相关的技术

深空探测是衡量世界各国综合国力和科技水平的一个重要标志，而且在深空探测中研发的新技术和新方法也带动了民用领域技术的发展和进步，该新技术与新方法具有重要的应用价值。随着以互联网为代表的第四次技术革命的兴起，新一轮的国际太空探索竞赛开启，并得以迅猛发展，不同于20世纪60～70年代苏联和美国单纯以工程、科技竞争为主的苏美太空竞赛，当前阶段的太空竞赛将焦点逐渐转移到重大科学问题的研究和解释上。NASA、ESA在天体生物学领域的研究开启较早，目前已具有较完备的教育和科研系统以及较清晰的远景规划，且在该领域具有深刻的国际交流合作。中国在天体生物学等行星科学研究领域与美欧等航天大国相比相距稍远，急需蓄力追赶，提高该领域研究的国际影响力。

自20世纪90年代以来我国开始实施载人航天工程和深空探测计划，已成功完成了5次月球探测任务。火星作为各国当前地外生命和宜居性探测的热点，也是我国深空探测的重点，我国已于2020年启动首次火星探测，后续将陆续开展火星取样返回、火星科研站建设等任务。我国首次火星探测任务“天问一号”已经于2020年7月23日成功发射，采用“绕”“落”“巡”探测方式，对火星进行联合探测。该探测任务将围绕五个科学目标展开：①火星形貌与地质构造特征；②火星表面土壤特征与水冰分布；③火星表面物质组成；④火星大气电离层及表面气候与环境特征；⑤火星物理场与内部结构。我国计划从研究火星表面土壤特征和水冰分布角度入手，利用环绕器搭载的火星矿物光谱分析仪、火星车搭载的次表层探测雷达，分别完成火星表面和地下水冰探测、火星土壤类型分布和结构探测、火星巡视区土壤结构（剖面）探测等，以此获取现今火星地质环境和内部结构信息，加深对其宜居性的认识。未来包括空间站建设以及木星及其卫星、近地小行星、主带彗星等在内的系列探测计划也将逐步实施。

深空探测已成为我国中长期发展的一项重要战略规划。寻找地外宜居环境和地外可能生命是我国深空探测中重要的基础科学问题，因此在现阶段加强和推进天体生物学的相关研究，特别是围绕我国已经规划探测的行星、卫星和小行星等开展，具有重要的现实意义和科学价值。此外，随着深空探测对相关人才的需求，在我国建设行星科学学科培养体系也迫在眉睫。天体生物学作为规划中行星科学一级学科中的一个重要方向，对行星科学学科培养体系具有重要的支撑作用。

第二章 发展规律与研究特点

第一节 极端地质环境的定义

极端地质环境微生物是指那些能适应并栖息在极端地质环境条件下的微生物。要理解极端地质环境微生物就需要理解“微生物”和“极端”这两个术语。微生物是借助显微镜才能看到或看得清的细菌、古菌和大部分单细胞真核生物和病毒等微小生物的总称。

这里比较重要的术语是“极端”。极端，是事物发展所达到的峰值点，是指一个事件偏离平均状态的严重程度。例如，极端气候是指气候的状态严重偏离平均态，在统计意义上属于不易发生的事件，也就是指一定地区在一定时间内出现的历史上罕见的气候事件，发生概率通常小于 5% 或 10%。通俗地讲，极端气候事件指的是 50 年一遇或 100 年一遇的小概率事件。世界气象组织规定，当某个（些）气候要素达到 25 年一遇时就称为极端气候。

极端地质环境主要是指那些已经或者正在发生地质作用（如风化、侵蚀、搬运、沉积、成岩、成矿、成藏等）的极端环境，不仅包括地质时期的极端环境，还涉及现代的极端环境。也就是说，那些正在发生地质作用的现代环境（如热泉、干旱环境、洞穴、深海底等）也是极端地质环境。除了极端地质环境，极端环境还包括极端生物体环境（如人体肠道）、极端物理环境（如闪电环境、核爆环境）、极端化学环境（如富含重金属的工业废水）等。因此，极端地质环境仅是众多极端环境类型中的一类。

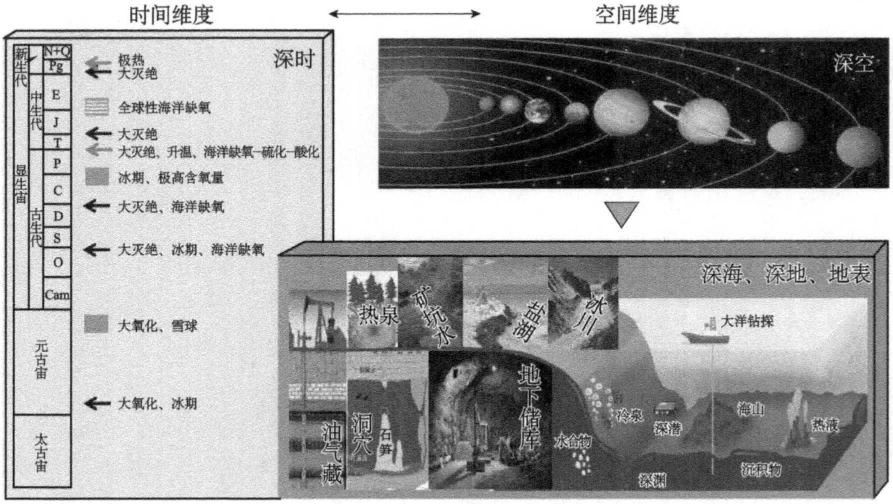

图 2-1　极端地质环境的时间和空间维度（文后附彩图）

资料来源：深空图来自 https://www.163.com/dy/article/DQSHR3B80525T0QI.html。

注：寒武纪（Cam）、奥陶纪（O）、志留纪（S）、泥盆纪（D）、石炭纪（C）、二叠纪（P）、三叠纪（T）、侏罗纪（J）、白垩纪（K）、古近纪（E）、新近纪（N）、第四纪（Q）。

对于极端地质环境，可以从时间和空间两个维度来理解（图 2-1）。对于当代的极端地质环境，可以从空间维度上去理解。也就是指偏离了当今地表环境总体状态的那些特殊环境条件。例如，地表环境总体是富氧的，那么缺氧环境就是一种极端地质环境。又如，地表环境总体处于中性 pH 条件，那么诸如酸性矿坑水、盐湖等 pH 很低或很高的环境都是极端地质环境。对应的极端地质环境有冰川及冻土、热泉 / 热液喷口、酸性矿坑水、碱湖、盐湖、深海、沙漠及地下深部等。这些极端地质环境都有一种或多种极端地质环境条件。这些极端地质环境条件对大多数高等复杂的生命形式往往是致命的，但对微生物可能会是最适宜的生活条件。

一、海洋极端环境

海洋极端环境包括如下一些类型（图 2-2）。

（一）寡营养盐海区

寡营养盐海区，通常是指营养物质（如氮、磷等生源要素）缺乏的海区（如热带的开阔大洋等），又称海洋荒漠，覆盖了约 40% 的海洋表面积，贡献了 30% 的海洋初级生产力，是表层海洋最大的生态系统。寡营养盐海区主要

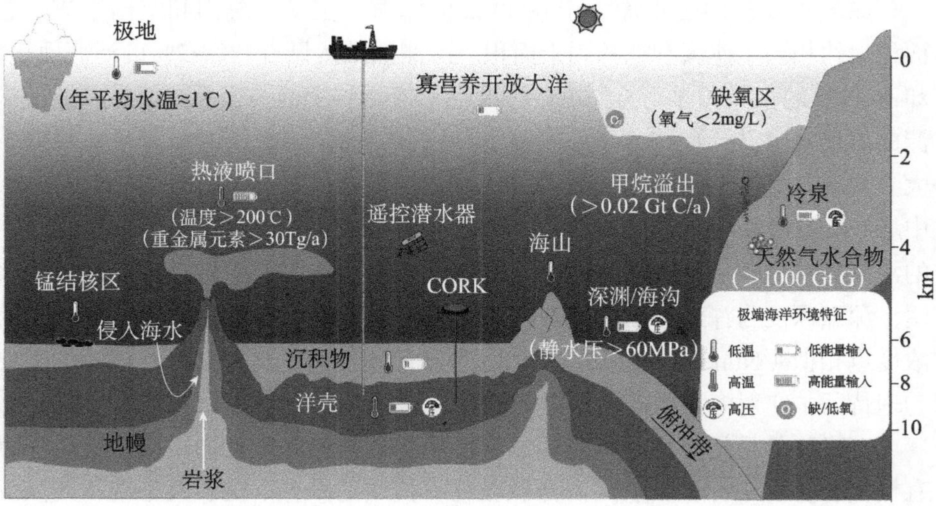

图 2-2　海洋极端环境示意图（文后附彩图）

位于副热带环流区，其重要特征是表层水体分层严重，阻碍了营养盐向真光层的补充，限制了浮游植物的生长。寡营养盐海区的营养元素有较快的周转速率，在寡营养盐海区，浮游动物消耗了约 80% 的初级生产，而在上升流海区，对初级生产的消耗大约为 66%。寡营养盐海区的新生产力主要受到大气沉降、垂直混合和生物固氮作用的影响。

（二）深海、深渊与海沟

深海一般指距海表 1000 m 以下的区域，是一个营养非常不均一的环境，总面积约为 3.2×10^{14} m^2，总体积为 1.3×10^{18} m^3，约占海洋总面积的 88%、海洋总体积的 75% 和全球生物圈的 62%。深海的温度通常为 −0.4～5.4 ℃，盐度为 34.3～35.1 ppt①，pH 为 7.5～8.0，溶解氧为 29～281 μmol/kg，深海中无机阴离子硝酸根（NO_3^-）、磷酸根（PO_4^{3-}）和硅酸根（SiO_3^{2-}）的浓度分别为 16～43 μmol/kg、1.0～3.2 μmol/kg、10～177 μmol/kg，压力范围为 10～60 MPa（Nagata et al., 2010）。深海具有永久黑暗、低温、高无机盐和寡营养等特点（Tamburini et al., 2013）。在这样一个极端的黑暗环境中，深海容纳着水生系统中最庞大的微生物类群，拥有全球海洋近乎 75% 的原核生物量（约 2.5×10^{29} 个细胞）和 50% 的原核生物产量，是生物圈中有机物的再矿化、碳的长期储存和埋藏的重要场所。

① 1 ppt = 1×10^{-12}。

深渊主要指的是水深大于6000 m的深海，主要由海沟和海槽组成，总面积大约相当于澳大利亚的国土面积。深渊区虽然只占1%～2%的海底面积，却占45%的全球海洋垂直深度，因此它是海洋科学的一个重要前沿领域。深渊的温度通常为1.0～2.5 ℃，盐度为34～35 ppt，与较浅的深海平原相差不大。现有的数据还表明，马里亚纳海沟的深渊区域（水深6500～11 000 m）中pH和溶解氧的范围分别为7.7～7.8和150～160 μmol/L，与6000 m以上的深海区域类似（Nunoura et al.，2015）。此外，来自马里亚纳海沟的数据表明，深渊深度与深海平原之间NO_3^-、亚硝酸根（NO_2^-）、PO_4^{3-}等无机阴离子的浓度也相似（Nunoura et al.，2015）。似乎静水压力是区分深海与浅海的主要物理特征，它对海沟中的生命影响很大。

有机物的可用性被认为是控制深海生物多样性和丰度的主要因素。深海有机质有多种来源，如表层海洋颗粒物有机质的垂直下沉、大型动物残骸沉降、陆源有机质和化能自养生物合成产物等。此外，海沟的特殊地形可能是导致有机质的另一个来源，即有机质通过斜坡的横向运输沉积到海沟底部（也称为海沟的聚集效应）。此外，诸如地震等构造活动可能将大量沉积物运输到海沟底部，这也塑造了深海系统中的有机质分布。在水文上，有报道称强的深水流，如绕极深层水和北太平洋深层水，流经西太平洋海沟。海底洋流、内部潮汐和振荡在几个海沟中都有记录。这些水文活动结合海沟的特殊地形，调节了深海的物理和化学条件。

深海与深渊都是高压环境。1949年，人们开始研究静水压对微生物活动的影响（Zobell and Johnson，1949）。随后，科学家逐渐发现，在深海等高静水压环境中存在着耐压微生物、兼性嗜压微生物和专性嗜压微生物等一系列高静水压环境下特殊的生命形式。耐压微生物是指可以在超过0.1 MPa（正常大气压）环境中生存，但是在高压条件下的生长速率比0.1 MPa时低的微生物。嗜压微生物则相反，即指最适生长压力超过0.1 MPa或需要高压条件维持正常生长（严格嗜压）的微生物（图2-3）。1995年，人们提出用piezophile来代替barophile作为嗜压菌的专属名词，piezo来自希腊语词根“压力”（Yayanos，1995）。嗜压菌是一种类群，而非某一特定的进化分支，广泛分布于各种微生物中。1979年，人们首次分离获得了一株纯培养的嗜压细菌，这种类似螺杆菌的菌株CNPT-3在50 MPa下可以快速生长，但是在常压下却数周都无法形成菌落，该细菌传代至少10次都未丢失嗜压特性（Yayanos et al.，1979）。随着一系列嗜压微生物的分离，高压下的生物活动形式等问题逐渐得到解答。

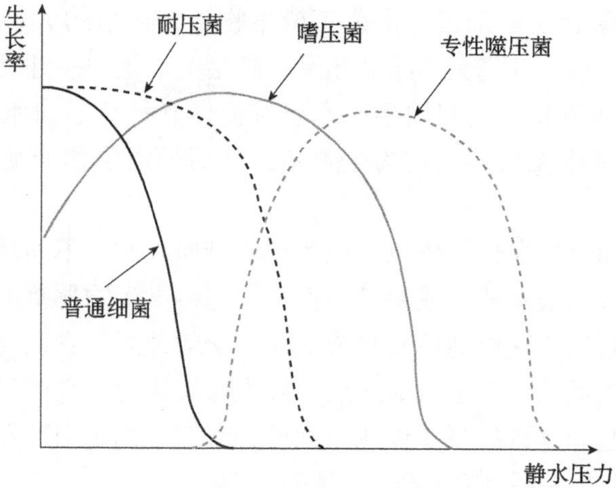

图 2-3 嗜压微生物

资料来源：Prieur D（1997）。

（三）热液区

1977 年，深海热液生态系统的发现改变了人们对深海生物学的看法。相较于其他深海地区黑暗、低温、缺氧的环境条件，热液区的环境条件更为苛刻，它还是一个温度变化急剧（4～400 ℃）、强酸（通常 pH 范围为 2.5～4）、强还原性的环境，其羽流中溶解了许多重金属元素（铝、铜、钴、铁、铅、镁、锌等）、含硫化合物等有毒化学物质。

截至 2021 年，已在全球海底发现 640 多处热液喷口，这些喷口集中分布在东太平洋隆起、北大西洋中脊、南大西洋中脊、西南太平洋的劳海盆、西南印度洋中脊、西北印度洋中脊、印度洋中脊三联点、西太平洋俯冲带等地（邵宗泽，2018）。海底热液由海水（2 ℃）沿着裂隙向下渗透到地壳的 1～4 km 深处，抵达地幔柱上方的岩浆房，岩浆房加热海水并与围岩发生水岩反应，生成的还原性物质与周边海水混合，沿裂隙上升至海底烟囱口喷出，形成的高温热液羽流温度高达 300～400 ℃。海底地幔柱、岩浆房驱动的热液循环，向大洋水体源源不断地输送着物质和能量，连通了地球深部与全球海洋。因此，深海热液喷口系统具有影响全球生物地球化学循环的巨大潜力。

尽管热液环境如此严苛，但这里的生态系统却极为活跃，孕育了多种多样生机勃勃的生命。其特征表现为动物种群的高密度性（生物量是附近深海环境中生物量的 10^3～10^4 倍，密度可达 50 kg/m^2）和物种的特殊性，其更为显著的特征是生长速率快，它们对热液的极端环境和毒性表现出了异乎寻常的适应性。除有大量细菌、古菌等微生物外，这里的大型生物种类也异常丰富，包括蠕虫、双壳类的蛤和贻贝、腹足类、蟹类、虾类、鱼等多种生物。化能自养细菌和古菌利用还原态热液流体和氧化态海水之间的化学不平衡构成的能量将无机碳固定为生物量，从而形成了热液喷口生态系统的基础。

针对共生和分布在热液喷口区域各种栖息地以及在不同环境下的全球热液系统进行研究的最新研究成果已经揭示了这些微生物群落的结构和相关生态功能，包括它们与底层地质和热液地球化学的关系，它们与动物的相互作用等。海底“黑烟囱”是最为活跃的海底生命聚集地，它的发现和探究为生命的海底热液起源学说提供了温床。此外，海底“黑烟囱”和微生物与大型动物群之间的共生关系因其新颖性而备受关注。

研究极端条件下的生命和栖息地（如扩散流区、海底含水层和热液羽流）对了解热液系统在全球元素循环中的作用具有重要意义。这些栖息地具有动态的、极端的和地理上隔离的性质，在物理和化学条件上存在着强烈的对比，表明热液喷口能够提供一个宝贵的窗口，让人们了解形成微生物群落的环境和生态力量，并由此深入了解微生物的极限、起源和演化。

尽管存在挑战，但在建模、原位生物地球化学和组学方法方面的最新进展为深入研究深海热液环境的生物提供了更完整的框架。尽管热液区的各种生境在物理和生态上是相互联系的，但它们的微生物群落却截然不同，这说明不同环境条件对微生物群落选择的重要性。深海热液喷口每个栖息地内的生物群落在很大程度上都是由流体化学和物理特性共同决定的。

（四）冷泉区与海洋沉积物

冷泉是由深部沉积物中甲烷、硫化氢及其他碳氢化合物流体向海底渗漏或者喷发而形成的独特环境。由于冷泉流体的主要成分是甲烷，冷泉也因此被称为甲烷渗漏，其渗漏的甲烷主要来源于深部沉积物中水合物的分解。冷泉主要分布在不同地质板块交界的大陆架边缘，目前已发现的冷泉主要分布于太平洋板块周围，是海底较为活跃的生态系统。

（五）海洋低/缺氧区

海洋低/缺氧区通常定义为水体溶解氧浓度＜2.0 mg/L 的区域。当水体中溶解氧的浓度＜2.0 mg/L 时，海洋中大部分水生生物将面临死亡，海底拖曳无法捕捉到鱼虾种群。广义上讲，一般在海水深度 500～1000 m，由于太阳光线不能到达，没有植物进行光合作用释放氧气，反因细菌和动物的呼吸作用及有机质的腐败分解而消耗氧气，致使该深度区域氧含量最少，称为最小含氧带。

（六）海山与多金属结合区

海底山（seamount）通常指从海底地面高耸但仍未突出海平面的山，简称海山。海洋学家定义海山的特征为在海底地面至少升高 1000 m。2011 年，全球海洋探测资料显示，全球共有 3 万多座海山，其中大部分位于太平洋地区。典型的海山由死火山形成，火山活动减弱或已静止的海山可作为地幔与海洋之间流体交换的通道，为深层或海底微生物群落提供营养物质。当海山达到一定高度后，对海流影响明显，从而影响到微生物等海洋生物栖息环境中营养物质和腐殖质的分布。

（七）洋壳岩石圈

海洋深部生物圈的研究取得的一个重要进展便是确定了海洋沉积物中栖息的微生物规模，即 2.9×10^{29} 个细胞，约为 4.1 Pg 碳，占地球生物碳总量的 0.6%。相比之下，对海洋洋壳岩石圈中生物量评估的难度要大得多，目前对海洋洋壳中微生物的规模还没有取得一致的认识。从岩石的角度来说，洋壳可以分成三个由不同岩石构成的圈层，分别包括最上部的玄武岩层，该圈层的厚度为 0.5～1.0 km；中部席状岩墙群复合体，以辉绿岩为主，主要分布于已有岩体的缝隙和裂缝中；下部的辉长岩层，平均厚度约 5 km。其中，上部玄武岩层通常发育一系列的孔隙，能够满足微生物生存所必需的空间需求，因而通常是洋壳微生物栖息和生命代谢过程的主要场所。

二、地表极端环境

（一）盐碱与干旱环境

盐碱环境通常包括（咸）盐湖泊和盐碱土壤两大类（图 2-4）。其中（咸）盐湖泊和淡水湖泊的区别是前者湖水的盐度大于 1 g/L。通常以海水盐度（约 35 g/L）作为咸湖和盐湖的截断值，盐度超过海水盐度的称为盐湖，而湖水盐度在 1 g/L 和 35 g/L 之间的则称为咸湖。盐湖的盐度范围一般为 35～400 g/L。全球范围内，（咸）盐湖泊主要分布于南北半球的干旱

图 2-4　常见陆地盐碱环境（文后附彩图）

注：（a）、（b）为盐碱土壤，（c）、（d）为盐湖泊。

带中，包括北非、北美以及亚洲的干旱地区以及南非、南美及澳大利亚等地的干旱地区；我国（咸）盐湖泊广泛分布，主要集中于北纬 30°～50° 的青藏高原、蒙新高原等干旱或半干旱地区。湖水盐度涵盖 1 g/L 到盐度饱和（300 g/L 以上）。盐碱土壤较为广泛的定义是盐度、碱度较高，影响植物正常生长的土壤，主要分布于干旱、半干旱的陆地环境。目前世界各个地区对盐碱土壤盐度、碱度上下限的界定尚未统一。经过初步总结，可以按照碱化程度将盐碱土壤分为盐土壤和盐碱土壤，前者盐度较高 [如含盐量＞0.5%～1% 或电导率（EC）＞4 mS/cm]，但碱度较低 [如碱化度（ESP）＜20%～30% 或 pH＜8.5～9]，后者的盐度和碱度均处于较高水平。盐碱土壤在世界各地均有分布，其中分布较为广泛的国家包括澳大利亚、俄罗斯、中国、印度尼西亚、巴基斯坦等，我国的盐碱土壤主要集中于北方的蒙新高原、河西走廊、松嫩平原等地区。

干燥指数（aridity index，AI）是指年平均降水量与潜在蒸发量的比值，常被用于定义干旱区。联合国环境规划署（UNEP）在 1992 年将干燥指数值在 0.05～0.65 的区域定义为干旱区，干旱区约占全球陆地面积的 39.7%；其中分为半湿润区（0.5～0.65）、半干旱区（0.2～0.5）、干旱区（0.05～0.2）和极端干旱区（＜0.05）。干旱区在地貌特征方面常以风沙地貌或荒漠地貌为主，形成沙漠、戈壁和雅丹地貌等代表性极端生境。沙漠广泛分布于世界各个地区，主要分布地区包括北非、北美、南美、澳大利亚、非洲南部、亚洲西南部和中亚干旱区，大部分处于副热带高压下沉气流控制的亚热带区域（北纬、南纬 15°～35°），以北非的撒哈拉沙漠和亚洲西南的阿拉伯沙漠最为著名。我国境内也有众多沙漠区域，主要集中在我国西北部新疆、青海等地，总面积为 130.8 万 km^2，约占我国国土面积的 13.6%。

（二）极端酸性环境

极端酸性环境通常具有极低的 pH（1～4），而且常常伴随着高重金属、高盐、高温等极端条件。常见的极端酸性环境分为天然酸性环境和人工酸性环境两大类。前者包括陆地或深海热液、热泉、洞穴、酸性硫酸盐土壤和酸性沼泽，以及自然暴露的矿床等；后者主要是极端酸性矿坑水（acid mine drainage，AMD）环境，这也是分布最广、研究最多，同时又与人类密切相关的极端酸性环境（图 2-5）。

图 2-5　各种类型的极端酸性环境（文后附彩图）

注：(a) 酸性矿坑水；(b) 深海热液；(c) 热泉；(d) 洞穴；(e) 酸性硫酸盐土壤；(f) 自然暴露的矿床。

（三）高温环境

高温环境是指温度超过人体舒适程度的环境，通常把 35 ℃以上的生活环境称为高温环境，在科学研究中重点关注 45 ℃以上的高温环境。最典型的自然高温环境主要来自地热。地热源可追溯到 46 亿年前的熔融状温度极高的原始地球，直至 44.5 亿年前才开始冷却形成地壳。目前的地表仍有大量的地热性自然高温环境存在，主要分布在板块生长、开裂的大洋扩张脊和板块碰撞、衰亡的消减带部位。环球性地热带主要有以下 4 个：①环太平洋地热带，是太平洋板块与美洲、欧亚以及印度板块发生碰撞的边界，主要分布在美国的阿拉斯加、加利福尼亚州，墨西哥，智利，新西兰，印度尼西亚，菲律宾，中国的东部沿海以及日本等地；②地中海—喜马拉雅地热带，是欧亚板块与非洲、印度板块的碰撞边界，主要分布在意大利和中国的云南、西藏等地；③大西洋中脊地热带，是大西洋海洋板块的开裂部位，该地热带上的国家和地区有冰岛以及葡属亚速尔群岛；④红海—亚丁湾—东非大裂谷地热带，主要分布在吉布提、肯尼亚、埃塞俄比亚等国家。中国位于环太平洋地热带与地中海—喜马拉雅地热带的交汇区域，处于全球构造的活跃地带，具有丰富的地热资源。

我国热泉系统主要分布在 4 个地热活动密集带：①滇西—川西—藏南密集带，该密集带的热泉数量占全国热泉总数的 33%，中国出露的沸泉、沸温泉、间歇喷泉和水热爆炸等高温热泉多集中分布于此区，记录到的最高温度可达 98～104 ℃；②东南沿海密集带，该区域主要包括东南沿海的闽、粤、琼三省，拥有占全国 21% 的热泉，记录到的最高温度可达 97 ℃；③胶东—辽东半岛密集带，该区域主要以低温热泉为主，仅有一处高于 80 ℃的热泉；④台湾及其邻近岛屿密集带，该区域有 103 处热泉，地热活动较为强烈，记录到的最高温度可达 120 ℃。

在热泉等高温环境中栖息着种类丰富的微生物，这些微生物在长期的进化过程中保留了喜好高温的特征，被称为高温菌（thermophiles）或嗜热微生物，一般认为高温菌生长温度在 45 ℃以上，是一类能在高温环境下生长并繁殖的极端微生物，被认为是地球上最古老的生命形式。根据其最适生长温度可划分为 3 类，即中度嗜热微生物（moderate thermophiles，45～70 ℃）、极端嗜热微生物（extreme thermophiles，＞70 ℃）、超级嗜热微生物（hyperthermophiles，＞80 ℃），它们不能在低于 50 ℃的温度下存活，主要是处于“生命之树”基部的部分细菌和古菌（图 2-6）。硫矿硫化叶菌（*Sulfolobus solfataricus*）和嗜酸热硫化菌（*S. acidocaldarius*）就是典型的超级嗜热菌。

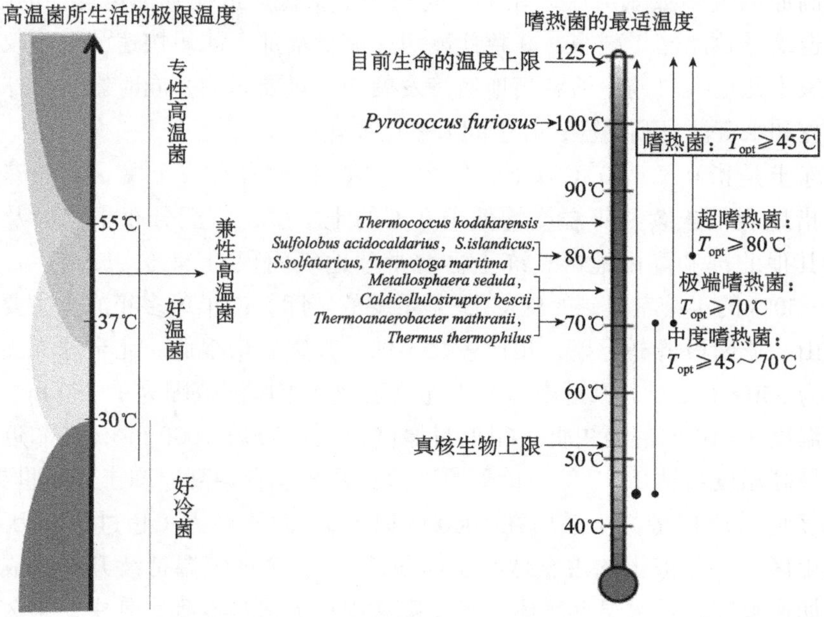

图 2-6　高温菌的生长温度（文后附彩图）

注：强烈火球菌（*Pyrococcus furiosus*）、柯达卡兰热球菌（*Thermococcus kodakarensis*）、海栖热袍菌（*Thermotoga maritima*）、热解纤维素菌（*Caldicellulosiruptor bescii*）、勤奋金属球菌（*Metallosphaera sedula*）、马氏热厌氧杆菌（*Thermoanaerobacter mahranii*）、嗜热栖热菌（*Thermus thermophilus*）。

（四）低温环境

当前的地球是一个寒冷的星球，大约 85% 的生境长久处于 5 ℃以下的低温中。这些极端低温生境的覆盖范围从南极到北极，从高山延伸到深海。地表的极端低温环境主要位于三极地区，包括南极地区、北极地区和被称为地球第三极的青藏高原及其周边地区，三极地区气温极低，养分贫乏，广泛分布着冰川（包括冰芯、冰雪和冰尘等）、冰川湖泊和冻土（季节冻土和多年冻土），其中冰川和冻土是面积最大的低温环境。

全球有冰川 168 311 条，总面积为 726 258 km²，冰储量为 113 915～191 879 Gt。南极地区有现代冰川 3274 条，总面积为 132 267 km²，冰储量为 27 224～43 772 Gt。北极地区分布有现代冰川 66 760 条，总面积为 434 143 km²，冰储量为 73 364～127 660 Gt。青藏高原是除极地之外最大的山地冰川区，冰川面积为 49 873 km²，冰储量约为 4561 km³。随着全球变暖，南极、北极和青藏高原的大多数冰川在近几十年间经历了从快速退缩到加速退缩的过程，冰川退缩导致冰储量减小，并致使冰川径流量增加。冰川快速消融不仅对周边国家十几亿人口的生活和当地经济发展产生重要影响，同时影响全球水循环过程和水资源可利用性，并引起海平面变化。

冻土是指 0 ℃或 0 ℃以下，并含有冰的各种岩石或土壤。永久冻土主要是指地表下连续两年温度均低于 0 ℃的土壤层，广泛分布于北极及北半球的其他北部和高山地区，冻土占北半球地表面积的 24%，却储存着全球 25%～50% 的土壤有机碳（SOC），是重要的碳库。南半球多年冻土主要分布在高山、亚南极群岛等地，相比较北半球，其碳储量很低。北半球冻土厚度范围为 350～650 m，靠近南部的不连续区域冻土厚度范围为 1～50 m。北极冻土温度（−10 ℃甚至更低）很低且厚度很大（500～1000 m），在靠近北极地区最高温度可达 1～2 ℃，最薄厚度只有几米。青藏高原冻土区是世界中、低纬度地带海拔最高（平均在 4000 m 以上）、面积最大（超过 100 万 km^2）的冻土区，多年冻土约占全球冻土面积的 6%。全球气温持续升高造成永久冻土加速解冻，导致温室气体二氧化碳和甲烷大量排放到大气中；永久冻土加速解冻释放出的二氧化碳和甲烷又使全球气温不断升高，导致积雪融化提前、冰川消融加速。

三、深地环境

深地环境主要是指大陆深部地下环境。深地环境有多种，有关生物圈的研究主要集中在地下水、油气藏、煤层、岩石与流体、洞穴等典型环境，这些生境以黑暗、低氧、寡营养为特征。与地球表面相比，油藏、煤田或矿井深部的温度通常较高，而洞穴中的温度一般与所在地的年平均温度接近，地下水温度较低且相对恒定。对深地生物圈的研究，最重要的是获取深地样品，不同生境需要采用不同的采样方法，地下水的样品往往用抽水泵就可以采集，油藏、煤层及岩石圈样品的采集主要依赖钻井与矿坑，而对于洞穴环境，可以直接进入采样或者进行原位实验。深地微生物功能与微生物所处的环境条件密切相关（图 2-7）。这些环境之间的微生物区别主要体现在碳源、电子供体、电子受体与能量代谢等方面。

（一）地下水环境

地下水环境中的微生物往往起源于地表，随着地下水的补给而进入含

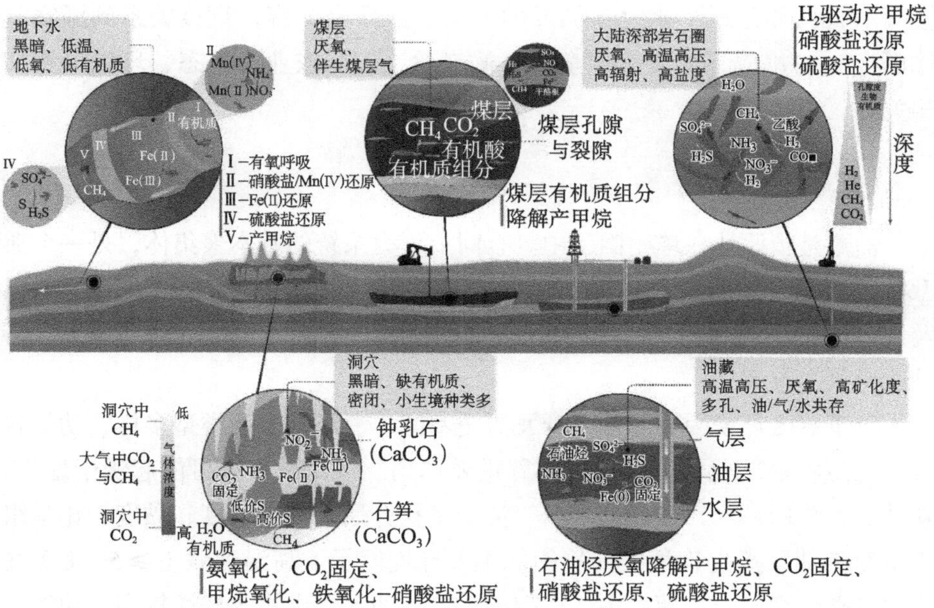

图 2-7　大陆典型生境和主要的生物地球化学过程综合示意图（文后附彩图）

注：硫酸根（SO_4^{2-}）、氨气（NH_3）、水（H_2O）、碳酸钙（$CaCO_3$）、氦（He）。

水层，并且随着地下水化学成分的演化而演替。地下水微生物主要以水岩反应产生的氢气为电子供体，其电子供体也包括少量从地表搬运至地下水的有机质。根据含水层的开放程度、溶解氧的高低与地下水年龄，电子受体可以是氧气、硝酸盐、Fe（Ⅲ）或者 Mn（Ⅳ）、硫酸盐与二氧化碳等。微生物活动能催化众多元素的地球化学元素循环，包括碳、氮、硫、铁、锰等。在厌氧条件下，还原菌可以把氧化态的元素还原成低价态；在有氧或厌氧条件下，氧化菌又把低价态的元素氧化成高价态，从而完成元素的地球化学循环。由于不同功能的微生物对电子供体的竞争，封闭含水层可以划分成几个不同的氧化还原带，从微氧到厌氧依次分成硝酸盐还原带、锰还原带、铁还原带、硫酸盐还原带、产甲烷带，每一个氧化还原带的类型可以依据反应物、产物类型与氢气的浓度来判断。地下水生境的一个典型特征是多元素的耦合，譬如甲烷的厌氧氧化与硫酸盐的还原，硝酸盐的还原与二价铁的氧化，甲烷的厌氧氧化与铁、锰的还原，甲烷的氧化与硝酸盐、亚硝酸盐的还原等。这一系列生物地球化学过程不仅会影响元素在地下水中的活动性与迁移、地下水的化学成分与水质，还会影响地下水的生态服务功能以及相关的地质过程，譬如矿物岩石的风化、元素的迁移转化及其生态毒理性等。碳循环是地下水元素循环的一个重点研究内容，直接关系到温室气体的产生与排放，从而影响全球气候变化，也与农业、生态、环境等密切相关。

（二）油藏环境

油藏是指原油在单一圈闭中具有同一压力系统的基本聚集体，是一个典型的集高温（约 180 ℃）、高压（数十兆帕）、高矿化度（约 20%）、厌氧以及油 / 气 / 水共存的多孔介质环境，这种特殊环境造就了地下深部一个独特的微生物生态系统。

由于油藏环境因子和油藏开发历史不同，油藏微生物类型多样、功能各异，在生物地球化学过程中发挥着重要作用。油藏微生物可能起源于地表，随着沉积物的埋藏，慢慢地寄居于油藏多孔介质之中。温度、黏度、化学组成是影响油藏微生物种类与功能的主要环境因子。高温油藏（≥50 ℃）往往含有嗜热细菌与古菌，多样性较低；低温油藏微生物多样性较高，功能丰富。由于油藏环境石油烃类物质丰富，因此烃类物质的厌氧生物降解是重要的生物地球化学过程之一。根据电子受体的不同，可以分成硝酸盐还原作用、硫酸盐还原作用、CO_2 生物固定等主要过程。

（三）煤层与煤层水环境

与油藏环境类似，煤层与煤层水的微生物可能也起源于地表环境，随着煤层的埋藏而进入地下环境。煤层也是一个富含有机质的环境，电子供体包括复杂的有机化合物，如苯甲氧基类化合物、甲基类化合物、乙酸、正构烷烃、类异戊二烯烷烃，以及简单的化合物，如 H_2、H_2S 和 CH_4。电子受体与地下水环境比较相似，包括硫酸盐、硝酸盐、CO_2、Fe（Ⅲ）与干酪根。厌氧发酵、产乙酸、产氢以及产甲烷等为主要生物地球化学过程。其中，产甲烷过程是低阶煤层典型的生物地球化学反应之一，也是低温煤层气的主要成因。生物成气是通过一系列的功能微生物，在厌氧条件下将煤炭中可利用的有机质组分解聚、转化，生成最终代谢产物甲烷的一系列复杂过程的叠加。在厌氧处理系统中，非产甲烷菌和产甲烷菌相互依赖，互为对方创造良好的环境条件，构成互生关系。同时，双方又互为制约，在厌氧生物处理系统中处于平衡状态。温度和 pH 可以很好地保持产甲烷菌活力，并且铁和镍元素的添加能促进浅层煤层气的生成。原位条件下注入营养液可以提高产气效率。

（四）洞穴环境

洞穴主要是由岩溶作用形成的一种喀斯特地貌，包括表生岩溶洞穴和深成岩溶洞穴，此外火山活动、冰川融化、海蚀作用等也能形成较大的地下空间，成为熔岩洞穴、冰洞和海蚀洞穴。洞穴环境普遍黑暗、相对封闭、缺少有机质，是一种寡营养的极端环境。表生岩溶洞穴中 CO_2 浓度高，CH_4 浓度低，而深成岩溶洞穴中则富含 CH_4、H_2S、NH_3 等还原性气体，小生境十分多样（图 2-7）。表生岩溶作用形成的洞穴中，微生物赖以生存的低浓度有机质主要来自风、流水 / 滴水或者蝙蝠粪便等外来有机质，有很大一部分微生物进行异养生活，生态系统的营养级较为简单；深成岩溶作用形成的洞穴生态系统与深海热液喷口的生态系统相似，以化能无机自养微生物为主，并孕育出原生生物、钩虾等真核生物，该生态系统是陆地深地环境中最复杂的生态系统。自养微生物利用 H_2S 等还原性物质进行生长，成为生态系统的初级生产者，并为真核微生物（如鞭毛虫、纤毛虫等）以及高等的生物（如钩虾等）提供了丰富的食物，养育了独特的生态系统。生物地球化学过程还包括硫的氧化、甲烷的产生与氧化、Fe（Ⅱ）的氧化、亚硝酸盐的还原、氮气（N_2）的固定、CO_2 的固定、氨的氧化等。这些过程相互耦合，往往产生自生碳酸钙（钟乳石和石笋等）和石膏等矿物。寡营养环境导致微生物之间的共生现象较为普遍。

（五）岩石与流体环境

基岩与地下流体环境通常具有厌氧、高温、高压、高辐射和高盐度等特征。在浅层地层，由地表带入的有机质作为微生物代谢的碳源与能量底物，但是随着深度的增加，有机质含量与其生物可利用性降低，因此异养微生物的丰度与活性也随之降低。在深部环境，岩石孔隙度减小，有机质含量低，微生物总量也随之降低，但是来自水-岩反应的 H_2 浓度升高，可为微生物的代谢提供能量。迄今，介导岩石圈金属异化还原和氧化过程的微生物也存在于地下深部，但是对其在原位的生长需求的报道甚少。

四、深时极端地质环境

对于地质时期的极端地质环境，可以从地质时间维度看，具体是指某个地质时期发生的环境事件严重偏离了当时的背景环境状态，这样，地质突变期对应的环境条件就是一些极端地质环境条件。例如，一些生物大灭绝时期的环境就是极端地质环境。又如，温度的大幅度升高或降低、大气-海洋氧含量的显著升高或降低、海水化学组成的显著变化等时期的环境均是极端地质环境。

需要注意的是，尽管前寒武纪的环境条件很恶劣，不适合人类或宏体真核生物生存，但不能把所有前寒武纪的环境都当作极端地质环境，只有前寒武纪的一些突变期环境才是极端地质环境。例如，第一次大氧化时期的环境相对于当时整体厌氧环境背景就是一个极端环境条件，“雪球地球”时期的环境相对于当时很温暖的地球来说也是一种极端地质环境条件。因此，对于地质时期的极端环境条件，需要从时间尺度去定义，即某个时期的环境条件偏离了当时的背景环境条件。也就是说，地质突变期的环境都可以归为极端地质环境（图 2-1）。

从这一层面理解，深时极端地质环境可以理解为地质历史上的重大地质转折期。在地球约 46 亿年的演化历史上，出现过非常多程度不同的重大地质转折期，包括但不限于：古元古代 / 新元古代大气-海洋自由氧含量显著升高的大氧化时期、缺氧富铁海洋向富硫海洋转换的时期、富氧海洋向富铁 / 硫海洋转换的时期、以新元古代“雪球地球”时期为典型代表的极端寒冷时期以及退冰过程的极端高温期、以早三叠世和古新世—始新世极热事件为代表的极端高温时期、海水 pH 显著降低时期以及显生宙后生动物大灭绝期等。

五、深空及临近空间环境

（一）深空环境

深空一般是指地球引力影响不到的区域，也是地球大气极限以外的空间，离地球的距离等于或大于 2×10^6 km。深空环境包括广袤的空间及分布于其中的各类天体所具有的环境。目前的研究范围主要局限在太阳系内。其中，与地外生命相关的研究，主要集中在火星及一些巨行星的冰卫星（土卫二、土卫六和木卫二）上。火星表面环境与地球有一定的相似度，但更加干旱和寒冷，大气稀薄且含氧量极低。巨行星的冰卫星通常被称为地外海洋世界，现有探测和研究认为，这些天体的表面覆盖有巨厚的冰层，而冰层之下有液态水存在，具备微生物形成和保存的条件。此外，位于小行星带内的 C 类小行星和彗星中，也存在大量氨基酸等构成生命的基本要素，是研究生命起源和在宇宙中分布的特殊样品，也是天体生物学研究的主要对象之一。

（二）临近空间

临近空间是海平面之上 20～100 km 高度的空间范围，包括平流层的大部分、整个中间层以及少部分热层（林巍，2020）。临近空间是地表环境与近地轨道空间的过渡区，是生物圈-大气圈-电离层的多圈层耦合区，也是保护人类和地表生物圈的一道重要屏障。临近空间的环境非常极端，具有高辐射、低温、干燥、低气压等复杂环境条件，可以类比现代火星的表面环境和早期地球的高辐射环境，是开展天体生物学研究的天然实验室。目前已经在临近空间 41 km 高度发现了微生物存在的确切证据，但也有报道称在 48～77 km 的高空可能有微生物，因此临近空间存在一个体积巨大的未知生物圈。目前在临近空间中发现的微生物均属于细菌和真菌，大多数类群可以形成孢子。

（三）天体生物学

深空及临近空间环境对生命来说也是一种极端环境。地外是否存在生命是一个开放性的话题，为了寻找太阳系内外可能存在的生命，国际上开发并使用了多种探测技术，并获得了初步成果。这就涉及天体生物学（astrobiology），也叫地外生物学（exobiology），它是一门涉及宇宙中生命的起源、早期演化、分布和未来的交叉学科。

第二节　极端地质环境微生物的内涵

在不同分学科领域，极端地质环境微生物的研究内容有不同的侧重点，但总体上可以归纳为以下四大主题的内容，分别是极端地质环境因子、微生物特征及其功能群、微生物对极端地质环境的适应和响应，以及微生物对极端地质环境的作用。具体内容介绍如下。

一、极端地质环境因子

极端地质环境有两种分类方案。一是按照单一的环境因子分类，如低温和高温环境、高压环境、酸性和碱性环境、缺氧环境、高盐环境、干旱环境、强辐射环境、富含重金属环境等。二是按照典型生境来划分，每种生境往往存在多种极端环境因子，体现了多种环境因子对微生物的综合影响。本书重点采用第二种方案，主要涉及深海极端环境（热泉、冷泉、深渊、缺氧区、海底沉积物），当代地表极端地质环境（陆地热泉、盐湖、冰川、酸性矿坑水、干旱地区等），深地环境（洞穴、油气藏、地下水、超深钻、矿坑等），深时极端地质环境（地质突变期的环境）以及深空环境。

重建重大地质转折期环境因子的变化类型和幅度是深时极端地质环境微生物研究中最基础的问题。只有明确了地质时期的各类环境条件参数，才能深入地研究当时微生物组成的影响因素以及微生物与环境相互作用的机理和过程。深时极端地质环境微生物研究中关注的环境因子主要包括温度、大气化学组成、古海洋氧化还原条件、古海洋水化学等。深时极端地质环境一般不是单一环境参数的变化，往往是多环境变量的协同变化，如海洋缺氧程度的加剧一般会伴随着海洋水化学条件的显著变化。

极端地质环境涉及各类环境因子具体数值的获取。现代的极端环境可以直接通过仪器的现场测量获得各类环境因子的具体数值。对地质时期的极端环境参数的获取，则需要采用各种地球化学、地球物理学和地球生物学方法通过代用指标对这些极端地质环境进行重建。例如，氧化还原敏感元素的富集程度可以反映局部海洋的氧化还原条件，磷灰石质生物壳体的氧同位素值、微生物的脂类分子等可以重建地质历史时期的地表环境温度，利用硼（B）同位素可以估算海水的 pH 等。然而，由于岩石样品都经历了长时间的成岩过程，原始的环境信息可能在不同程度上遭受了改变。在使用代用指标讨论环境变化时需要综合多方面的因素。

二、微生物特征及其功能群

微生物方面的研究主要涉及微生物的时空分布、丰度、多样性生态学和生物地球化学功能，以及随时间的演化。无论是现代的还是地质时期的微生物，最重要的是各类微生物功能群，特别是与碳、氮、硫、氧、铁等循环有关的微生物功能群。地质微生物主要通过代谢过程的功能群对地质环境起作用，因此，功能群可以直接将微生物的特定功能与环境变化紧密联系起来，也将微生物代谢与生物地球化学联系起来。现代极端环境和深时极端地质环境所涉及的微生物有所差异。

（一）现代微生物的起源、生物量、生存方式与代谢活性

现代极端环境主要涉及微生物的丰度、多样性、种类、生态学和生物地球化学功能。得益于分子钟的不断发展，微生物的地质演化研究也在不断

深入。自1998年以来，科学工作者采用多个数学模型对深地生物圈的量进行了估计，但是这个估计还一直在修订之中。据最新估计，深地生物量为 $0.7\times10^{30}\sim1.1\times10^{30}$ 个细胞，占全球生物量的3%～4.5%，其中大陆深部与海洋深部生物量大体相当。与地表环境相比，大陆深部微生物的多样性比较低，以变形菌门（Proteobacteria）与厚壁菌门（Firmicutes）为主。与地表环境中的微生物相比，深地微生物的代谢速率极其缓慢，至少要慢100倍。根据深地微生物的代谢速率和最新的模型预测，死亡微生物氨基酸在深地的平均周转时间达数千年之久，并且深地微生物营养细胞平均每数年到数十年分裂一次。深地微生物缓慢的代谢是适应深地低能量的结果。

有研究者曾使用幂函数将细胞计数与深度相关联，估算出在海底沉积物平均厚度为500 m范围内微生物的丰度为 8.7×10^{29} 个细胞（Parkes et al.，1994）。随后，人们统计了沉积物深至2000 m范围内微生物的丰度，得出的微生物丰度则要高一些，为 3.6×10^{30} 个细胞（Whitman et al.，1998）。有人将细胞脂类浓度与海底沉积物的总有机碳（TOC）进行关联，得出 5×10^{30} 个细胞的细胞丰度（Lipp et al.，2008）。如果将低生物量区添加到海底细胞计数数据库中，并考虑了沉积速率和距陆地的距离后，得出海底以下4000 m左右的海底沉积物的细胞丰度则减至 2.9×10^{29} 个细胞（Kallmeyer et al.，2012）。这与近年得出的数据（4×10^{29} 个细胞）较为接近（Bar-On et al.，2018）。在沉积物的特殊生态系统中，近岸、冷泉、热液等环境附近的沉积物往往拥有相对更高的微生物丰度。在垂直深度上，微生物丰度的分布大致遵循如下的经验公式，lg（细胞）=7.98−0.57 lg（深度），随沉积物深度增加，微生物密度呈对数线性减低，由表层1 m的 10^9 个细胞 /cm^3 降低到深度500 m的 10^6 个细胞 /cm^3（Parkes et al.，1994）。

深海微生物的呼吸速率［10^{-6} mol CO_2/（cm$^2\cdot$a）］要比上层海洋浮游植物的光合作用 CO_2 的消耗速率［10^{-3} mol CO_2/（cm$^2\cdot$a）］低1000倍。以硫酸盐还原过程为例，沉积物浅表［10^{-11} kJ/（细胞·a）］微生物代谢的吉布斯自由能也是深部微生物［10^{-15} kJ/（细胞·a）］的1000～10 000倍，显示深海深部生命代谢活性差异巨大。

现代微生物功能群可以借助分子生物学方法，从分子水平上直接通过功能基因来确定。借鉴现代微生物代谢组学新技术与环境微生物大数据是深化我们对地质微生物在地球系统以及重大地质历史事件中作用认识的一个重要环节。随着代谢科学技术的发展，突破地质微生物功能群及其地质过程的时机目前已经部分具备，研究对象包括与碳、氮、硫和铁这些重要元素的地球化学循环有关的微生物代谢等，研究内容涵盖相关特征、起源、地质识别标志及其地质作用等。近年来高通量基因组测序技术极大地推动了微生物学研究，特别是微生物组研究。通过采用建立微生物“群落构建单元”（community assembly unit，CAU）的概念，可将不同的微生物功能群整合到现代基因组大数据系统中，有利于克服每项研究对基因组数据采用不同的分析流程，但分析结果往往存在不可忽视的异质性，难以汇聚成大数据以获得鲁棒性结论的困难。只有通过对微生物或微生物组研究获得的基因组数据（基因型），与相关的生物学特征数据（表型、代谢产物）以及环境数据（包括早期地球环境）在统一的标准下存储和分析，才能实现数据集的有机整合，得到普适性的结论或推断。

例如，深海高压条件下不同生境中的微生物群落往往由数个不同种类的微生物相互作用，在代谢过程中形成特有的网络，并形成较为紧密的联系。这些微生物群落通常具有活跃的代谢能力，吸收并转化其生长所需的营养物质及不同的电子供体/受体，进而驱动着环境中碳、氮、硫元素循环。以甲烷丰富的冷泉生态系统为例，全球深海冷泉中通常是由以甲烷厌氧氧化古菌（anaerobic methane-oxidizing archaea，ANME）中的 ANME-2a/c 及硫酸盐还原细菌为特征的微生物组成，进行甲烷的厌氧氧化和硫酸盐的还原，同时释放二氧化碳及硫化氢。然而，在热液甲烷渗漏区，ANME-1 类群会取代 ANME-2a/c 形成新的特征微生物群落组成。此外，通过将深海冷泉和淡水生态系统和热泉生态系统比较，发现这两类生境中存在的电子受体 / 供体类型明显区别于深海环境，因此，其关键微生物群落组成已经由 ANME-1 或 ANME-2a/c 与硫酸盐还原菌，转变为 ANME-2d 与亚硝酸盐还原菌，以及韦斯特拉古菌门（*Ca.* Verstraetearchaeota）和深古菌门（*Ca.* Batharchaeota）古菌特征微生物群落，并伴随代谢网络的变化。随着研究的深入，全球高压生境的组学数据也在持续积累，为发现有类似的环境特征及关键微生物功能群落，以及其形成的代谢网络驱动环境中元素循环及生态系统演化现象提供了重要的数据基础。

（二）地下生物圈的能量来源、全球分布与生存边界

地表环境的能量主要来自光合作用产生的有机质，但是深地环境以化能自养型微生物为主。深地环境一系列地质过程包括水岩反应、构造活动等会产生氢气、甲烷等地质气体，深地微生物可以利用这些物质进行氧化还原反应来获取能量。热液水岩相互作用（如蛇纹石化）可合成有机物并迁移至地

下生境，为地下生态系统提供能量与碳源。另外，地下深部的岩石经历放射性衰变而产生氧化还原活性物质，从而进行氧化还原反应，释放能量，也为地下生态系统提供能量来源。为了进一步理解这些机制的重要性，深地微生物研究必须分析地球化学背景并了解微生物代谢途径，确定微生物生长及个体间相互作用所需要的最小能量。目前有许多尚未解决的问题，如地下微生物群落规模和碳与能量之间的不匹配性。也就是说，深地环境虽然能源极其匮乏，但是仍然有微生物活动。

深海环境的微生物代谢策略多样，种类丰富，规模庞大，近年来相关研究取得了重要进展。以沉积物生物圈为例，微生物细胞数量与海水中的细胞数量相当；对其分布范围的研究发现，在海底超过 2 km 下的沉积物中仍存在尽管代谢缓慢但仍然具有活性的微生物，其平均代增时间可能为几年到几十年，甚至百年至千年；在沉积物中微生物的类型方面，发现了与微生物数量相当的孢子，发现了沉积物中一些重要古菌，在生命演化和元素循环中发挥了重要作用（He et al.，2016）；H_2 和乙酸等小分子也成为深部生态系统重要的物质能量来源；发现了洋壳生态系统，可能在洋壳风化和元素循环中发挥了重要作用（Santelli et al.，2008；Li et al.，2020）；海底通过热液和冷泉等系统向上释放的碳量远超过以前的认知。

温度可能是决定生物圈生存边界的一个最重要因素。虽然已知微生物的最高生长温度在常压下为 122℃，但是在油藏环境中，往往温度在 85℃ 以上就没有微生物了。因此，深地微生物的最高承受温度可能由生物维持代谢所需要的能量与化学反应能提供的能量相对大小所决定。温度越高，氨基酸外消旋作用（racemization）的速率越快，微生物需要不断地合成新的蛋白质才能维持生命，所需要的能量也就越高，因此需要那些高能的化学反应才能维持这样的生态系统。在寡营养的环境，化学反应所产生的能量比较低，因此只有低温微生物才能生存，但是在富营养环境，孕育高温微生物的概率就会比较高。目前嗜热微生物（90℃）都来自富能量的热泉环境，因此深地微生物生存的最高温度可能随着环境条件的变化而变化。

深地微生物有着许多共性，如细菌以变形菌门和厚壁菌门为主，但是其中的原因尚不清楚。在深地环境由于岩石的不连通性，微生物扩散应该是比较困难的，也许地下类似的极端环境造就了类似的微生物，也许处于不同深地环境的微生物开始都来自同一个地表环境，因此它们具有相同的群落结构，尽管已经搬运到地下，但是它们进化速度极慢，因此保留了原先的物种。或者地下极端环境病毒非常活跃，能够杀死宿主释放 DNA，加速基因的重新组合，从而使得深地微生物有足够的机会发生水平基因转移（horizontal gene transfer，HGT）。深地微生物有可能在物种上类似，但是在功能上差别巨大，因此将来需要从功能上对深地微生物进行全球对比研究。

（三）代谢网络多样性

代谢网络不仅关系到微生物的功能和地质作用，还涉及微生物的演化，连接着现代与深时微生物之间的关系。要研究代谢网络的演化，首先需要了解不同微生物代谢网络组织方式（metabolic network organization）的多样性。这种多样性主要体现在以下三个层次。

（1）利用不同前体合成相同终产物，这些前体即代谢网络中的不同关键节点（key nodes）。关键节点的不同将影响整个代谢网络的构架，是代谢网络多样性中最重要的部分。以芳香族氨基酸合成为例，经典合成途径中使用赤藓糖-4-磷酸（erythrose 4-phosphate，E4P）和磷酸烯醇式丙酮酸（phosphoenol-pyruvate，PEP）作为合成芳香族氨基酸的前体，E4P 产生于磷酸戊糖途径，而 PEP 产生于糖酵解的后期步骤，因此经典芳香族氨基酸的合成与磷酸戊糖途径紧密关联；在一些产甲烷古菌中，E4P 被 L-天冬氨酸半醛（L-aspartate 4-semialdehyde，A4SA）和 6-脱氧-5-酮-果糖-1-磷酸盐（6-deoxy-5-keto-fructose-1-phosphate，DKFIP）代替。A4SA 来自天冬氨酸，是合成赖氨酸、苏氨酸、甘氨酸等多种氨基酸的重要前体，而果糖-1，6-二磷酸（fructose-1，6-dephosphate，F16BP）来自糖酵解的初期步骤，因此这一类芳香族氨基酸合成途径与经典途径和磷酸戊糖途径紧密相连不同，它与天冬氨酸的浓度密切相关，并与其他氨基酸的合成呈现竞争关系，并且其中另一个前体 DKF1P 由 F16BP 和果糖-1-磷酸（fructose-1-phosphate，F1P）合成而来，来源于糖酵解前期步骤，因此在能量代谢上也与经典途径中来自糖酵解后期步骤的 PEP 有所不同。然而目前就此两种途径在代谢网络中造成的差异仍不清楚，尚有待研究。

（2）从相同的前体出发合成相同的终产物，中间产物有不同的可选择性代谢途径（alternative pathway）。这些可选择性代谢途径因经过不同的中间产物而在一些环境理化因子下可能具有不同的热动力学偏好（thermodynamic favorability）。与此同时，不同途径之间存在不同的辅因子转换，尤其是与能量或还原力相关的辅因子，因此在不同途径下可能在能量和还原力消耗上有所差异。以赖氨酸的合成为例，其以草酰乙酸（oxaloacetate，OAA）为前体，目前已知有 6 种可选择的代谢途径合成赖氨酸，其中涉及不同的辅因子转化，如 ATP、NADPH、NADH、琥珀酰辅酶 A（succinyl-CoA）、乙酰辅酶 A（acetyl-CoA）等。研究者对这六种途径进行热动力学计算得出其中一条 M00031（LysW 酶参与的赖氨酸合成途径）在高温下具有热动力学偏好，而对嗜热菌的基因组分析也证实，这条代谢途径在嗜热菌中被广泛应用。

（3）从相同的前体出发经过相同的中间产物合成相同的终产物，但选择不同的辅因子。这类代谢差异往往不如前两者对代谢网络组织方式的影响大，但关键辅因子的差异可能与整个代谢网络中能量和还原力的平衡直接相关。如超嗜热古菌中特有的糖酵解途径（modified EM pathway）与经典糖酵解途径中的中间产物完全相同，但关键酶——磷酸果糖激酶（phosphofructokinase，PFK）利用二磷酸腺苷（ADP）生成一磷酸腺苷（AMP）供能与经典途径中利用 ATP 生成 ADP 供能不同，而另一个关键酶——铁氧还蛋白氧化还原酶（glyceraldehyde-3-phosphate:ferredoxin oxidoreductase，GAPOR）用铁氧化还原蛋白（ferredoxin）做辅因子而非经典途径中的 NAD 或 NADP。

对上述三个层次的代谢网络多样性进行研究时发现，不同研究方法具有不同的优劣势。例如，可以通过分析转录或表达差异来研究代谢网络中可能的变化趋势，但通过转录组或蛋白质组得出的结果并不直接反映代谢网络的变化。虽然可以通过测定碳源和能量源的吸收以及代谢产物（或中间产物）的产生来判断代谢流的走向，但多数代谢网络整体仍然是个黑箱而很难被定量解析。可以通过遗传操作敲除或插入关键代谢途径的酶来研究关闭或开启特定代谢途径时生物体的变化，但此法依赖成功率，耗时耗力，无法进行高通量测试，而且有时外源基因并不能成功表达，或者即便成功表达也不能很好地行使功能（在极端生物材料中尤为突出）。

代谢网络建模在研究代谢网络多样性与演化中具有较好的优势。它可以整合现有数据（单菌 ≥ 富集微生物群落 ≥ 原位微生物群落），包括组学数据、生理生化数据、代谢质谱数据等，帮助拆解代谢网络的黑箱。可以通过数学模拟进行实验无法进行或很难进行的分析，包括：①测定不同代谢途径下的能量及还原力消耗，以及对整个代谢网络的能量和氧化还原度的影响；②模拟敲除或插入特定代谢途径，分析碳流、能量流和还原力的变化，预测可能的生理现象，解读代谢网络的鲁棒性，指出特定代谢网络组成形式下相对薄弱的环节；③模拟分析不同代谢途径在特定环境因子下的偏好性和在整个代谢网络中的兼容性，从代谢网络的层面解读微生物的环境适应性。它还可以通过整合反应物的热动力学参数，进行不同热动力学条件下的模拟。

可应用于代谢网络建模研究的生物材料有两类。一类是模式菌株建模。

例如激烈火球菌已有30年的研究基础，有大量实验数据可用于对模型的校正，可以模式菌株为基础，模拟通过应用不同的前体、可替换的代谢途径和不同的辅因子会如何影响这类菌株的生理性能，以此获得对此类超嗜热微生物代谢网络的基本认知（节点偏好、热动力偏好、辅因子偏好，以及途径之间的协同与制约）。另一类是与模式菌株近源的、互相之间基因组差异不大的且在进化上有明显代谢途径差异的菌株类群，如火球菌（*Pyrococcus* sp., LMO），以模式菌株激烈火球菌作为模板，进行批量建模，并逐个分析代谢途径差异带来的优势和代价，与演化研究相联系，从代谢的层面解读微生物的环境适应。

（四）深时极端地质环境微生物特征

揭示重大地质转折期微生物组成的变化特征是深时极端地质环境微生物学研究的重要内容。由于个体微小、缺乏硬体骨骼，微生物很难以化石的形式保存下来。同时，即使有少部分微生物以化石的形式保存下来，但由于微生物形态简单，也很难从形态方面进行鉴定。因此，与地质历史时期动物和植物可以通过化石进行属种的分类学研究不同，深时极端地质环境微生物的研究很难开展属种等门类水平的分类学研究。

由于微生物与环境相互作用主要取决于微生物的生理代谢类型，如自养微生物通过其生理代谢将CO_2转换成有机碳进而作用于碳循环，而无论其是蓝细菌还是藻类，因此在无法开展微生物属种水平研究的情况下，地质微生物功能群是开展深时极端地质环境微生物研究的重要手段。目前的研究主要涉及典型功能微生物的种类、丰度、生态学和生物地球化学功能以及随时间的变化等内容。

常见的与C、N、S循环密切相关的微生物功能群是深时极端地质环境微生物研究的重要对象，如与N循环有关的主要微生物功能群有固氮微生物功能群、硝化微生物功能群、反硝化微生物功能群和厌氧氨氧化微生物功能群等；与S循环有关的主要微生物功能群有硫酸盐还原微生物功能群、非产氧光合硫氧化微生物功能群、中间价态硫的歧化代谢微生物功能群等。深时极端地质环境微生物研究的主要内容就是综合多种手段，揭示环境因子的巨变所伴随的微生物功能群的变化特征。

地质时期的微生物功能群不可能通过功能基因确定。深时极端地质环境微生物重点关注微生物代谢在沉积物和岩石中留下的一系列标志，如分子化石、微生物的各种地球化学指纹、微生物成因矿物、微生物沉积组构等。各种分子和同位素技术、微区分析技术等为研究这些宏观、微观标志提供了重要技术手段。

例如，在甲烷代谢微生物功能群方面，产甲烷古菌有特征性的类异戊二烯结构，能够产生古菌醇化合物。ANME-1 能够合成具有 0～3 个五元环的四醚脂类，ANME-2 的生物标志物为 2,6,11,15-四甲基十六烷，而 ANME-3 的生物标志物为 2,6,10,15,19-五甲基二十烷。I 型好氧甲烷氧化细菌含有 3-甲基藿烷类化合物以及 $C_{16:1}$ 单不饱和脂肪酸这些特征性化合物，而 II 型好氧甲烷氧化细菌含有较多的 $C_{18:1}$ 单不饱和脂肪酸。在氮代谢微生物功能群方面，固氮微生物往往具有 2-甲基藿烷类化合物，部分固氮蓝细菌的异形胞膜脂含有特征性的糖脂化合物。好氧氨氧化奇古菌的细胞膜为单层结构，区别于细菌的磷脂双分子层，细胞膜主要由甘油二烷基甘油四醚组成，主要为泉古菌醇（crenarchaeol）。厌氧氨氧化细菌的生物标志物为梯烷类化合物。在硫代谢微生物功能群方面，光合硫细菌能够形成带芳环的 2-烷基-1,3,4-三甲基苯系列化合物，而单烷基链甘油醚类可能是目前指示硫酸盐还原菌最好的化合物。

三、微生物对极端地质环境的适应和响应

（一）微生物对极端地质环境的适应

极端地质环境条件强烈地影响地质微生物，导致地质微生物通过各种方式来响应这些环境条件。微生物主要通过调节或改变细胞膜（孔道大小、脂肪酸组分、色素等）、细胞质（溶质组分构成）和基因（G+C 含量、抗逆基因、稳定性、特殊修复系统等）成分构成，以及表达特殊蛋白质（抗逆酶表达）等方式适应不同的极端环境条件。特别重要的是，极端地质环境微生物往往采用共生的方式生存，以减轻极端环境条件带来的生长压力，这导致许多生物地球化学过程出现耦合关系。

同时，微生物是通过信使分子感应这些外界环境条件并产生适应的。核苷类分子、ROS 等可以作为信使分子，感应到特定化学成分或环境因子后，通过合成酶或分解酶改变自身胞内浓度，调控下游相关代谢基因转录水平，改变胞内初级与次级代谢反应速率，使细胞实时响应化学及环境因子波动，使得微生物能够适应外界环境条件。为了加深对信使分子代谢调控功能的了解，以微生物功能群中关键类群为对象，研究信使分子种类特异性、合成与水解机制，以及对营养限制、DNA 损伤、温度等环境压力的相应调控机制，

并厘清信号分子的进化历程，从而也有助于微生物演化的研究。

（二）地质微生物功能群对重大地质环境变化的响应

生物对重大环境变化的响应在宏体生物层面更加直观。例如，显生宙出现的数次生物大灭绝事件就是宏体生物对环境变化响应的结果。在这方面已经开展了大量的工作，并取得了非常好的认识。

微生物对重大地质环境变化同样有很好的响应，而且可能更加敏感。微生物功能群对重大地质环境变化的响应主要表现在两个层面。一是微生物功能群类型的变化。例如，当环境从氧化变成缺氧状况时，由好氧代谢占主导的地质微生物功能群转变成以厌氧代谢占主导的地质微生物功能群。二是微生物的生化组分和代谢途径或代谢产物通量方面的变化，而微生物功能群的总体类型并未发生显著变化。例如，反硝化代谢的最后一个阶段是将 N_2O 在 N_2O 还原酶（NOS）的作用下还原成 N_2。该酶需要大量 Cu 作为辅因子，当环境中 Cu 浓度过低的时候，NOS 的活性会失去或明显减弱，进而影响反硝化微生物功能群的代谢。关于以上这两个层面的变化，第一个层面的变化更容易发现，第二个层面的变化则需要更加深入的工作，包括实验室的模拟、稳定同位素和脂类生物标志物等方面的综合研究。

四、微生物对极端地质环境的作用

微生物不是被动地适应极端地质环境，而是可以通过生物地球化学过程对地质环境产生很强的改造能力。例如，产氧光合微生物功能群释放的氧气改变了环境的氧化还原条件；硫酸盐还原微生物功能群代谢过程产生硫化氢进而形成缺氧的硫化环境；固氮微生物功能群将大气中惰性的氮气转换成生物可利用的活性氮，提高了其初级生产力，进而作用于碳循环。这些都说明微生物不仅仅是简单地响应环境的变化，更会对环境条件有很明显的改造作用。

微生物对地质环境的作用，主要通过生物地球化学循环来实现，包括与微生物代谢密切相关的碳、氮、硫、铁、氧的循环。这进一步体现了微生物功能群的重要性。微生物在极端地质环境的生物地球化学过程往往与其他环境存在明显的差异。特别是，由于极端地质环境微生物代谢速率极其缓慢，所以对它们的地质作用检测极其困难。目前的认知局限于实验室可培养的少数菌株，检测到的地质作用只是在室内理想条件下，缺少原位检测。将来需要在微生物群落水平上，通过追踪微生物对地质作用的长期累积效应，来反推微生物功能群在特定时期和特定过程对地质作用的反馈。

（一）深地和深海环境的元素循环

与浅部光合作用有关的有机质可以通过沉积和成岩作用直接保存在地下深部，成为地下深部生物圈的一种碳源。然而，这种复杂有机质难以利用，只能作为地下微生物非常有限的碳源。地下微生物可能通过固定溶解无机碳形式或甲烷氧化产物来获得碳源以供异养微生物利用。产甲烷菌、产乙酸菌分别耦合溶解无机碳（DIC）还原和氢氧化形成的 CH_4 或乙酸盐。无机成因的 CH_4 和烃类化合物也可以为嗜甲烷菌和异养菌提供碳源来维持地下生物圈的稳定。

在地球表层，自养微生物光合作用形成了初级生产力，这些有机质被异养微生物降解并矿化，异养微生物首先利用氧气作为电子受体。当沉积物表层以下几毫米到几厘米范围内的氧气消耗完后，微生物便利用其他电子受体将有机质在厌氧条件下继续氧化。其中，大部分的有机质在硫酸盐还原带被消耗掉。烃类的厌氧微生物代谢是地下生物圈有机质矿化的一个重要特征。在厌氧条件下，降解芳烃与烷烃的微生物主要与反硝化细菌、铁还原细菌、硫酸盐还原细菌以及产乙酸细菌和产甲烷古菌的共生体有关，降解过程可以在这些不同的氧化还原带发生。

人们对地下深部诸如氮等营养物质的来源了解得很少。这些营养物质的来源与生物地球化学循环密切相关。在氮循环涉及的所有作用中，发生在缺氧和厌氧条件下的主要是反硝化作用和氨的厌氧氧化作用，地下生物圈的氮循环因此主要涉及反硝化作用和氨的厌氧氧化作用。当然，近年发现，地下环境也出现了固氮作用、硝化作用等。因此，通过氮循环过程，微生物不仅为地下环境提供营养物质，还有助于改变地下环境条件。

硫循环主要涉及硫酸盐与硫化物之间的转化，以及部分有机硫化合的形成与分解。由于地下深部环境主要以缺氧环境为主，因此硫酸盐的微生物还原作用是地下深部的主要硫循环过程。近年来，硫循环的一个重大进展是发现了隐形硫循环（cryptic sulfur cycle）。这种硫循环主要出现在水体中（如最小含氧带），由微生物把硫酸盐还原成的 H_2S 很快又氧化成硫酸盐，从而监测不到 H_2S 的产生过程，故称之为隐形硫循环。但该循环最近也在沉积物深处（甲烷生成带）发现。而且，这种隐形的硫循环可能在沉积物深部普遍存在。这个过程很像水体最小含氧带的隐形硫循环，只是水体中产生的 H_2S 主要被别的氧化剂氧化成硫酸盐。

（二）深时极端地质环境的元素循环

深时极端地质环境元素循环的研究较为困难，往往包含生物因素和非生物因素方面的信息，很难定量化地将两者剥离开来。目前的工作主要是从与微生物代谢活动密切相关的 C、N、S、Fe 同位素角度开展的，因为不同的微生物代谢途径对这些元素的同位素的分馏效应有较为明显的差异。

地层中碳同位素组成特征很好地反映了深时极端地质环境碳循环方面的信息。例如，在奥陶纪—志留纪之交和晚泥盆世弗拉期—法门期之交，碳酸盐岩碳同位素组成都表现为正偏移，但在二叠纪—三叠纪之交和三叠纪—侏罗纪之交的碳同位素值出现负偏移，在前寒武纪的两次“雪球地球”时期也出现非常强烈的碳同位素值的负偏移。这些碳同位素值的正、负偏移都反映了大气碳库 ^{12}C 丰度的变化，但具体机制过程（有机碳埋藏、生产力变化、沉积物甲烷释放等）还有待深入研究。

地质时期氮循环过程往往与特定的环境条件和微生物作用有关。硝化代谢需要氧气的参与，一般发生在氧化环境中。在前寒武纪大气中的氧气含量增加之前，硝化过程未能发生。反硝化作用和氨的厌氧氧化作用主要发生在缺氧环境中，同时将环境中的生物可利用氮转换成惰性的 N_2。这些微生物过程都伴随着特征性的氮同位素分馏，因而，氮同位素记录可以较好地反映深时极端地质环境的氮循环信息。例如，在二叠纪—三叠纪之交，后生生物的主灭绝伴随着显著的 $\delta^{15}N_{org}$ 负漂移，而且主灭绝界线之上的 $\delta^{15}N_{org}$ 与大气氮气的同位素组成类似。这表明该时期海洋缺氧程度加剧，导致海洋中的生物可利用氮非常缺乏，微生物固氮作用成为当时生物可利用氮的主要来源。

硫循环很好地反映了海水硫酸盐和硫化氢等不同价态化合物之间的转换情况，这与硫代谢微生物的作用密切相关。石膏和重晶石，以及碳酸盐晶格硫酸盐（carbonate associated sulfate，CAS），可以很好地记录古海水硫酸盐的硫同位素组成。同时，沉积成因的黄铁矿可以很好地记录沉积—成岩期硫酸盐还原过程的微生物代谢特征。这两者综合起来可以很好地重建深时极端地质环境的硫循环特征。此外，海水硫酸盐和黄铁矿的硫同位素组成也可以反映古海水化学成分的变化特征。例如，在二叠纪—三叠纪之交 $\delta^{34}S_{CAS}$ 的快速变化揭示了当时海洋硫酸盐库的储量很小；黄铁矿的硫同位素、分子化石和铁组分则记录了晚太古代至寒武纪初期长达 22 亿年的前寒武纪硫化海洋的发育，以及显生宙一些极端地质环境时期的硫化海洋成分，这与细菌硫酸盐还原作用有关。

第三节　学科领域的发展规律和特点

不同极端地质环境微生物具有一些共性的发展规律，它们的发展都受到工程技术创新、工农业生产应用前景、国际重大科学计划等多方面的推动。

一、受工程技术创新的推动，体现了前沿科学与工程计划的结合

极端地质环境微生物的研究充分体现了科学与技术的结合。没有关键核心技术的创新，极端地质环境微生物的重大科学突破就无从谈起。实际上，人们开始关注极端地质环境微生物，也是从分子技术的突破开始的。与常规微生物研究不同，极端地质环境微生物的研究涉及一系列能够适合极端环境条件下开展微生物研究的特殊技术，包括保真取样、原位培养、微区分析、大数据构建等核心技术。如果说一系列的分子和同位素技术、微区分析和示踪技术是地球科学与生命科学经常用到的技术，那么极端环境的保真取样、极端环境的原位分析、极端条件的分子和同位素技术则需要特色的技术支撑。

例如，深部生物圈的提出主要源自极端地质环境微生物的研究，后者得益于大陆超深钻技术以及深海海底的原位观察技术的突破。又如，从高压环境分离的微生物只能在高压条件下才能很好地培养出来，只有高压培养技术取得突破，才能研究这些微生物在高压条件下的生物地球化学过程，才能理解它们对极端环境的适应能力及其地质作用。一方面，这些技术的突破为开展极端地质环境微生物的研究及其科学突破提供了可能；另一方面，极端地质环境微生物的研究需求推动了一些技术的革新。

（一）与深地、深海与深空计划的结合

极端地质环境微生物研究需要从野外现场取样到实验室分析等一整套相关技术的突破（图 2-8）。这些技术环环相扣，形成一整套的系列技术，而不只是一个环节的技术突破。例如，如果突破了野外保真取样技术，但实验室高压培养技术没有实现，那也很难得到真实的研究结果，因为许多高压环境的微生物只能在高压条件下才能培养出来。微生物在高压条件下的生物地球化学过程与常规环境明显不同，如对碳同位素的分馏明显有很大的差异。因此，极端地质环境微生物要取得重大突破，需要整建制的技术突破，因而需要与深地、深

海、深空等重大工程计划相衔接。近几年来，人们特别关注深渊微生物的研究，这主要得益于深渊保真取样技术和实验室高压培养技术的双重突破。“科学要突破，技术要先行”在极端地质环境微生物领域得到了最充分的体现。

图 2-8　极端地质环境微生物研究的三大技术体系（文后附彩图）

资料来源：http://news.sina.com.cn/c/2010-12-13/070321629707.shtml；https://baike.baidu.com/item/%E7%A7%91%E9%92%BB；https://5b0988e595225.cdn.sohucs.com/images/20170628/10fd2c9458644480afa8b3c53b2fc3f7.png；https://zhidao.baidu.com/question/1927524789956804747.html；http://s12.sinaimg.cn/orignal/4be5cbafte10f301a2e8b；http://www.hnhxnh.com/upload/201905/09/201905091725395352.jpg；http://www.ioz.ac.cn/xwzx/kyjz/201401/W020140116345537641504.jpg。

由于需要技术创新的推动，极端地质环境微生物的研究越来越需要与深地计划、深海计划和深空计划等国家重大工程计划结合。我国在深地、深海、深空等领域推出了一些涉及国家战略的重大工程计划，在国际上产生了重要影响，也为科学前沿研究提供了难得的发展机遇。同时，这些前沿科学研究极大地提升了国家重大工程的经济社会效应和文化效应。

事实上，极端地质环境微生物研究与深地计划结合，可以通过研究微生物对深地环境的改造作用而为地下人居环境工程提供服务。极端地质环境微生物研究与深空计划结合，可以为探索地外生命、研究生命适应极端环境条件提供素材，进而为拓展人类的生存空间和了解人类自身的过去和未来服务。对深时地质微生物的研究，则可以了解生命的起源与演化，探索早期生命对恶劣地球环境的改造作用等。

以上这些科学与工程的结合，需要人们转变科学研究范式。人们不仅需要实验室规模的模拟研究、野外现场的原位研究，更需要形成与重大工程衔接的大科学研究体系。这就决定了研究组织形式的变化。第一，需要相关的自然科学家参与一些重大工程的决策、布局和实施，以便为科学研究提供重要窗口。第二，需要工程技术专家不断与自然科学家进行交流，完善现场施工，以便更好地服务科学家，提升重大工程的多方面效应。

（二）受重大科学工程技术、保真取样技术和原位观测分析技术的推动

观测是海洋学研究的重要方法学，海洋工程技术与海洋科学的交互发展促进了海洋学的发展。人们对海洋痕量金属及其海洋学意义的认识过程就是一个典型的科学与技术交互进步的案例。2020 年是人们提出“铁假说”的 30 周年。在此之前，人们并不关注采样设备的专属设计和实验室分析的质量控制等方面对数据准确程度的重要性。国际上先后组织了从地球化学海洋断面研究计划（GEOSECS）到痕量元素及其同位素海洋生物地球化学研究计划（GEOTRACES）等多个国际合作项目，以促进工程技术的发展，使样品采集设备逐步机械自动化，室内的样品分析体积也从 10 L 降低至 1 mL。通过组织国际的方法互校，最终使痕量金属的样品采集和实验室分析逐步标准化，极大地促进了该学科的发展。在 GEOTRACES 项目之后，国际的科学家希望通过参照 GEOTRACES 的工程技术研发、国际合作和组间互校的方法，提出新的全球海洋尺度生态系统与痕量元素相互作用的项目（BioGeoSCAPE），以不断推动海洋科学的发展。

压力对微生物多样性（含功能群）的影响随水深的增加而加强（Tamburini et al.，2013），深海调查取样与实验室模拟培养装备的发展为高压下微生物功能群的研究提供了基础条件。1977 年，科学家在东太平洋中脊加拉帕戈斯裂谷首次发现了深海热液口，随后，美国载人潜水器“阿尔文号”首次在该地区观察到了深海热液口繁荣的生态系统，并采集了生物样品。科学家从深海热液口中发现了大量的嗜热细菌和古菌，深海热液口的发现同样也为研究生命起源和进化提供了一个窗口。“阿尔文号”在样品采集方面的贡献大大推进了科学家对深海热液口的认识，使得深海热液口成为 20 世纪后期的研究热点，并取得了一系列成果。我国研制的“蛟龙号”载人潜水器在 2014 年 12 月 18 日首次赴印度洋下潜。1968 年开始的深海钻探计划（Deep Sea Drilling Program，DSDP，1968～1983 年）及其后续的大洋

钻探计划（ODP，1985～2003年）、综合大洋钻探计划（IODP，2003～2013年）和当前的国际大洋发现计划（IODP，2013～2023年）是地球科学领域迄今规模最大、影响最深、历时最久的大型国际合作研究计划，也是引领当代国际深海探索的科技平台。半个世纪以来，科学大洋钻探在全球各大洋钻井3600余口、累积取芯超过40余万m，所取得的科学成果验证了板块构造理论，揭示了气候演变的规律，发现海底"深部生物圈"和"可燃冰"，形成了地球科学一次又一次的重大突破，这些成果的取得同样也是基于深海取样技术的发展。目前，已知的嗜压菌均来自海洋环境，其中仅有热自养甲烷球菌（*Methanococcus thermolithotrophicus*）DSM2095一株菌分离自浅海环境，其余绝大部分菌株来自水深超过4000 m的深海区域，深海采样技术直接决定了能否获得生物样品以进行高压环境下的微生物研究，深海微生物研究的发展也直接受深海取样技术的推动。

深地微生物研究受到深地钻探与掘进技术、样品保真取样技术、原位观测和模拟培养技术等的推动。深地生物圈研究最重要的是获取深地样品，而钻井取样是最方便、直接的方法之一，利用已有坑道直接采样则是获取深地生物圈样品的一条重要途径。全封闭式的保真采样器可以确保在采样过程中维持原有的极端和无菌环境，不至于导致微生物及其生活的环境条件发生变化。原位观测则可以通过检测与微生物代谢活动相关的指标来推测深地微生物的活动，模拟培养技术可以利用钻井处的地层流体进行微生物的模拟培养。

深空环境的遥感探测技术、原位探测技术、地外样品的研究技术等都推动了天体生物学的研究。例如，人们第一次看到地球之外的流水作用遗迹是"水手6号"（Mariner 6）传回的火星照片。又如，"火星快车"（Mars Express）轨道器首次原位探测到了很少量的甲烷，美国NASA的"好奇号"还在盖尔撞击坑内探测到大气中甲烷含量随季节发生变化的现象。此外，围绕类太阳恒星运动的系外行星的发现获得了2019年诺贝尔物理学奖。

（三）受微生物培养和检测技术的驱动

显微镜、代谢产物、第一代测序技术、高通量测序、同位素标定等微生物分析与检测技术极大地推动了极端地质环境微生物的研究。

1. 现代微生物

这里以嗜（耐）高温和低温微生物为例。1897年，Bradley Moore Davis

报道称，在美国黄石国家公园 85 ℃热泉中存在生命。6 年之后，生命的温度上限推至 89 ℃（Setchell et al.，1903）。20 世纪 60 年代，从美国黄石国家公园发现的水生栖热菌（*Thermus aquaticus*），可以说是掀起了高温微生物研究的热潮（Brock et al.，1969）。1976 年，从水生栖热菌中分离获得 *Taq* DNA 聚合酶，该酶的高温特性使其很快代替聚合酶链反应（PCR）技术中之前使用的来自大肠杆菌的 DNA 聚合酶，进而使 PCR 技术自动化（Chien et al.，1976）。*Taq* 酶是目前公认的最具通用性和一致性的 DNA 聚合酶，广泛应用于生物技术相关领域，产生了巨大的商业价值。随后，人们发现了越来越多的高温微生物，可供使用的高温酶也越来越丰富。高温微生物还蕴藏着新颖的代谢途径，譬如目前已报道的 6 种生物固碳途径中，3-羟基丙酸 /4-羟基丁酸循环和二羧酸盐 /4-羟基丁酸循环两种固碳途径是在热泉极端嗜热古菌——硫化叶菌（*Sulfolobales* sp.）中首先发现并证实的，而 3-羟基丙酸双循环是在高温细菌橙色绿弯菌（*Chloroflexus aurantiacus*）中被首先发现的。之后随着分子生态学技术的发展进步，研究人员在陆地热泉高温环境中探测到了大量未培养的微生物类群，被喻为"微生物暗物质"。新近快速发展的宏基因组技术为人们探究这些未培养类群的代谢潜能提供了可能性，但基于纯培养的验证仍然是关键。

微生物分析和检测技术的发展也大大地促进了低温环境微生物的研究。以往主要依靠传统的分离培养方法研究可培养微生物的形态和生理特征，近年来，分子杂交、高通量测序等免培养方法的发展，加快了在分子水平上对低温环境微生物的基因组信息和生态功能的研究进度，使对低温环境微生物的研究越来越深入，极端微生物资源的挖掘越来越全面。对极端低温环境中古老生命形式的探索，一方面，针对某一特定低温环境中微生物的研究，从有利于微生物群落长期保存的环境中提取并分离出优势微生物的纯培养物，探索古老生命信息；另一方面，是以空间代替时间，分析不同年代序列的低温环境样品的微生物，发掘新基因，研究微生物多样性，揭示过去环境的变迁以及生物进化规律，如可以通过研究冰芯中的微生物特征来反映古气候的特征。微生物是冰雪和冻土环境中的重要生命体，近年来，关于全球变化导致低温环境微生物群落结构及其所发挥的生态功能改变的研究，既有通过对微生物与环境因素之间关系的探索，来预测未来全球变化下微生物对环境的响应，又有以古 DNA 作为重要信息载体，诠释微生物进化和迁移过程及它们对历史环境和气候变化的响应。

2. 地质微生物

对肉眼可见的生物大化石研究的快速发展离不开分析测试技术的提高，如扫描电子显微镜、透射电子显微技术、同步辐射技术、激光拉曼光谱技术等，这些分析测试技术极大地提高了古生物大化石的微细结构、内部结构和化学组成方面的分析。同样地，深时极端地质环境微生物学的发展也受到了分析测试技术提高的推动。一方面，一系列与分子生物学研究相关的分析技术的提高，如高通量测序技术等，极大地推动了演化生物学的发展，估算了重要微生物功能群的进化时间；另一方面，一系列测试技术的进步，极大地提高了地质样品分析的精度和灵敏度。例如，气相色谱-质谱技术和液相色谱-质谱联用技术的发展，推动了从脂类生物标志物角度开展深时极端地质环境微生物的研究；气体同位素质谱技术和多通道接收电感耦合等离子体（MC-ICP）质谱技术的发展，从 C、N、S 等传统稳定同位素和非传统稳定同位素角度促进了深时极端地质环境微生物的发展；二次离子质谱（SIMS）、纳米-二次离子质谱（Nano-SIMS）以及飞行时间-二次离子质谱（TOF-SIMS）实现了原位同位素、化学组成和脂类生物标志物的原位分析，更进一步促进了深时极端地质环境微生物学的发展。

3. 模拟培养技术

模拟培养技术的发展同样对微生物研究起到了巨大的推动作用。深海微生物原位研究的技术难度大、易受环境波动影响、定量不精确等缺点使得科学家对模拟培养产生了强烈的需求。模拟培养更有利于对实验变量的控制、实验的重复，从而得到更准确、更可靠的结果，同时，便捷的模拟培养装置也更有利于深海嗜压微生物的分离工作。深海微生物的模拟培养技术从早期的简易式低温静置高压釜逐步更新迭代，发展到现在的操作更简便，安全性更高，覆盖 0.1～200 MPa、0～200 ℃的静置压力培养装置和 0.1～40 MPa、0～150 ℃气液混合高压连续流动培养系统，能够更好地实现对深海环境的模拟，更有利于深海微生物科研工作的开展。

二、受工农业生产应用前景的推动，体现了基础研究与应用服务的结合

极端环境微生物与常规微生物存在很大的差异，它因要适应特殊环境而形成了很多特殊的功能基因、重要功能的酶和特殊的蛋白质。这决定了从极

端环境分离出来的微生物是一类极其重要的生物种质资源，具有广阔的应用前景。实际上，从极端地质环境分离的微生物已经在环保、冶金、能源、电子、材料、医药、食品等各类工业上得到了很广泛的应用，很好地体现了极端地质环境微生物的科学前沿研究成果的知识转移问题，其广泛的工业应用促使人们从自然界分离出越来越多具有不同功能的微生物。

同时，一些重要微生物的特殊功能是在工业生产应用时才发现的，从而促进了这些微生物功能基因和生态学功能的研究。例如，著名的氨的厌氧氧化现象就是在工业应用的生物处理系统中发现的，随后从环境中分离出具有这种功能的微生物，并查明了它们的生态学功能和生物地球化学功能，以及在自然界中识别它们的标志。

（一）受工农业生产与生活等方面应用的驱动

由于极端地质环境微生物具有特殊的生理生态特征，譬如嗜（耐）热、嗜（耐）冷、嗜（耐）酸、嗜（耐）碱、嗜（耐）压、嗜（耐）盐等，因此这些微生物本身或其丰富的基因和代谢产物都有重要的应用价值，在工业、农业、食品、医药、环保、冶金等领域有着重要的潜在应用前景。例如，嗜酸微生物可以应用于生物冶金和 AMD 环境修复，嗜盐微生物可以应用于腌制食品，耐干旱微生物可以应用于干旱环境绿化、微生物方法培育耐旱植物，高温微生物可应用于湿法冶金，低温微生物可应用于食品领域，洞穴微生物可提供抗菌药、用于太空环境育种等。

1. 嗜（耐）酸微生物

早期的 AMD 微生物学研究主要基于利用嗜酸微生物进行湿法冶金以及 AMD 污染防治两方面的动机。大量研究主要关注嗜酸微生物催化的硫化物矿物氧化和 AMD 生成过程。20 世纪 90 年代中期以前，AMD 微生物的研究主要依赖于传统的培养方法。由于极易从这些酸性环境中分离出嗜酸氧化亚铁硫杆菌（*Acidithiobacillus ferrooxidans*）以及氧化亚铁钩端螺旋菌（*Leptospirillum ferrooxidans*），它们成为研究微生物催化 AMD 生成过程的主要模式生物，并由此建立了关于其能量代谢及其对硫铁矿氧化溶解贡献的多种重要模型。

自 1998 年以来，相关领域的研究更多是以 AMD 作为模式系统 / 群落进行微生物生态学的理论探索和验证研究。加州大学伯克利分校的 Jill Banfield 小组对此做出了重要贡献。*Acidithiobacillus ferrooxidans* 尽管被普遍认为在

AMD 的形成过程中扮演着关键角色，但其在美国里士满（Richmond）矿内的产酸过程中只起着有限的作用（Schrenk et al.，1998）。后来发现了一种新的铁氧化极端嗜酸古菌（*Ferroplasma acidarmanus*，Fer1），并确定该物种是 AMD 的重要贡献者（Edwards et al.，2000）。2004 年，人们首次用宏基因组方法，从 AMD 生物膜中重建了包括Ⅱ-类钩端螺旋菌（*Leptospirillum* group Ⅱ）和Ⅱ-型热原体属古菌（*Ferroplasma* type Ⅱ）在内的 5 个基因组，确定了这些微生物种群的遗传多样性，揭示了这些微生物在 AMD 系统生物地球化学过程中扮演的关键角色（Tyson et al.，2004），被《科学》（*Science*）评为 2004 年十大科技突破之一。人们还利用宏蛋白质组方法确定了遗传重组是 AMD 生物膜中微生物适应特殊生态位的关键驱动力，利用宏基因组结合群体遗传学方法阐明了 AMD 生物膜中重要微生物物种的种群遗传多样性结构，并揭示了潜在的微进化过程。

2. 嗜（耐）干旱微生物

干旱区营养贫乏、盐碱化严重、植被稀少等，致使干旱区形成了特殊的微生物类群。土地干旱和荒漠化问题不仅限制了人类的生产生活，也对整个地球的生态系统造成了威胁。研究发现，干旱区蕴藏着丰富的微生物资源，其多样性和群落结构能够反映地区的环境变化。此外，微生物的抗逆基因是宝贵的基因库资源，在恢复干旱区生态系统中具有重要的研究价值。

近年来，随着高通量测序技术和培养组学的快速发展，干旱区微生物生态学研究得以广泛开展，研究区域涵盖了全球主要干旱区。干旱区微生物的分离和培养对未来该领域的研究工作具有极其重要的作用。此外，利用微生态系统重建技术，人为地进行生态干预，可明显缩短微生态重建时间，迅速提高干旱土壤肥力，改善干旱区的微生态环境，增加地区生物多样性，进而实现干旱区土壤的恢复，防止土地沙化。

尽管干旱区微生物丰富多样，但功能微生物类群在其中仅占很小一部分。在不同的干旱土壤中，参与碳固定、分解和氮代谢的微生物基因各不相同，这表明了干旱区微生物在功能上的差异。了解微生物参与干旱区生物地球化学的过程，对预测全球环境持续变化的生态后果和更好地理解生物地球化学循环机制至关重要。

沙漠作为极端环境的典型代表，一直是科学研究的热点，其中不乏具有重要科学意义的发现。例如，通过对智利北部的阿塔卡玛（Atacama）沙漠样品进行宏基因组测序发现，该环境中存在着大量的未可培养细菌，并且推测这些微生物还具有重要的生态学功能（Idris et al.，2017）。蒙古沙漠中的未可培养细菌可占总体细菌群落的69%（Lu et al.，2009）。这些研究从某种程度上改变了人们以往对沙漠贫瘠荒凉的印象。以往沙漠微生物的研究主要集中在微生物群落多样性调查及其生态修复功能研究方面，而在菌株资源培养方法学和应用开发方面的研究都相对薄弱。

沙漠生境蕴藏着极为丰富的放线菌资源。随着高通量测序技术的快速发展，从环境基因组中直接获得微生物生命“暗物质”的遗传信息，为人们开辟了一条认识微生物生命“暗物质”的新途径。研究发现，甚至在被称为“微生物生命的干旱极限区”的阿塔卡玛沙漠，也有大量放线菌门的物种分布，在低海拔和高海拔干旱区丰度可达75%和98%，在部分核心极端干旱区占比常在94%以上。该地区占主导地位的放线菌类群为酸微菌科（Acidimicrobiaceae）、地嗜皮菌科（Geodermatophilaceae）、微杆菌科（Microbacteriaceae）、微球菌科（Micrococcaceae）和小单孢菌科（Micromonosporaceae）等（Connon et al.，2007）。人们在腾格里沙漠和巴丹吉林沙漠的微生物多样性研究中发现，其放线菌类群相对丰度分别占到了35%和29%，且蕴藏着大量稀有放线菌类群，如地嗜皮菌科、红色杆菌属（*Rubrobacter*）、诺卡氏菌属（*Nocardia*）和类诺卡氏菌属（*Nocardioides*）等（Sun et al.，2018）。此外，沙漠中存在大量的放线菌未分类单元亟待人们去挖掘和开发利用。这些研究表明，相比于普通土壤或其他极端生境，沙漠蕴藏着更为丰富的放线菌资源，是新放线菌资源的可靠来源。

可培养的极端生境放线菌资源是放线菌系统学和应用研究的基石。通过免培养技术，研究者可以发现沙漠环境中存在的许多未知类群，并基于宏基因组信息进行功能注释和进化分析。但如果没有纯的、可传代培养的菌株，其分类地位无法得到确认，其蕴含的独特代谢途径及新结构天然产物也无法得到有效开发。因此，分离培养是了解和利用沙漠微生物的关键步骤。

国内学者对沙漠微生物的可培养研究已有初步探索。人们从塔克拉玛干沙漠土壤中分离出了13个属的27株细菌，优势属为链霉菌属（*Streptomyces*）、考克氏菌属（*Kocuria*）和芽孢杆菌属。从内蒙古自治区鄂尔多斯南部的毛乌素沙地的生物结皮中分离得到13株菌，优势属为链霉菌属。整体而言，从沙漠样品分离培养放线菌的技术手段相对滞后，仍停留在对普通环境放线菌分离培养方法的模仿阶段，沙漠放线菌难分离、难培养的问题依然严峻，亟待开发建立针对沙漠样品的放线菌高效分离培养方法体系。

3. 嗜（耐）盐微生物

盐湖沉积是特定自然地理和地质环境的产物，它对气候和环境变化有灵敏的记录。盐湖在各个发展阶段详尽地保存着周围环境变化的信息，包括一般湖沼相所缺乏的成盐阶段的信息，因而具有湖泊环境变化记录的全息性。盐湖和盐碱地对于一般动植物来说，是一种极端环境，由于其含盐度较高，一般的生物不可在其中生存。因此，盐湖、盐碱地通常给人的印象似乎是死寂的、没有生物活动的水域。其实不然，在这类高盐环境中，还有少数生物种群与其种群的多数成员不同，能够适应高盐环境而得以生长、繁衍。

根据国际粮农组织不完全统计，目前全球有盐碱土壤 9.5×10^8 hm^2。中国的盐碱土壤面积达到了 9.9×10^7 hm^2，且有不断扩张的趋势（如耕地盐碱化等）。随着人口增长及粮食需求增加，合理开发、改良盐碱土壤，使其恢复耕作能力变得尤为重要。在恢复盐碱土壤肥力的过程中，微生物发挥着重要的作用，合理利用相关耐盐 / 嗜盐微生物有利于水土保持、增加土壤有机质以及提高植物耐盐性（Jiang et al.，2019）。微生物本身的群落结构也会在修复过程中发生变化，因此，盐碱土壤环境中的耐盐 / 嗜盐微生物研究具有显著的生态指导意义，也一直是土壤微生物学的研究热点。除此之外，部分盐碱土壤环境同时具有高盐、干旱及强辐射等特性［如柴达木盆地的大浪滩干盐滩及美国俄克拉何马州的大盐湖平原（Great Salt Plains）国家公园］，这与现代火星地表环境类似，因此在探究火星生命形式的研究中往往被当成理想的类比模型。除了环境、农业及地外生命等学科的飞速发展，微生物培养、测序技术及组学研究的不断革新也极大地促进了盐碱土壤环境中耐盐 / 嗜盐微生物的研究工作。

（二）受能源资源开发、环境保护、气候变化应对等实际需要的推动

随着经济的发展与社会的进步，人类对油气、水资源的过度开发利用，带来了一系列的环境污染问题。人类活动对生物圈产生了干扰，但也为研究人类活动与生物圈的相互作用提供了机遇。极端地质环境微生物在推动地下水污染治理、核废料处置、油气资源开采、新生物制剂开发、气候变化应对等方面均有重要的应用。

1. 极端地质环境微生物与环境治理和气候应对

极端地质环境微生物在环境治理方面有广阔的应用前景，其发展深受环境治理这一实际需求的推动。早在20世纪80年代初期，世界范围内地下含水层受到不同程度的污染，功能丰富的微生物则成为优先的治理工具。与化学与物理方法相比，微生物方法能使污染物在原位降解，并且没有副产物的产生，同时适用于有机和无机污染物的治理。近几年出现的新型污染物［如砷（As）、汞、氟、全氟化合物等］，对污染地下水的生物治理提出了更高的要求。

极端地质环境微生物也对人类活动产生了负向反馈。一个典型例子是核废料的处置。核能的开发产生了许多核废料，这些废料往往储存在地下岩层，但是深地微生物会影响核废料储存库的可靠性和安全性。例如，金属的厌氧腐蚀会释放氢气，从而增加储存库的压力，危及其结构完整性；氢气能为许多微生物提供能量和电子，增加其活性，从而对核废料储存库的长期稳定性产生影响。这些影响应该是多方面的，包括参与核素的氧化还原反应、络合作用，对存储容器的风化、腐蚀作用，对周围环境介质孔隙度、渗透率的改造作用等。

将超临界态的CO_2注入地下环境是降低温室气体浓度的一种有效途径，但是目前对捕获的CO_2在深地环境的归宿还认识不足。对CO_2如何进入地层，如何与矿物、有机物及微生物相互作用尚不清楚。CO_2在不同时间尺度如何影响深部碳循环及生物圈，科学家对此还知之甚少。这些问题的提出强有力地推动了深地微生物的发展。

2. 极端地质环境微生物与资源能源

对许多沉积型矿产资源的研究发现，成矿过程与微生物的地质过程有着密切的联系，如前寒武纪的铁建造、新元古代成冰系间冰期的锰矿、陡山沱组的磷矿等。前寒武纪铁建造作为研究前寒武纪地球表层环境和生物演化的重要载体，对其成因目前已有较为深入的认识，揭示了一系列铁代谢微生物功能群在铁建造型矿床中的作用，同时也揭示了伴随的其他微生物功能群在调节前寒武纪气候环境中的作用。

新元古代晚期是深时重大地质事件的频发期，发生了“雪球地球”冰期事件、新元古代大氧化事件等，同时也是地质历史上重要的成矿期之一，发育了大量的沉积型锰矿和磷矿。对这些大型矿床的成矿过程和成因机制的探讨，推动了深时极端地质环境微生物的发展。以磷矿资源为例，1978 年起，国际地质科学联合会发起国际地球科学计划 156 项计划，开展了近十次全球磷矿对比研究，并出版了“世界磷矿床”系列丛书共三卷。我国在同一时期也开展了大规模的磷矿勘察工作，发现了一批大型、超大型磷矿。磷矿资源勘查对基础地质和成矿理论研究提出了新要求，在总结找矿经验、开展实验研究的基础上，提出了上升洋流、物理富集、化学淋滤各种过程来解释高品位磷矿的成因。与此同时，生物活动在成矿中所起到的作用逐渐受到关注。在磷块岩中，常见各种微体化石，而大规模磷矿的开采为开展地球生物学研究提供了很好的研究对象。研究显示，硫代谢细菌可能在不同时代（包括古元古代）的磷块岩形成过程中都起到了重要作用。显然，对地质时期尤其是重要成矿期海洋磷循环机理的深入理解还依赖于未来深时地质微生物的研究。

三、受国际重大科学计划的推动，体现了学科领域的国际前沿性

极端地质环境微生物的研究涉及地球科学与生命科学这两大自然科学领域，因而同时受到这两大领域重大国际科学计划的推动。其中，国际大洋钻探计划和国际微生物组计划对推动和发展极端地质环境微生物的研究尤其关键。

（一）深地微生物研究计划

美国能源部（United States Department of Energy）1985 年启动的“地下科学研究计划”（Subsurface Science Program）是美国最早从事深地微生物研究的项目，主要目的是治理有机与无机污染物，在美国一些白垩纪和三叠纪盆地建立野外场地，聚集了一批来自国家实验室的顶尖地学和微生物学家，开发了一整套深地微生物的采样方法，初步探索了生物量、多样性与地质环境之间的关系，分离培养了一批具有潜在应用价值的菌种。

紧接着，美国国家科学基金会和美国国家航空航天局联合资助“极端环境微生物”计划。这个项目于 2001 年正式启动，总体目标是研究极端环境的微生物及其与环境之间的相互作用。这个计划以后，美国国家科学基金会相继出台了“环境中的生物复杂性”（Biocomplexity in the Environment）、“微生物观察”（Microbial Observatory）、“微生物作用过程”（Microbial Interaction

Processes）等一系列研究计划，其中有许多与深地或者极端环境微生物有关。特别是在 2009 年左右，有私人基金会艾尔弗·斯隆基金会（Alfred P. Sloan Foundation）资助美国卡内基科学研究所开展深碳观测的工作，深地微生物（包括陆地与海洋）是其中四个方向之一，内容包括深地微生物的多样性与分布、深地微生物的边界和深地微生物与碳循环的关系。它和国际大陆钻探计划一起资助过钻探项目，极大地推动了国际深地微生物的研究。

国际大洋钻探计划近年来越来越重视深地生物圈的研究，近 10 年以来，一共资助过 5～6 个与深地生物圈有关的项目，内容包括由水库蓄水或开矿诱发的地震和断层活动带来的脉动式氢气爆发及其与深地微生物的关系，蛇绿岩、蛇纹石化与深地微生物的关系，大火成岩省的氢气和微生物活动的关系，超临界 CO_2 注射到地下以后深地微生物的响应，地下页岩气开发与页岩微生物群落、功能以及地表微生物的相互作用等。

（二）深海极端环境微生物计划

探测深海微生物的计划主要由 IODP 承担，历史上第一次致力于探究沉积物深部生物圈的大洋钻探航次为 201 航次，现在微生物学家被邀请参加 IODP 航次已经成为常态，绝大部分是采集海洋沉积物岩心（王风平和陈云如，2017），真正打到洋壳岩石（辉绿岩）的是 2010 年实施的 IODP 304 航次、305 航次［北大西洋的亚特兰蒂斯高地（Atlantis Massif）］与随后的 301 航次与 327 航次（胡安·德富卡岭侧翼的洋壳玄武岩）。

以深部岩石圈研究为例，IODP 极大地推动了海洋深部生物圈的研究。在海床以下沉积物和岩石以及其中的流体中生活的生物被统称为海洋深部生物圈。人们对这些深部生物圈的探索是从大陆架沉积物开始的。在早期的海底勘测中，人们偶然发现在海床下数厘米至数米的深海沉积物内存在细菌。随后，科学家依托国际大洋发现计划（IODP，2013～2023 年）及其前身大洋钻探计划（ODP，1985～2003 年）和综合大洋钻探计划（IODP，2003～2013 年）组织的科学考察航次获取大洋沉积物样品，并对深部生物圈开展了系统性的研究。

另外一个有组织的研究计划是美国的暗能量生物圈调查中心（Center for Dark Energy Biosphere Investigations，C-DEBI），C-DEBI 于 2011 年左右启动，受美国国家科学基金会资助，一直运行到现在，主要目的是研究海洋沉积物与洋壳微生物的量、分布规律与功能，以及与地质条件之间的关系。

（三）国际微生物组计划

当前，极端地质环境微生物研究因微生物组学技术的发展而迎来了新的机遇。美国在能源部支持下，于2007年率先启动了一项微生物组地球计划，旨在对当时所有纯菌进行全基因组测序与功能预测。美国又在2010年启动了地球微生物组计划，旨在对所有环境样品进行微生物测序与功能预测。在2015年底，一批科学家联名在《自然》（*Nature*）和《科学》上发表文章，呼吁启动联合微生物组计划（Dubilier et al.，2015）。随后，美国白宫于2016年宣布启动国家微生物组计划，这是与脑计划等同等量级的大科学计划。这一计划旨在促进不同生态系统微生物组的整合研究。截至2016年，有资料可查的国际微生物组研究计划有22项，其中13项与人类健康相关，9项与环境相关。

在国内，早在21世纪初，中国科学院微生物研究所有关专家就开始推动微生物地球研究计划。2017年，国家自然科学基金委员会正式启动“水圈微生物驱动地球元素循环的机制”（简称水圈微生物）重大研究计划，也推动了深地（譬如地下水）生物圈相关研究。2017年，科技部推出了“中国微生物组计划”，其中环境微生物组是主要研究内容之一。2016年，由地球科学家发起、中国科学院资助的“深部地下生物圈”学科发展战略研究项目开始启动，经过近三年的努力，项目组完成了一系列“深部地下生物圈”丛书，2020年春季由科学出版社出版。该项目系统探讨了深地生物圈的起源，物质营养与能量来源，生物圈和岩石圈-水圈之间的相互作用，深地生物圈的多样性、代谢特征、进化历史和环境适应特征，深部和表层生物圈之间的联系以及深部生物圈与气候变化等内容。由此可见，这一系列的研究项目与计划深入推动了深地微生物学的发展。

四、具有多学科交叉和共性科学难题等特点而相互促进地发展

（一）具有多学科交叉的特点

极端地质环境微生物的研究具有明显的多学科交叉特点，涉及地球科学与生命科学的交叉，还涉及与化学等学科的交叉。也就是说，它涉及最基础的自然科学之间的交叉联合，而不仅仅是地球科学内部二级学科之间的交叉联合。首先，极端地质环境的研究需要地球科学领域的基础理论和技术方法的支撑，尤其是地质时期极端环境的研究，只能依托地球科学才能完成；其

次，微生物的研究需要生命科学领域的基础理论和技术方法的支撑，无论是微生物的多样性还是生态功能的研究，都离不开生命科学；最后，一些生物地球化学功能的研究涉及许多化学反应动力学，有待化学知识的支撑。

这里以氨的厌氧氧化过程为例，它的发现最能体现化学、生命科学和地球科学三大自然科学交叉联合的特点。首先，化学家进行了预言。根据化学反应热力学，理论化学家预言，自然界存在以硝酸盐或亚硝酸盐为氧化剂的氨氧化反应，这是因为这些反应所释放出的自由能不逊色于以氧为氧化剂的氨氧化反应。随后，这一预言得到了生物化学家的证实。生物化学家在生物脱氮试验中发现，除人们所熟知的反硝化外，还进行着未知的某个反应使氨消失了，同时硝酸盐和亚硝酸盐也有一半以上不明去向，而气体中的最主要成分为 N_2。紧接着，培养实验证明了氨确实与硝酸盐同步转化，气体的主要成分是 N_2。由此认为，生物脱氮试验中的氨和硝酸盐转化是按化学家所预言的方式进行的，理论值与实测值非常接近，并将其称为厌氧氨氧化。之后，主要微生物纯培养获得成功，发现了关键的酶和脂类，并用 ^{15}N 的示踪实验研究表明，厌氧氨氧化是通过生物氧化的途径实现的，过程中最可能的电子受体是羟胺（NH_2OH），并推测出其代谢途径；^{15}N 同位素标记把氨的厌氧氧化与反硝化区分出来。最后，微生物群落的宏基因组学揭示了氨的厌氧氧化的生物化学与基因基础。从此，生物地球化学家把氨的厌氧氧化作用应用到地质研究中。现在人们已经知道，厌氧环境中有一种比较普遍的与氮有关的生物地球化学过程。从化学家到生命科学家再到地球科学家的不断接力，才使得氨的厌氧氧化作用大白于天下，它是自然界的一大重要而普遍的现象。

正是由于这种多学科交叉的特点，极端地质环境微生物的研究需要培养大量具有学科交叉知识背景的学者。研究者既要有地球科学的背景知识，也要有生命科学的基础知识，还要有化学方面的知识。这样，一方面，要求建立多学科交叉的科学共同体，促进多学科交叉和碰撞；另一方面，要求资助机构改变资助管理模式，特别需要资助具有学科交叉特质的前瞻性研究。

（二）具有共性科学难题

极端地质环境包括地表极端环境、深地环境、深海极端环境、深空环境和深时极端地质环境，这些极端环境的微生物存在密切联系。它们具有一些共性特征：①生物的生活环境条件比较恶劣乃至极端；②生物代谢缓慢乃至处于休眠状态；③生物多样性比较低，代表了生物比较原始的类型。极端地质环境微生物因而具有类似的重大科学难题，包括微生物在极低能量、极低营养等极端环境中是如何获得足够的能量以维持生命及其演化的，古菌、细菌和病毒等生命系统是如何相互关联的，又是如何改善这些恶劣环境条件的，这些都是急需解决的重大共性科学前沿问题。

1. 受到了现代微生物学基础科学问题研究的推动

现代微生物学的发展在不同层面上突出了微生物的重要性。其一，微生物学的研究揭示了微生物数量庞大，是生物圈的主体，包括所有的细菌、古菌和绝大部分真核生物，而且这些微生物中的绝大部分都具有非常久远的演化历史，是地球早期生物圈的主导者。其二，微生物的生态位非常多样，除正常生态系统中繁盛的大量微生物之外，在许多相对于后生生物来说的极端环境，如前文所述及的地表、深地、深海等极端环境中都有大量微生物的分布。部分极端环境在早期地球上广泛分布，如缺氧海洋、热液喷口和热泉系统等。对这些极端环境微生物的研究推动了深时极端地质环境微生物的发展。其三，微生物的生理代谢类型丰富多样，在获取能量和碳源的过程中推动生源要素的生物地球化学循环，进而作用于环境。微生物也被称作生物地球化学循环的引擎。这些认识过程推动着深时极端地质环境微生物学的发展。

特别是，微生物的地质作用体现在微生物在不同环境条件下代谢产物对环境的影响，因而代谢科学的发展对极端地质环境微生物具有重要的推动作用。代谢科学关注生物体代谢物组成、代谢途径及代谢网络的作用和调控，也包括生物体代谢与环境之间相互作用的规律。代谢组学作为系统生物学最下游的数据类型，可以体现出生物体系整体状态和功能的最终结果。其中，代谢流量组（fluxomics）是细胞内代谢反应速率的集合，反映基因、蛋白质、代谢物之间各种相互作用导致的代谢网络的最终功能。相应地基于稳定同位素示踪的代谢流量分析技术、流量平衡分析技术、代谢可视化技术、网络代谢模拟技术，都将有助于地球微生物研究从定性走向定量，并从被35亿～40亿年地质历史所覆盖、扭曲的基因组/膜脂等有限信息中还原可能的早期生命过程与环境效应，为前寒武纪早期某些重要代谢方式和生物类群的起源提供新的探索思路，在研究生命的起源、辐射、灭绝和复苏方面对古生物学继承并发展。

如何依据地质历史记录设计合理有效的模拟策略？在全面了解微生物代谢机制的基础上，设计科学合理的代谢通路，再现早期生命的若干代谢途径并与地质历史记录进行对比和溯源，是解析与验证微生物功能群特征、起源、地质识别标志及其地质作用等的必由之路。

代谢流量组是细胞内代谢反应速率的集合，代谢流量分析（metabolic flux analysis）能够用于识别代谢通路，解析代谢调控规律，目前发展出了基于稳定同位素示踪的代谢流量分析技术和流量平衡分析技术。其中比较典型的是基于 ^{13}C 标记的代谢流量分析技术，即在细胞的培养基中加入 ^{13}C 标记底物，经过一段时间后收集细胞，提取出胞内代谢物，利用质谱或核磁共振技术检测细胞内中间代谢物的 ^{13}C 标记异构体分布，基于代谢反应中碳原子转移特征建立代谢物 ^{13}C 分布的数学模型，这样结合实验测量信息和数学模型来计算代谢网络中的流量分布。

全基因组规模代谢网络重构与分析能够发现基因敲除、插入、异常表达、环境变化等对生物系统表型的影响。随着现代基因测序技术的高速发展，已构建出 130 多个物种的代谢网络，覆盖原核、真核和古生生物，其中 22 种得到了实验数据的验证。《自然-生物技术》（*Nature Biotechnology*）期刊报道的最完整的大肠杆菌 1515 菌株的代谢网络模型，已经将基因和蛋白质的对应关系细化到结构域层面，更精细地模拟催化过程，揭示基因型与表型之间的关联。利用该模型宏基因组测序数据也构建了人体肠道微生物菌群中的大肠杆菌菌株代谢模型。

此外，代谢网络分析和模拟也在基因工程改造、代谢疾病研究和药物靶标识别上都发挥了重要的作用。在代谢途径设计方面，使用传统的方法在宿主细胞中构建一条新的代谢途径，通常要过量表达来自异源的编码代谢途径关键酶的基因。这种代谢工程的操作主要集中在通过实验分析，鉴定宿主细胞内代谢途径的限速步骤，确定关键节点，然后进行基因敲除和过表达，消除限速步骤。这种策略在理论上非常有效。然而，尽管这些操作在实验室中有很多成功例子，但主要集中在乙醇、丁醇、甘油、有机酸、氨基酸等代谢物中，且这些化合物的结构和代谢途径简单，仅通过一到两步反应便可以实现。因此，对于更多复杂化合物的大规模生产和工业化改造，代谢网络分析和模拟都有着不可替代的作用。

2. 受到地质学重大科学难题研究的推动

在地球 46 亿年的演化历史上，发生过一系列重大的生物演化事件和环境突变事件，对这些事件的研究推动了深时极端地质环境微生物学的发展。以显生宙的动物大灭绝事件为例，在探索生物大灭绝过程和原因的过程中发现了很多与微生物有关的现象。一方面，许多研究发现，后生生物的大灭绝伴随着微生物的繁盛，如二叠纪末的动物大灭绝伴随着大量微生物岩的广泛沉积，以及生物标志物指示的蓝细菌和绿硫细菌等的繁盛（Grice et al.，2005）。又如，晚奥陶世赫南特期的生物大灭绝也伴随着海洋生态系统中原核微生物所占比例的显著升高以及好氧甲烷氧化代谢微生物丰度的显著升高。是什么因素导致这些地质微生物的繁盛？这些地质微生物的繁盛对地表环境有什么反馈作用，对后生生物的灭绝和复苏又有什么重要的影响？对这些问题的探索极大地推动了生物大灭绝期地质微生物的发展。

另一方面，重大地质转折期的研究发现，后生生物大灭绝期一般都伴随着 C、N、S 同位素组成的显著异常，代表着 C、N、S 循环的波动。对这些 C、N、S 循环波动原因的研究揭示了其与微生物功能群变化之间的密切关系。例如，氮循环的异常一般受控于各种类型的氮代谢微生物功能群，包括固氮微生物功能群、硝化微生物功能群、反硝化微生物功能群和厌氧氨氧化微生物功能群等。这些微生物功能群的变化与环境条件的变化密切相关，对认识动物大灭绝的机制有重大的帮助，进而推动了深时极端地质环境微生物的发展。

3. 不同极端地质环境的类比研究日益受到重视

极端地质环境微生物因具有共性的科学难题而促使不同领域的研究可以相互借鉴，它们不是相互割裂而是相互关联的，因而可以开展很好的类比研究。

1）当代与深时极端地质环境的对比

无论是当代还是深时，无论是海洋还是陆地，都存在许多类似的极端环境条件，如高温、缺氧、酸化、盐碱化等，而且这些极端地质环境中的微生物具有一些共性特征，如代谢缓慢、多样性低、丰度低、代表一些原始类型等。鉴于以上两方面的原因，进行这些极端地质环境微生物的对比研究是很必要的。

由于深时微生物与深地微生物在组成、多样性、环境适应等方面有可能有高度相似性，现在一般都认为，深地生物圈代表着地球最早期极端环境下的生命。诸如深海热泉环境微生物等深地微生物可以为探索地球早期的生命起源提供素材。极端地质环境微生物研究与深地计划结合，可以了解这些微生物对深地环境的改造作用，了解一些特殊微生物的生态环境功能，为人类移居地下深部环境提供服务。

2）地球极端环境与外太空环境的类比

地球上有多种环境可以在某些方面类比火星和冰卫星环境，通过实地勘察、样品采集和实验室分析，结合空间探测数据，有助于探索在未知天体上存在生命宜居环境的可能性。极端环境中的主要生命形式是微生物，包括真菌、细菌和古菌等。目前已经发现了数十个不同的类比研究地点或环境，包括我国柴达木盆地类火星环境和地貌、临近空间类火星环境等。这些环境可以类比早期地球环境、现代火星环境、现代冰天体环境、金星大气环境等。该方面的研究内容包括极端环境中微生物的多样性、极端环境微生物的生命边界、微生物在极端环境中的代谢过程和适应策略等。研究极端环境中的特殊生命形式可以为我们探测地外生命提供借鉴。

3）行星环境的实验室模拟

行星环境的实验室模拟包括在地面实验室和空间环境（如空间站、临近空间）模拟行星环境进行生物学实验，研究生命物质在模拟环境中的变化。地外空间环境（如空间站、临近空间）本身具备多种天然的极端环境因子，如高 / 低温、缺氧、干旱、微重力、不同辐射源和强度等，是地外生命生存极限探测的良好试验场。

4）系外宜居行星

利用地基和空间望远镜对太阳附近（<100 光年）的类太阳恒星和小质量恒星（红矮星）开展类地系外行星搜寻；利用掩星事件观测系外行星大气透射谱，发展大口径地基和空间望远镜直接观测系外行星发射谱以了解系外行星大气和表面成分（这些光谱观测用以研究系外行星宜居性和目前的物理化学状态）。此外，有可能在未来几十年对环比邻星的类地行星开展就位探测。

第三章 国际发展现状与发展态势

第一节　学科整体的发展现状与态势

一、不同极端地质环境存在微生物的共性规律

不同极端地质环境微生物的研究都经历了从初期的纯培养模式，发展到随后的基于 16S rRNA 基因克隆文库方法，以及近年来的全基因组测序、宏基因组测序、新一代高通量测序和单细胞测序等方法的不断革新，引发了许多新的发现，而且有许多共性特征。这些共性特征都是由极端环境特征以及微生物的适应能力决定的。

（一）代谢缓慢

起初，人们认为海洋沉积界面之下的微生物代谢速率超低，相比于表层生态系统几乎可以忽略不计。后来发现，从海洋沉积物表面至海洋深处 1626 m 处都存在大量微生物细胞和微生物活动，微生物有能力参与细胞运动、分裂，氨基酸、脂质和碳水化合物的合成和转运，以及 DNA 修复等（Orsi et al., 2013）。然而，与正常环境的微生物相比，极端地质环境微生物往往具有代谢缓慢或活性比较低的特征，这是由于微生物受到获取能量的限制，是微生物适应低能量环境的结果。

例如，在深海沉积物中，有机质含量和生物可利用性都随沉积深度的增

加逐渐减少和降低，导致微生物代谢速率随沉积深度增加变得越来越低。在全球尺度，边缘海深部生物圈的微生物年呼吸速率大约为上覆海洋表面真光层年生产率的 1%，而在大洋区，该比例下降到 0.01% 甚至更低（Parkes et al.，2000；D'Hondt et al.，2002）。所以，与表层海洋微生物相比，海底深部生物圈中的大部分微生物都是不活跃的或只具有极低的代谢速率。

与大陆地表环境微生物相比，大陆深地环境微生物的代谢要慢得多。大陆地表环境微生物碳代谢率通常为 0.1～10 fmol 碳 /（细胞·d）（即每克细胞每天代谢 10^{-3}～10^{-1}g 碳），而深地微生物碳代谢率仅为 10^{-5}～10^{-3} fmol 碳 /（细胞·d）（即每克细胞每天代谢 10^{-7}～10^{-5}g 碳）（Jørgensen，2011；Morono et al.，2011）。可见深地环境微生物的代谢速率通常要比地表环境微生物的代谢速率至少慢至 1/100。如果按 10^{-5} fmol 碳 /（细胞·d）的代谢速率推算，深地微生物细胞平均每 1000 年才分裂一次（Jørgensen，2011）。估算也证实，地下微生物的碳和蛋白质周转十分缓慢，可达 1000 年（Jørgensen and D'Hondt，2006；Onstott et al.，2014）。模型预测也表明，死亡微生物氨基酸在深地环境的平均周转率达数千年之久（Braun et al.，2017）。

为了承受极端环境并在其中生存，一些细菌形成了具有特殊结构的内生孢子。内生孢子是细菌在抵抗极端恶劣环境时形成的用于保护其遗传信息的一种存在形式，以深度休眠、代谢率极低并且可以抵抗恶劣环境为特征。据估计，在有 1000 万年之久的海洋沉积物中，每平方厘米有 10^7 个内生孢子（Lomstein et al.，2012）。

（二）多样性低

极端地质环境微生物需要适应一种乃至几种极端环境因子，只有那些适应这些极端环境因子的微生物才能获得能量来维持生命。在深海海底，随着沉积物深度的增加，大部分微生物因无法抵抗极端环境而被淘汰，仅有少量幸存者，如嗜高压生物。例如，从 IODP 337 航次采自海底之下 2406 m 深处的沉积物中发现了产内生孢子的革兰氏阳性嗜高压菌，从下北泽（Shimokita）煤层沉积物中分离培养出革兰氏阳性厚壁菌门嗜高压菌，这显示了深部极端环境的微生物群落结构及其物种多样性与表层生物圈差异较大，多样性低得多。

大陆深地环境尽管已经发现超过 30 门的微生物，但这些微生物丰度普遍较低。在大陆基岩环境，细菌主要以厚壁菌门（Firmicutes）和变形菌门（Proteobacteria）占优势，而古菌是以广古菌门（Euryarchaeota）占优势，尤其以产甲烷菌为主（Li et al.，2017）。在洞穴系统，细菌主要类群为变形菌门、放线菌门（Actinobacteria）、厚壁菌门和拟杆菌门（Bacteroidetes），古菌类群主要是奇古菌门（Thaumarchaeota）和广古菌门。油气藏环境因受温度和流体化学组成的影响而主要出现包括变形菌门、厚壁菌门等在内的 15 门的细菌。

在海洋系统，表层沉积物比深部沉积物的微生物多样性更高一些。表层沉积物主要是海洋古菌类群Ⅰ（Marine Group Ⅰ）、奇古菌、ε-变形菌（ε-Proteobacteria）、浮霉菌（Planctomycetes）和绿弯菌（Chloroflexi）等。深部沉积物中深古菌、海洋底栖古菌类群 D（MBG-D）、洛基古菌（Lokiarchaeota）、暗黑菌（Atribacteria）和厚壁菌等类型的丰度相对较高。

（三）耦合的生物地球化学过程与微生物种间的互养共生

与正常环境相比，极端地质环境的生物地球化学过程往往出现许多耦合特征（图 3-1），这既与极端地质环境有限的能量但丰富的化学物质有关，也与化学反应的热动力学有关。这些耦合的生物地球化学过程使得许多微生物之间出现互养共生。极端地质环境微生物就以互养共生的形式建立了一个相互交织的代谢网络，通过这个代谢网络，它们在有限的能量条件下有效地将碳、氮、硫和铁等循环耦合在一起。

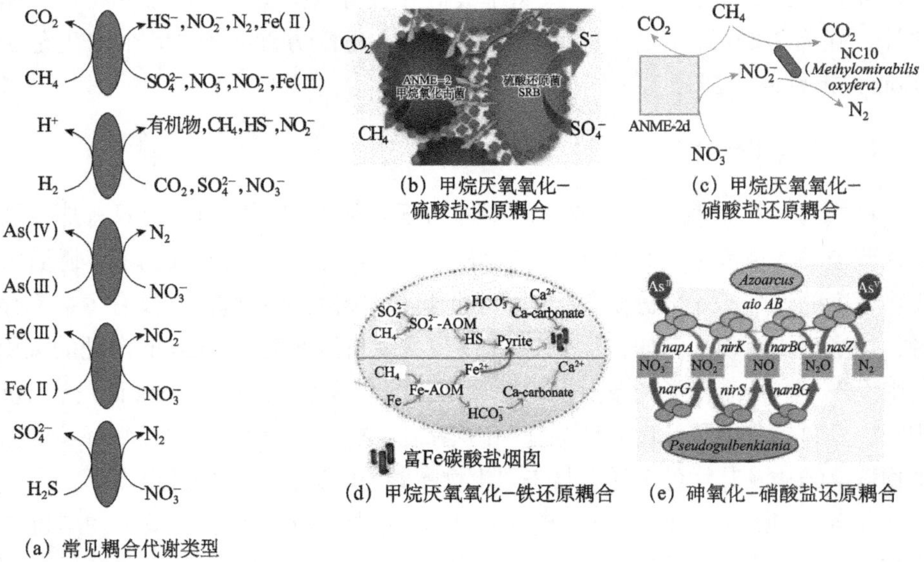

图 3-1　极端地质环境的常见耦合代谢类型及部分典型生物地球化学过程（文后附彩图）

耦合过程在理论上是可行的。根据热力学上最有利的反应，即微生物氧化有机碳和还原电子受体所产生的吉布斯自由能大小，氧化有机碳可按次序先还原氧气，然后依次还原硝酸盐、铁、硫酸盐，最后进入甲烷发酵阶段（Stumm and Morgan，1996）。这样，微生物能量产生过程就受制于这些电子传递反应。而且，在一些极端地质环境中有丰富的硝酸盐、亚硝酸盐以及含铁和锰的矿物。所以，从热动力学和环境物质分布的角度，将有机碳氧化为二氧化碳是可以和还原氧气、硝酸盐、氧化锰、氧化铁和硫酸盐耦合起来的。这方面最典型的例子是甲烷厌氧氧化作用与其他过程的耦合。

首先，在厌氧条件下，微生物介导的甲烷氧化可以和硫酸盐还原耦合在一起，这是极端地质环境（如深海沉积物、陆地深部环境等）比较普遍存在的生物地球化学过程。甲烷厌氧氧化古菌先氧化甲烷，将释放的电子传导给硫酸盐还原细菌。这一电子传导过程与细胞内的电子传递链不同，涉及细胞内和细胞外的电子传送，称为微生物的胞外电子传导。细胞色素 c 和微生物纳米导线很可能将电子由甲烷厌氧氧化古菌直接传导给硫酸盐还原细菌。这一氧化还原反应过程也可由一种微生物完成。一个来自大陆深地、属于 ANME-2d 的 *Methanoperedens nitroreducens* 拥有几乎完整的甲烷厌氧氧化和异化硫还原所需要的功能基因（Haroon et al.，2013）。

其次，甲烷厌氧氧化微生物也可利用氧化甲烷后释放的电子来还原硝酸盐和亚硝酸盐（Timmers et al.，2017）。例如，ANME-2d 可利用硝酸盐为末端电子受体氧化甲烷。这种微生物富含细胞色素 c，这些细胞色素部分可参与硝酸盐的还原。亚硝酸盐也可作为间接的末端电子受体，在这一反应中神奇甲烷产氧菌（*Methylomirabilis oxyfera*）先利用亚硝酸盐产生氧气，再用产生的氧气来氧化甲烷。

最后，甲烷厌氧氧化还可以与锰和铁的还原相耦合（Beal et al.，2009）。甲烷厌氧氧化古菌可以利用水溶性的三价铁为电子受体氧化甲烷，甲烷厌氧氧化古菌 *Ca. Methanoperedens nitroreducens* 能够将甲烷厌氧氧化过程与三价铁和四价锰矿物的还原相耦合（Ettwig et al.，2016）。这说明甲烷厌氧氧化古菌和三价铁异化还原细菌希瓦氏菌和地杆菌等微生物都拥有用于胞外电子传导的通道，这些通道主要由细胞色素 c 构成。

又如，一些嗜热硫酸盐还原微生物可以耦合氢气氧化与硫酸盐还原反应。这一过程从热力学上来讲也是可行的，并且氢气和硫酸盐广泛存在于一些极端地质环境中（Colman et al.，2017）。在南非威特沃特斯兰德（Witwatersrand）盆地 2800 m 深处，发现了一个属于厚壁菌门的自养硫酸盐还原菌金矿菌（*Ca. Desulforudis audaxviator*），该菌在那个深度的微生物群落中占主导地位（Chivian et al.，2008）。在那个深度，氢气和甲酸盐氧化与硫酸根还原相耦合所产生的能量很可能用来进行无机碳和氮的固定。*Ca. D. audaxviator* 有相当多的水平基因转移和病毒性感染等特征。这不仅说明氢气在其新陈代谢中的必要性，还说明水平基因转移在其进化上的重要性（Colman et al.，2017）。

（四）微生物的巨大空间差异性

在不同的极端地质环境，无论是微生物的代谢、种类、多样性还是生态功能，都存在很大的空间差异性，这是由环境因子的空间梯度分布决定的。在横向空间上，以海底热泉、冷泉和天然气水合物为代表的高能区微生物的丰度比较高，而在深海盆寡营养区的沉积物中，微生物的丰度相对较低。在不受海底热泉、冷泉和天然气水合物等地质化学过程影响的大部分海域，不同水深的沉积物表层微生物的丰度随水深增加而降低，这主要与真光层新生产的有机质输出和水深等因素有关。

在纵向空间上，微生物在沉积物中的垂直分布大致遵循 Parkes 等提出的经验公式，即 lg（细胞数）= 7.98−0.75lg（深度）。也就是说，随着沉积深度的增加，微生物丰度呈对数线性减少，在表层 1 m 的沉积物中，微生物细胞丰度为每立方厘米 10^9 个细胞，而在深度 500 m 的沉积物中，细胞丰度降低到每立方厘米 10^6 个细胞。海洋沉积物、陆地沉积物与岩石的微生物量都会随着深度增加而降低（Kallmeyer et al.，2012）。沉积物的埋深温度、氧化还原条件、有机碳含量和生物可利用性、沉积物孔隙度、地质年龄和地球化学特征都会影响微生物的空间分布。

如果同时考虑横向和纵向上的变化，那么这种空间差异性更大。例如，在岩石或者沉积物最表层的 1～2 m 处，微生物的生物量与总有机碳表现出负相关关系。2 m 之下，随着温度和盐度的升高，微生物的生物量快速降低，这表明温度、盐度的重要性在加强。在 300 m 以上，每立方厘米地下水中的生物量要明显低于每克岩石中的生物量。但是，在 300 m 以下，两者的生物量却相当。与岩石中的生物量不同，地下水中的生物量只与深度、温度相关，而与 pH、盐度或者可溶性有机碳含量无关。

除了丰度，微生物的多样性也存在很大的空间差异性，不同深海环境就有显著的微生物多样性差异。例如，在深海表层沉积物中，广泛存在着绿弯菌门、γ-变形菌纲（γ-Proteobacteria）、δ-变形菌纲（δ-Proteobacteria）、 α-变形菌纲（*α*-Proteobacteria）和放线菌等细菌类群。但在大陆架边缘沉积物、冷泉和热液，以及深部沉积物中，则出现了大量的古菌，主要是深古菌、深海底栖古菌类群 B（Marine Benthic Group B，现在叫 Thorarchaeota）、海洋底栖古菌类群 D、奇古菌、广古菌和乌斯古菌（Woesearchaeota）等（Orcutt et al.，2011）。深海环境微生物的这种空间分布差异与沉积物类型，以及环境中的理化因子（硫酸盐浓度、氧气含量、甲烷浓度、无机碳浓度、水温、水深等）有关。

微生物的功能群也出现很大的空间差异性。沉积物从表层到深部，由于氧化还原电位（Eh）的变化而分别出现了有机碳的有氧氧化、铁锰氧化、硝酸盐还原、硫酸盐还原和甲烷形成等具有不同功能的微生物群落，这是典型的微生物种类和功能在纵向上的差异。如前所述，尽管极端地质环境微生物的代谢总体很缓慢，但在沉积物深部有可能出现以甲烷为能量的微生物代谢活跃带。这也是空间差异性的一个典型体现。又如，在一个海底热泉系统中，从喷口到外围，环境温度逐渐降低，化学物质也存在很大的空间梯度，那么微生物的多样性、种类和生态功能都会发生变化，表现出巨大的空间差异性。

（五）微生物丰度评估结果的较大不确定性

微生物的一系列特征存在很大的空间差异性，导致了对极端地质环境微生物特征的评估出现很大的不确定性。最典型的代表是有关微生物丰度的估算争议很大。

人们早期估计海洋和陆地深地微生物细胞数量分别为 3.5×10^{30} 个和 2.5×10^{29}～2.5×10^{30} 个（Whitman et al.，1998）。随后，人们估算海洋深地微生物细胞数量仅为 2.9×10^{29} 个，总碳量为 4.1×10^{15}g，约占全球生物量的 0.6%（Kallmeyer et al.，2012）。不过，一些学者又推测陆地深地微生物细胞数量应该在 5.0×10^{29}～5.0×10^{30} 个，总碳量为 1.4×10^{16}～1.35×10^{17}g，占全球生物量的 2%～19%（Sean and John，2014），这与早期估计的结果基本符合。后来，科学家收集了 3800 个深地生物量的数据，其中 1/3 的数据来自地下水，2/3 来自岩石。通过对生物量与深度、温度关系的拟合并且假定生物圈的下限是 122 ℃，最终得出大陆深地的总生物量为 2×10^{29}～6×10^{29} 个细胞。如果采用同样的拟合公式，对洋壳深度 0.5～6.5 km 进行积分，得出海洋深地生物量为 5×10^{29} 个细胞（Magnabosco et al.，2018）。这样全球海洋和陆地的深地生物量总和为 7×10^{29}～1.1×10^{30} 个细胞。这个估计值不包括土壤和海水中的生物量。不同学者对微生物丰度如此大差别的评估结果，就是这些微生物空间分布差异的体现。

在海洋系统中，海底沉积物的厚度可能超过 10 km。在这种极大深度下，温度很可能成为影响微生物分布的最终要素。如果按照 30 ℃ /km 的地温梯度计算，113 ℃相当于深地 4 km 的温度，但截至 2021 年人类还未获得如此深的样品来开展微生物研究。在海底热泉、冷泉和天然气水合物富集区，生物地球化学过程格外活跃。由于在不同沉积深度，生物获取能量的方式不同以及地球化学特征分布不均匀，这些特殊地区微生物的分布一般不遵循 Parkes 归纳的对数线性变化规律，而是呈现出离散的局部最大或高度压缩的梯度分布特征（Parkes et al.，2000）。这些特征都决定了微生物各方面的评估结果均存在很大的不确定性。

二、微生物与极端地质环境的相互作用亟待全方位深入研究

（一）微生物地质作用巨大

任何生物与所在环境都是存在相互作用的。微生物在适应这些极端地质环境的同时，又是如何改变这些环境条件的？目前相关的研究还非常薄弱。这方面的主要工作还是聚焦在微生物参与的碳、氮、硫、砷、铁等元素的循环上，一些重要的微生物功能群不断被发现，一些重要的新的耦合过程不断被发现。极端地质环境微生物对大气圈、水圈、生物圈和岩石圈均具有多方面的地质作用（图 3-2）。特别是微生物对油气藏、酸性矿坑水、地下水等极端地质环境，以及对沉积物的天然气水合物等资源的形成均具有重要的地质作用。

在全球尺度上，人们对极端地质环境微生物的生物量评估还存在很大的不确定性，微生物对极端地质环境的地质作用还有待准确评估。可喜的是，在一些深时地质环境，微生物的地质作用相对比较明确。例如，在大约 23.3 亿年前，大气圈出现的第一次大氧化事件就与产氧光合生物的作用有关。在大气第一次大氧化事件之前，生物界主要是进行厌氧代谢的微生物。产氧光合生物出现之后，产生的氧气先消耗表层系统的还原性物质，之后氧气才在大气中积累下来，使得当时大气的氧含量达到现在的 1% 左右。产氧光合生物的出现和大发展，使得中元古代的海洋表层出现氧化，促进了真核生物的起源和演化。在新元古代，产氧真核生物的演替使得海洋整体出现氧化，动物等开始起源。这是在地球演化的长河里，极端地质环境微生物的长期地质作用的体现。

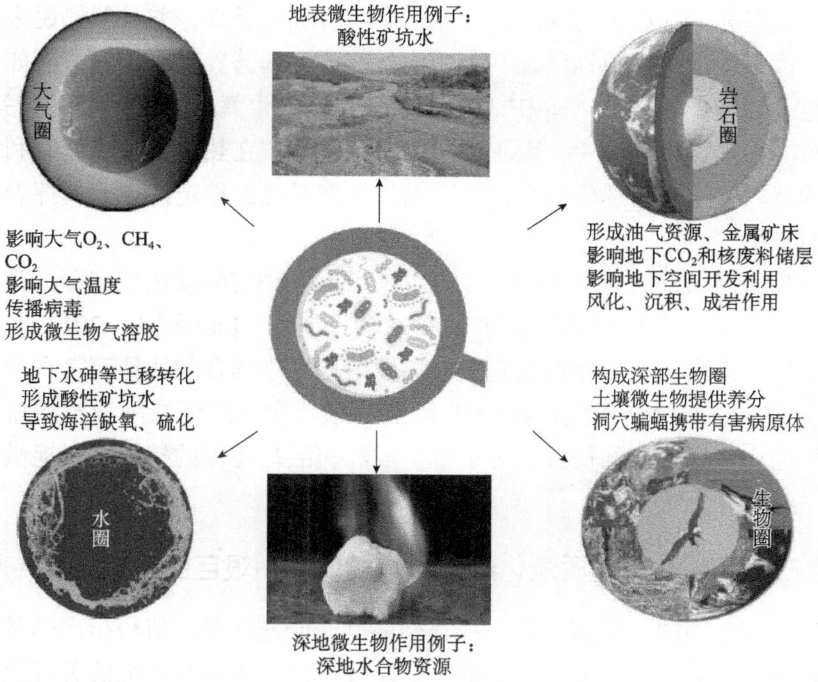

图 3-2 极端地质环境微生物对大气圈、水圈、生物圈和岩石圈的影响（文后附彩图）

资料来源：https://huaban.com/pins/1581316228；https://www.sohu.com/a/379569294_194423；http:// www.ah.xinhuanet.com/2014-12/04/c_1113512649.htm；https://bbs.le.com/thread-2107630-1.html；http://news.cnr.cn/native/gd/20170605/t20170605_523785210.shtml；https://www.21cnjy.com/H/11/93151/3912884.shtml；https://www.sohu.com/a/424813303_120012589。

又如，在当前全球变暖的背景下，在许多海洋水体的一定深度（200～300 m）出现了最小含氧带或者低氧区。海洋缺氧的加剧，将进一步导致海洋出现硫化事件（水体中富含 H_2S），称为硫化海洋。在地球历史上，最著名的硫化海洋出现在第一次大氧化事件之后的中元古代和新元古代，位于氧化的表层海水和缺氧铁化的深部水体之间，呈现一个楔状的硫化带。很显然，海洋缺氧水体与硫化水体的出现都与极端地质环境的微生物作用有关。

（二）微生物与微生物之间的密切相互作用

值得指出的是，极端环境条件下微生物与微生物之间的相互作用特别重要与密切。微生物之间的相互作用包含种间的互生、共生、拮抗、互作、竞争等诸多方面。

对极端地质环境的微生物来说，共生关系极其重要。微生物共生指的是代谢伙伴互相依赖以创造对各自代谢活动都有利的条件，并且严格说来，相互依赖性不能通过简单地添加共基质或任何类型的营养素来消除。例如，在海底地下环境中发现的 ANME 和硫酸盐还原菌共生就是一种典型互作关系。在寡营养大陆深地环境中，由于许多生物化学反应在热力学上略微有利，微生物的互养作用在组建生物多样性和保持生态系统的稳定性方面发挥着至关重要的作用。

科学家揭示了三对微生物互养作用：ANME 和硫酸盐还原细菌、产甲烷菌和 ANME，以及硫酸盐还原细菌和硫氧化细菌（Lau et al.，2016）。在后两种互利关系中，产甲烷菌和硫酸盐还原菌相对于其合作伙伴而言在产能上不太有利。它们的合作伙伴 ANME 和硫氧化细菌把产甲烷菌和硫酸盐还原菌的代谢产物迁出，从而分别为甲烷生成和硫酸盐还原的持续进行提供了驱动力。

（三）微生物的地质演化研究还是一项长期艰巨的研究任务

有关微生物的地质演化是一项长期艰巨的研究任务。值得指出的是，非常有限的研究显示，许多微生物都是在以极端地质环境为特征的关键地质突变期开始起源的。实际上，在“生命之树”的根部，许多微生物都是具有适应极端地质环境特征的嗜极微生物。微生物的地质演化来自两方面的证据，一是现代微生物的基因依据，二是各类地质记录。

首先，微生物的地质演化主要来源于分子生物学的研究。例如，分子生物学发现了氨氧化古菌（ammonia-oxidizing archaea，AOA）在 23 亿年前的第一次大氧化时期起源，并分别在 8 亿年前和 6 亿年前左右扩张，后者对应于表层海洋的氧化向深部海洋扩展（Ren et al.，2019）。这不仅反映了前寒武纪时期的大气含氧量变化直接关系到 AOA 的起源和演化，而且显示了 AOA 的起源与演化都出现在极端地质环境时期。

其次，地质记录可以提供一些关键微生物的起源信息。例如，微生物过程可以造成非常强烈的硫同位素分馏，使得地质记录中的硫同位素可以用来识别硫酸盐还原菌的起源。已经证实，这类微生物早在 34.7 亿年前就已经起源了（Shen et al.，2001）。

最后，地质记录与分子生物学的结合可以提供更加可靠的微生物起源信息。分子钟可以提供微生物起源的大致年代范围，而地质记录不仅可以提供地质年代和环境条件信息，还可以提供微生物的分子化石证据。例如，分子生物学证实，一些产甲烷古菌在距今 2.5 亿年前后的二叠纪—三叠纪之交出现（Rothman et al.，2014），而地质证据显示，当时是一个极端地质时期，出现许多极端环境事件，包括缺氧海洋、大量温室气体的释放造成的温室效应等。地质证据和分子生物学结果高度吻合。

三、微生物资源与生物安全

（一）微生物以多种方式适应极端地质环境而有广阔的工业应用前景

在微生物与极端地质环境相互作用方面，更多地开展了微生物如何适应这些极端环境条件的研究，包括微生物如何从分子水平通过特殊的基因、酶、蛋白质、细胞膜等来应对环境的极端性（图 3-3），如何通过耦合的生物地球化学过程提高能量利用效率，如何通过形成内生孢子和降低代谢速率等

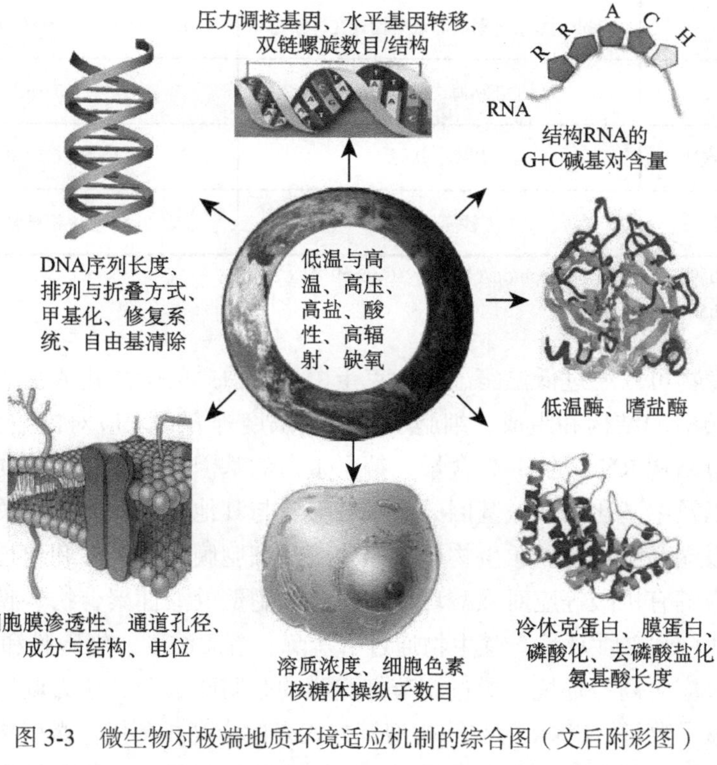

图 3-3　微生物对极端地质环境适应机制的综合图（文后附彩图）

资料来源：http://www.dashangu.com/postimg_16512606.html；https://max.book118.com/html/2016/1228/77790533.shtm；https://wenwen.sogou.com/z/q339795956.htm；http://www.dashangu.com/postimg_13440518.html；http://www.dashangu.com/postimg_13955397_12.html；https://www.zhihu.com/question/60239234/answer/174064750；https://www.biowolf.cn/m/view.php?aid=219；https://www.photophoto.cn/show/09695293.html。

一系列策略来延长生存时间等。已经发现，许多微生物的生存条件超出了人们的想象（表 3-1）。

表 3-1　地球生命的边界条件

环境因子	环境类型 / 源	极限条件
高温	海底热液通道	110 ～ 121 ℃
低温	冰盖、冰川	−20 ～−17℃
碱性环境	碱湖	pH ＞11
酸性环境	火山泉水、废矿淋滤液	pH= −0.06 ～ 1.0
辐射	宇宙射线、X 射线、放射性衰变、阳光	1500 ～ 6000 Gy 紫外辐射 5000 J/m^2
高压	马里亚纳海沟及地球深部	1100 bar
高盐 / 低水活度	盐湖、盐壳	水活度 0.5～0.7
干旱	沙漠地区	约 60% 相对湿度

资料来源：Gilmour 和 Sephton（2004）、Cockell 等（2015）。

注：1bar=10^5Pa。

微生物可以通过低温酶、冷休克蛋白、蛋白质的磷酸化或去磷酸化、细胞膜脂肪酸的结构和组成、细胞质内溶质浓度等方式来应对极端低温环境，通过调节结构 RNA 的 G+C 含量、编码蛋白的基因长度、氨基酸的排列及折叠方式以及甲基化方式、基因的水平转移、与其他微生物的共生关系等方式来适应极端高温环境。微生物可以通过调节细胞膜的渗透性和通道孔径、膜电位，降解有机酸等应对强酸环境，通过细胞壁的酸性聚合物、膜蛋白运输系统等适应强碱性环境。微生物通过嗜盐酶、溶质浓度、细胞膜和色素及 H^+ 泵作用等应对高盐环境，通过调节细胞膜的脂肪酸成分、压力调控基因、核糖体操纵子数量、DNA 的空间构型等方式来适应高压环境。微生物通过排出或绑定金属、将金属转成无毒形式，以及调节细胞膜渗透压和细胞某些成分来应对重金属的毒性。微生物通过 DNA 修复系统、自由基清除、细胞色素产生等应对高辐射环境。

也正因为对微生物适应极端地质环境的基因、生理和代谢产物的深入研究，才在各类工业上具有广阔的应用前景，并且已经很好地应用到环保、冶金、能源、电子、材料、医药、食品等各类工业的生产上。例如，源于嗜碱微生物的相关酶被广泛用于洗涤剂，因为碱性酶在洗涤剂中有长时间的稳定性，能降低洗涤温度，从而节省能源和减少后期污水处理。由中度嗜盐或耐盐细菌生产的四氢嘧啶是一种最为常见的渗透压平衡溶剂，可加在防衰老或抗干燥的化妆品中，使生物酶或核酸免受不利影响，并可修复由紫外线照射诱发的皮肤问题。一些嗜盐微生物能产生紫膜蛋白，紫膜蛋白可以作为光驱动的质子泵，将太阳能转化成电能，人工合成 ATP 等，作为光敏材料应用于全息光学、信号存储和处理以及光学检测领域。耐辐射菌对放射性废物的富集与处理技术具有重大科学意义。异养嗜热微生物可以在高温条件下快速降解多种天然高分子物质，包括淀粉、纤维素、几丁质等，在生物工程领域具有广阔的应用前景。

（二）病毒的潜在作用及其危害

与地球上细胞生物的分布一致，病毒存在于地球上的各种环境中。病毒与其宿主相伴，是自然界中微生物的“巨人杀手”。它的作用之一是“杀死胜利者”（kill the winner），因而是物种进化的重要推手。尽管人们对环境中病毒的分布、丰度、多样性、与宿主的相互作用、生态学功能等问题了解得很少，对极端地质环境的病毒则几乎属于完全未知，但少量研究发现，病毒的存在可能有利于宿主适应极端地质环境。来自极端地质环境的病毒对人类具有潜在的危害。

实际上，自 20 世纪七八十年代以来，人们就在深海、盐湖、热泉、极地等极端地质环境发现了病毒。目前，环境病毒的研究主要集中于海洋、湖泊、热泉等生态系统，极端环境病毒的形态多种多样（图 3-4）。从秘鲁沿岸到赤道东太平洋开阔海域的深海沉积物，原噬菌体占全部病毒的 15% 左右，这些病毒的存在可能有利于宿主适应极端环境（Engelhardt et al.，2014）。在深海热液口，发现了与维生素代谢、辅因子代谢以及铁硫簇形成等相关的病毒噬藻体辅助代谢基因（AMG），这些 AMG 可能有利于宿主适应热液极端生境（Anantharaman et al.，2014）。AMG 被认为是病毒通过水平基因转移的方式从宿主获取的，而其也可能再通过水平基因转移将这些基因传递给其他宿主。病毒介导的水平基因转移加速了宿主和病毒的共进化，增强了两者的环境适应性。

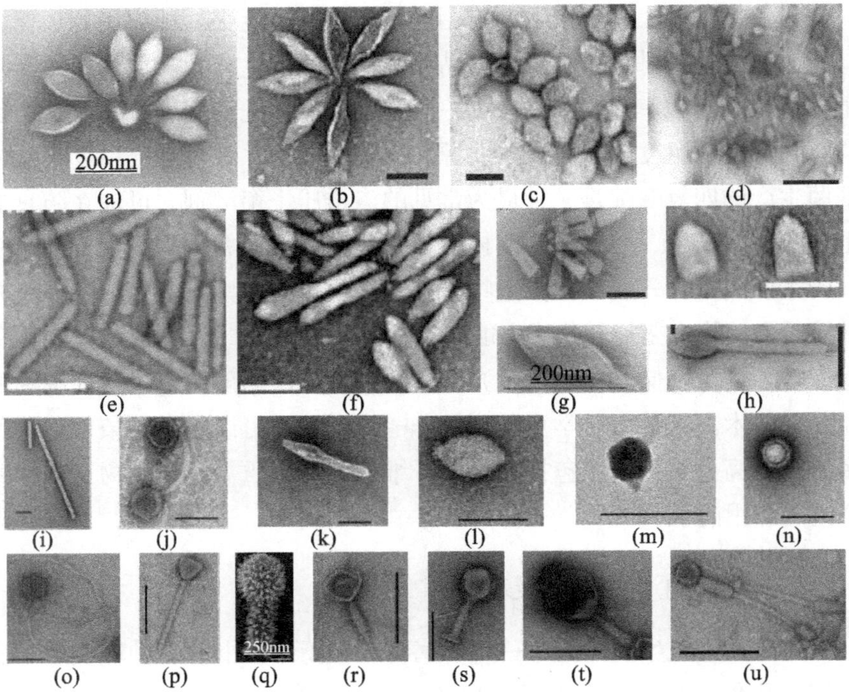

图 3-4 热泉（a～h）以及海底洋壳与沉积物（i～u）的病毒形态

资料来源：Abrahão et al.（2018）；Nigro et al.（2017）；Prangishvili（2013）；Jakubowska-Deredas et al.（2012）。

注：各比例尺，第 1～3 行除标记外均为 100 nm，第 4 行除标记外均为 200 nm。

四、分学科之间的对比与国际平台

（一）各分学科领域发展很不均衡

由于人类的可到达性以及研究技术的差异，各类极端环境微生物的研究程度差别很大。各类极端环境微生物研究程度从高到低依次为地表极端环境微生物、深海极端环境微生物、陆地深部环境微生物、深时极端地质环境微生物、深空环境微生物。

由于在地表极端环境中很容易获得研究样品，也不需要依靠钻探技术等，对地表极端环境微生物的研究程度最高，许多在工业上应用的微生物种质资源都是从地表环境中获得的。无论是陆地热泉、盐湖、酸性矿坑水还是雪冰和冰川等都开展了大量的微生物研究，并取得了许多进展。酸性矿坑水的微生物由于与环境污染和治理有关，研究程度特别高，在微生物的丰度、多样性、生态学功能及其与矿物之间的相互作用等方面都开展了大量的研究。雪冰和冰川是全球变化研究的一大关键载体，得益于全球变化这一热点领域的发展，雪冰和冰川微生物也受到人们的关注，以查明这些微生物对气候变化的响应以及通过碳循环等对气候的影响。热泉微生物受到微生物对高温适应能力及其工业应用的驱动，开展了许多研究。

深海微生物虽然很难获取，但IODP大洋钻探工作的全面开展以及诸如热泉和冷泉群落的发现，极大地激发了人们的研究热情，研究程度也比较高。深海微生物的研究涉及沉积物和玄武岩区、热泉和冷泉区、深渊等关键区域的微生物，在微生物多样性、丰度、分布，以及生态和生物地球化学功能方面已经有了初步的认知，产生了许多令人激动和兴奋的发现，甚至颠覆了我们之前对生命的一些认知。例如，海底存在深部生物圈，颠覆了人们的许多认知；暴露在海水中的玄武岩微生物丰度比上覆的底层海水中的微生物浓度要高出3～4个数量级，远远超乎人们的想象；微生物通过许多耦合生物地球化学过程参与深部碳、氮、硫和铁的循环，与表层系统有很大的差异；沉积物中存在许多高温厌氧化能自养氢氧化菌、小分子烷烃氧化微生物等，极大地影响了深部环境。

20世纪早期，人们就对地表以下生物圈开展了研究。但与深海微生物相比，陆地深部地下微生物的研究程度要低得多，主要集中在油气藏、煤层、洞穴、地下矿坑流体、超深钻和断裂带等环境。其中，油气藏的微生物因为与油气的勘探、开采等有关，其研究程度是相对较高的，在油气藏原位环境微生物活动、代谢途径及转录表达水平等基础研究上得到了许多新的认识，生物采油的一些技术方法也已经进入工业化试验阶段。许多用于驱油的微生物是直接从油气藏中分离出来的。洞穴微生物因与全球变化、微生物资源有关而受到人们的关注和重视。已经发现，洞穴放线菌是抗菌药最重要的来源之一。

受研究技术手段的限制和微生物保存的影响，深时极端地质环境微生物的研究程度很弱。无论是微生物的丰度、种类、多样性和功能研究还是这些微生物随时间的演化研究都比较弱。在地质时期，微生物研究面临的最大难题是种类的鉴定问题，不能利用分子生物学来区别各种不同的微生物属种，只能依靠脂类分子和同位素推测一些关键功能群，包括固氮微生物功能群、氨氧化微生物功能群、产甲烷微生物功能群、嗜甲烷微生物功能群、硫氧化微生物功能群、硫酸盐还原微生物功能群等。在了解这些微生物功能群的基础上，可以定性评估与这些微生物有关的生物地球化学过程及其环境条件，但很难定量评价它们对地质环境的作用大小。一个比较有优势的地方是，深时地质微生物的演化有助于人们了解微生物随时间的演化问题，同时可以了解微生物对极端地质环境的快速响应能力。

深空微生物是研究程度最低的领域，该领域的研究聚焦在两个方向。一是地外生命存在的可能性探索，主要根据水等环境条件以及甲烷代谢等可能的生活活动迹象开展探索；二是聚焦在地球极端环境的相似性研究，主要集中在沙漠等恶劣环境的对比性研究。所有这些工作都基于人们对地球生命的认识。

（二）国际平台

国际上对极端地质环境微生物做出最重要贡献的研究平台是IODP。DSDP最早提供了海洋沉积物中细菌活动性的证据。历史上第一次致力于探究沉积物深部生物圈的ODP航次为2002年启动的201航次，主要目标是探索洋底以下生物的生活习性、分布范围和细菌活动所造成的生物化学影响，以及在不同埋深环境的原核生物群落。钻探集中在秘鲁的盆地和大陆架以及赤道太平洋地区，钻探到的沉积物深达420 m，年龄达到40 Ma，也钻到玄武岩洋壳。ICDP近年来越来越重视深地生物圈的研究，近10年以来，一共资助过5～6个与深地生物圈有关的项目。可以说，没有这些钻探平台，深海和陆地深部极端环境微生物的研究就无从谈起，一些颠覆性的认识也不能获得。

同时，美国也启动了许多科学计划，开展相关研究。1985年启动的“地下科学研究计划”（Subsurface Science Program）是美国最早从事深地微生物研究的项目。2001年又启动了“极端环境微生物”计划。在2009年前后，艾尔弗·斯隆基金会资助美国卡内基科学研究所研究深碳观测，其四个研究方向之一就是深地微生物（包括陆地与海洋）。2011年美国建立了暗能量生物圈调查中心。

国际重要学术组织包括国际地下微生物学会（International Society for Subsurface Microbiology，ISSM）、国际生命起源研究学会（International Society for the Study of the Origin of Life，ISSOL）、地球生物学会（The Geobiology Society）、美国地质学会地球生物学和地质微生物学分部（Geobiology & Geomicrobiology Division，The Geological Society of America）、美国地球物理学会（American Geophysical Union）、生物地球科学（Biogeosciences）学部等。

与极端地质环境微生物最密切的国际学术刊物有《极端微生物》（*Extremophiles*）、1962 年创刊的《地质微生物期刊》、2003 年创刊的《地球生物学》和 2010 年创刊的《微生物前沿》等。比较系统的专著有 1997 年出版的《陆地深部地下微生物学》（*The Microbiology of the Terrestrial Deep Subsurface*）、2014 年出版的《深部生物圈的微生物生命》（*Microbial Life of the Deep Biosphere*）、2016 年出版的《深处生命：地球、火星及其他天体隐秘生命的探索》（*Deep Life：The Hunt for the Hidden Biology of Earth，Mars，and Beyond*）等。

第二节　海洋极端环境微生物

海洋不同的极端环境赋存了不同的生态系统，栖息在这些生态系统中的生物具有多样的代谢潜能与代谢途径（图 3-5），在全球生物地球化学循环中扮演了重要角色。

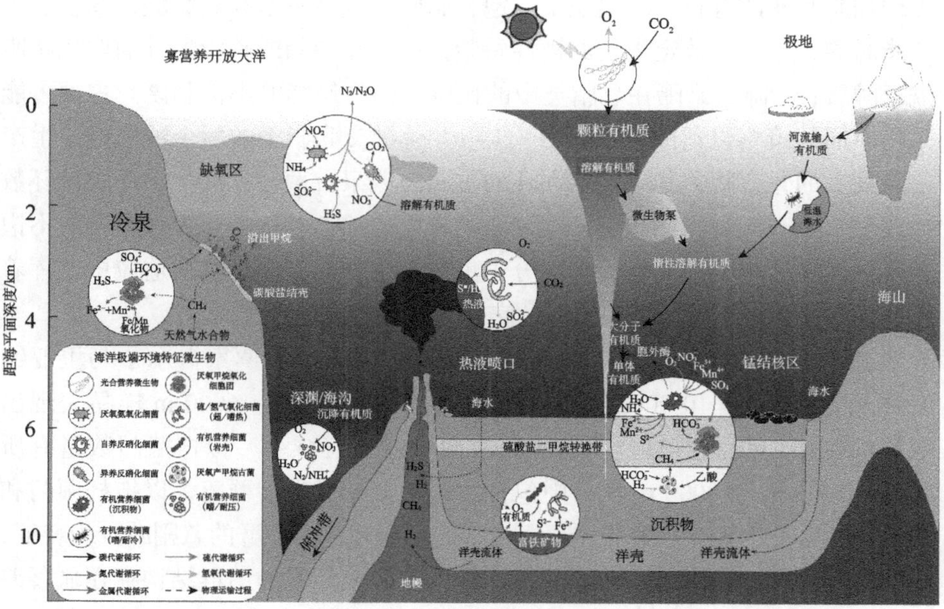

图 3-5　海洋极端环境特征微生物的主要类型与代谢途径示意图（文后附彩图）

资料来源：改自 Xiao et al.（2021）。

一、微生物研究进展

（一）深海、深渊与海沟

深海区具有极端的环境条件，最显著的是极高的静水压力，从而诞生了微生物学的一个新分支——深海高压微生物学。历史上一直认为，从深海采集的微生物种群在高压条件下的代谢活动可能低于常压条件。然而，最近对不同深海系统的研究普遍得出了相反的结论。根据适应压力的不同，可将嗜高压细菌分为压力敏感菌（0.1 MPa）、耐高压细菌（0.1～10 MPa）、嗜高压细菌（10～50 MPa）和嗜超高压细菌（＞50 MPa）4 类；根据适应温度的不同，一般将嗜高压细菌分为嗜冷嗜压菌（＜15 ℃）、嗜温嗜压菌（15～45 ℃）、嗜高温嗜压菌（45～80 ℃）和嗜超高温嗜压菌（＞80 ℃）4 类。

Jamieson（2015）对深海动物群应对静水压力的机制已经进行了系统总结。简而言之，过去几十年的研究强调了合成氧化三甲胺（trimethylamine N-oxide，TMAO）等小分子的重要性以及细胞膜和蛋白质结构的调控作用。为了提高细胞膜在高压环境下的通透性和稳定性，细菌可以采取以下策略：①增加脂肪酸的链长；②累积单不饱和脂肪酸或多不饱和脂肪酸；③积累支链脂肪酸；④积累特定类型的极性磷脂，如以卵磷脂或磷脂酰甘油取代磷脂酰乙醇胺；⑤降低蜡酯在总脂类中的比例，提高蜡酯的不饱和度。古菌可能采用不同的策略以适应静水压力：①以共价键交联脂类的两个酰基链，生产一系列大的化合物，其中包括大环的古菌醇或泉古菌醇；②增加环戊烷环数或改变二醚与四醚脂质比例；③调整类异戊二烯类结构的非极性脂质的不饱和水平。除了脂类，其他的生物分子也在微生物对极端环境的适应中发挥着重要作用，包括核酸、蛋白质以及它们在细胞内的相互作用。

关于深海微生物群落空间变化的报道少得多。科学家揭示了马里亚纳海沟从 6000 m 到 10 000 m 微生物多样性的逐渐变化。6000 m 样品表现出明显的深海向深渊群落过渡特征，古菌数量略有减少，拟杆菌门数量有所增加；7000 m 和 8000 m 样品中包含结构相似的微生物群落，以拟杆菌门和 SAR406 脱铁杆菌门（Deferribacteres）的增加和古菌种群的急剧减少为特征；极端深度 9000 m 和 10 000 m 的菌群结构与较浅深度的菌群结构存在显著差异，γ-变形菌数量显著增加，古菌数量进一步减少（Nunoura et al.，2015）。最近一项关于马里亚纳海沟微生物多样性的研究也证实了 γ-变形菌在水深 10 000 m 以上的深渊水域中占优势（Tarn et al.，2016）。海沟内微生物群落的变化可能主要受有机质输入的驱动以及沟内生境异质性的影响。

海沟之间在微生物群落丰度和多样性方面也存在较大差异。例如，日本海沟从表层到深渊的微生物群落由异养种群如拟杆菌门和 SAR406 所主导，而马里亚纳海沟表层水中的主导微生物群落主要是无机化能营养种群如 SAR324 和奇古菌门（Nunoura et al.，2015，2016）。这一结果可能与两个海沟浅水区富营养化水平直接相关：日本海沟地区相对富营养，而马里亚纳海沟通常被认为是寡营养的（Nunoura et al.，2016）。

（二）深海热液区

深海热液区是典型的化能合成生态系统（chemosynthetic ecosystem），维持有机物与能量来源的初级生产者是化能自养微生物。该类微生物氧化热液喷发带来的地球深部还原性物质，获取能量、固定二氧化碳合成有机物，从而形成热液生态系统的初级生产力。热液区存在的化能合成作用主要包括硫氧化（sulfur oxidation）、氢氧化（hydrogen oxidation）、甲烷氧化（methane oxidation）和甲烷产生（methane production）。通常这些过程发生在微好氧和厌氧条件下。多样的电子供体［H_2S、H_2、CH_4、铵根离子（NH_4^+）、Fe^{2+}］和电子受体（O_2、NO_3^-、SO_4^{2-}、S_0、Fe^{3+}）为化能自养微生物生长繁殖提供了无限可能。这些氧化还原反应的耦合成为微生物获取能量的主要途径，它们在热液区从无机到有机的物质能量转化、元素生物化学循环以及生态系统的形成稳定中发挥着至关重要的作用（邵宗泽，2018）。

热液生物固定碳的途径同温度和氧浓度之间具有重要关联，高效节能的氧敏 rTCA 循环通过取代三种关键酶来反方向运行三羧酸（TCA）循环，通常在 20～90 ℃时占据主导地位。TCA 循环可以通过双向酶（与特定的 rTCA 循环酶相反）反向运行，其方向性（异养与自养）可能取决于环境条件。能量消耗大但耐氧的卡尔文循环在温度低于 20 ℃时占优势。在 90 ℃以上，大多数生长的微生物是嗜热的古菌，它们分别通过 Wood-Ljungdahl 途径和二羧酸-4-羟基丁酸盐途径进行甲烷生成或硫酸盐还原来固定碳。

热液微生物具有多样的环境适应性机制，比如弯曲杆菌门（Campylobacterota）菌株可以利用丰富多样的电子供体和受体；一些菌株如γ-变形菌门热液口盐水球形菌（*Salinisphaera hydrothermalis*）、冰岛热棒菌（*Pyrobaculum islandicum*）既可以自养生长又可以异养生长；一些高温菌，如热球菌目菌株也可以在低温环境中存活，当环境温度合适时则大量繁殖生长。热液区的烃类物质是可以供热液区微生物氧化利用的电子供体，既可以有氧氧化，也可以厌氧氧化。因此，深海深层生物圈的能量支撑可能部分来自烃类物质。

深海热液喷口周围的生态系统可以对周围的海洋产生更大范围的影响。生长在热液羽流中的微生物能出现在离喷口几十千米远的地方，这些无机营养所富集的微生物群落至少能迁移数百千米远的距离。考虑到在海洋微粒、OMZ 和溢油环境都存在无机营养的电子供体，以及在这些生境中的微生物群落与热液羽流有相似之处，因此推测在热液喷口和非喷口之间，微生物群落可能存在相互作用。

（三）冷泉区与海洋沉积物

人们发现，深海沉积物中蕴藏着数量庞大、种类繁多、性质独特的微生物资源。其中，冷泉作为沉积物中一个特殊的生态系统更是备受关注。

表层沉积物具有丰富的有机质以及含有少量的氧气，其微生物的多样性也相对较高，分布着大量能够利用氧气或硫酸盐来降解大分子有机质的微生物类群。洛基古菌在很多沉积物表层作为优势类群被发现，是连接原核与真核微生物演化关系的一个重要桥梁（Spang et al.，2015），具有重要的演化意义。表层沉积物中另一个主要的古菌类群是海洋古菌类群Ⅰ（MG-Ⅰ），其最早是在海水中被发现的（DeLong，1992）。沉积物中的MG-Ⅰ类群，可能是由海水中的MG-Ⅰ逐渐渗入进化而来的。

从海洋沉积物整体范围来看，分布最广泛的古菌可能是海洋底栖古菌类群D与深古菌。目前这两类古菌还尚未有报道被成功分离进行纯培养的菌株。不过已有一些被成功富集的深古菌（Yu et al.，2018），并通过基因组研究与培养发现，其具备降解一些大分子有机底物的潜能，如木质素、胞外蛋白，此外它们还具备能够通过产乙酸获得物质能量的潜能（He et al.，2016）。

在较深的沉积物中，有机质逐渐被降解为小分子化合物，如 CO_2、乙酸、小分子甲基化合物等，这些小分子物质又会被产甲烷菌用来产甲烷。产甲烷菌常在沉积物深部寡营养层位作为优势种群出现，但由于其能利用的底物较为广泛，因此也能在表层沉积物中被发现。目前的产甲烷菌通常属于广古菌门，而最近的研究基于基因组层面的分析表明，深古菌门与韦斯特拉古菌门（Verstraetearchaeota）也具备利用甲烷的潜在可能性。海洋沉积物中普遍存在硫酸盐还原和产甲烷微生物代谢过程的竞争（Zhuang et al.，2019）。在有机质含量低、微生物代谢活动弱的大洋沉积物中，沉积物在较长的深度内都可保持较高的硫酸盐浓度，在此浓度下，微生物以硫酸盐还原与发酵为主要的代谢活动。在大陆边缘区域，硫酸盐下渗的深度则要大大降低。在海洋沉积物中，硫酸盐浓度通常随沉积物深度的增加而逐渐降低，甲烷浓度则逐渐升高。在一定深度的沉积物中会同时存在硫酸盐浓度降低与甲烷浓度升高的独特的化学离子梯度环境［硫酸盐-甲烷转化带（SMTZ）］，此区域主要进行微生物参与的硫酸盐还原与甲烷厌氧氧化作用。

甲烷同时也是微生物的一个重要物质能量来源，尤其是在冷泉环境中，为冷泉生态系统提供能量物质来源。甲烷的消耗被一类 ANME 通过 AOM 所消耗。ANME 往往与其他细菌（如硫酸盐还原细菌）形成共生关系，共同执行 AOM 过程，这也代表了深层微生物一种重要的能量代谢模式。在海洋沉积物中，SMTZ 代表了深部的氧化还原交界区域，通常表现出较高的微生物活性，AOM 通常就发生在这一区域。尤其在冷泉区，由于其上渗的甲烷浓度较高，对硫酸盐的消耗较大，SMTZ 通常只有几厘米到几十厘米，其 AOM 的活性较高，也是冷泉环境中最具代表性的微生物代谢活动。

冷泉中微生物参与的甲烷厌氧氧化机制一直是科学家研究的重点，尤其是围绕着该过程中的电子受体与传递做了很多的研究。目前对 ANME 和硫酸盐还原菌之间的物质能量传递体的研究有很多，分别推测 AOM 反应的中间物质可能是氢气和乙酸盐，而甲基硫、零价硫作为中间物质。也有研究指出，ANME 与 SRB 之间通过细胞色素传递进行直接的电子传递，这也是一种重要的细胞间电子传递方式。大量直接证据表明，ANME 和 SRB 之间以直接电子传递为主。在冷泉生态系统中，微生物通过转化冷泉喷涌出的甲烷和硫化物等物质为其他动物提供物质和能量，从而支持整个生态系统。从生态意义上看，在全球海洋环境中，每年有大量的甲烷产生，在海底沉积物中有约 5500 亿 t 碳以甲烷水合物形式储存在深部沉积物中，但只有一小部分，约 2000 万 t 碳以甲烷气体形式释放到上层海水中，其余约 90% 的甲烷在沉积物被 AOM 途径所消耗。

真菌是海底沉积环境中真核生物的主体，其数量占总真核生物数量的比例随沉积深度的增加而增加（Orsi et al.，2013）。海底沉积物真菌与陆源真菌具有高度的相似性，几乎所有的海底沉积真菌在陆地生境均有分布。其中，子囊菌和担子菌是海底沉积环境的优势种群。

海洋沉积物中的微生物往往需要进化出自己独特的代谢特征以适应不同的极端环境。沉积物中的深部微生物往往体积较小、呈圆形，这可能是其适应极端环境的结果。微生物的适应性还体现在细胞膜上，由于建立跨膜离子的浓度梯度需要消耗能量，相对于 H^+、Na^+ 的跨膜扩散速率会比较低，为了节约能量，细胞在产生跨膜电位的时候，会优先采用 Na^+，而不是常见的 H^+。在代谢调控机制方面，深部沉积物中的微生物往往会选择简单的代谢调控策略与缓慢的生长速率。海洋沉积物的微生物还是科学家研究地球生命起源和进化的理想材料。

（四）海洋低/缺氧区

关于海洋缺氧研究的第一篇文章发表于 1983 年，2000 年以后其研究热度增加得尤为显著，至 2020 年共计发表海洋缺氧相关研究文章 457 篇。最高引论文为 2008 年发表于 *Science* 上的题为“死亡带的扩张及其对海洋生态系产生的后果”（“Spreading dead zones and consequences for marine ecosystems”）的论文。

“温室气体排放量增加导致的全球变暖”被认为是造成开放海域持续脱氧的主要原因，然而河口和近海的营养物质（氮和磷）与有机物质增加会导致微生物呼吸作用加强，又促进了脱氧，而全球变暖则会进一步加剧脱氧。溶解氧的下降将改变生态系统结构，并急剧影响区域内微生物参与的氮、磷和微量金属等元素循环。低氧下反硝化微生物将亚硝酸盐转化为温室气体一氧化二氮或在更低氧浓度下将一氧化二氮进一步还原为氮气，而厌氧氨氧化细菌将亚硝酸盐和铵根离子直接转化为氮气。在最小含氧带中观察到大量的氮损失，约占全球反硝化的 10%（Wright et al.，2012）。最小含氧带被发现存在一个隐藏的硫循环，硫酸盐还原产生的硫化物可以加强硝酸盐还原（Canfield et al.，2010）。海洋脱氧最初可能会提高初级生产者对磷和铁的可利用率，然而生产力的提高导致有机物浓度提高，会促进微生物的呼吸作用，因此，生产力、脱氧和磷、铁释放形成了正反馈回路（Wright et al.，2012）。

（五）海山与多金属结合区

除了铁氧化深海菌（*Mariprofundus ferrooxydans*），还有很多细菌和古菌从海山中被分离得到，包括极端嗜热产甲烷菌（*Methanocaldococcus bathoardescens*）、能利用氢气作为能源的新型化能自养细菌以及在厌氧条件下能利用氢气和硫代硫酸盐的古菌等。海山火山口微生物群落主要由 ε-变形

菌纲或 ζ-变形菌纲细菌所组成，二者可具有时空上的不一致性，ζ-变形菌纲主要是罗希海山周围或里面喷口地带的菌群类型，而 ε-变形菌纲则主要出现在温度更高的火山出口处。

海山从浅到深的特征决定了其垂直连续分布的生境，形成了生物、固相栖息地、海水流动相之间的复杂关系。国际社会正逐渐认识到深海海山生物资源的重要性，海山成为研究微生物生态功能和地理分布的热点，如海山之间的微生物基因交换。国际上对海山矿产资源的日益关注，促进了海山调查以及海山微生物群落在矿物形成或风化中的功能研究。富钴结壳区微生物作为海洋特殊生境中重要的生物类群，其多样性及其在海山生态系统中的作用将成为未来研究的重要方向。

（六）洋壳岩石圈

上洋壳玄武岩与流经裂缝的海水发生化学反应，释放出的能量可以被化能自养微生物利用来支持自身的生长，特别是玄武岩中还原态的硫和铁与海水中的溶解氧以及硝酸盐之间的氧化还原反应支撑着较高的微生物生物量和多样性。通过近几十年的研究，已经有一系列的间接证据表明了在洋壳中栖息着活跃的微生物群落，这些证据包括微生物蚀变作用形成的特征性显微结构、$\delta^{34}S$ 与 $\delta^{13}C$ 的亏损、洋壳观测孔中获取的 DNA 序列等。这些研究主要集中于玄武岩质的上洋壳中。相比之下，由辉长岩组成的下洋壳，平均厚度为 5 km，其体积占整个洋壳体积的 2/3，近期的研究表明，其内部也栖息着独特的微生物系统（Li et al.，2020）。总的来说，地球深部生物圈研究取得的重要进展之一就是洋壳生物圈的发现，多种证据表明它们在洋壳风化和地球元素循环中发挥着重要作用。

尽管上洋壳玄武岩的体积巨大，高达 10^{18} m^3，但在玄武岩表面的玄武玻璃中栖息的微生物密度通常较低，约 10^3 个细胞 /cm^3。假定微生物细胞与元素碳之间具有 60 fg/ 细胞的转换关系，且海底玄武岩中大约有 2% 属于玄武质玻璃，估算出海底玄武质玻璃中栖息的微生物的细胞总数为 2×10^{25} 个细胞，生物碳的总量为 6×10^{11} g，仅占地球生物总碳规模（8×10^{17}g）的很小一部分，不到 1%（Fisk et al.，1998）。最近一项针对下洋壳微生物的研究也发现，洋壳中微生物的细胞丰度非常低，为 10^2～10^3 个细胞 /cm^3（Li et al.，2020）。然而，在东太平洋隆起（East Pacific Rise，EPR）9°N 海底玄武岩区，细菌和古菌的丰度在 3×10^6 ～ 1×10^9 个细胞 /g，其中细菌的总数远高于古菌，占微生物总量的 88%～96%（Santelli et al.，2008）。在某些古老（33.5 Ma～104 Ma）的洋壳玄武岩蚀变的区域，局部微生物的细胞丰度可以高达 10^{10} 个细胞 /cm^3，表明玄武岩的蚀变能够支撑起局部繁茂的微生物群落，它们经常以微生物菌席的形式存在（Suzuki et al.，2020）。

上述研究表明，洋壳岩石圈中微生物的分布具有很强的异质性，微生物细胞的丰度变化很大，差别可以超过 8 个数量级。目前在对洋壳这一地球最大的微生物栖息环境的评估中，通常按照接近微生物所能够耐受的温度上限 120 ℃的等温线来对洋壳进行划分，在靠近大洋中脊新形成（1 Ma 左右）的热的洋壳附近，这一界限大致在 500 m 的深度范围，而在老且冷（180 Ma）的洋壳中，这一深度可以达到 5 km。最新的估算表明，洋壳深部生物圈中微生物的细胞总量为 4×10^{29} 个细胞。该研究还进一步估算，在海洋洋壳深部生物圈中，20%～80% 的微生物，或者说 8.0×10^{28}～3.2×10^{29} 个细胞是以生物膜的形式黏附于岩石表面的（Flemming and Wuertz，2019）。微生物以生物膜的形式聚集，而不是分散于岩石中，这被认为是一种适应寡营养环境的生存策略。

洋壳中栖息的微生物广泛参与洋壳的各种地球化学过程，在洋壳元素的淋滤、聚集以及矿物的转化过程中起着重要作用。微生物可以从玄武质玻璃中获取代谢所需能量或者营养物质。玄武岩中存在的还原态铁、锰和硫均可以作为电子供体，Fe^{3+} 可作为电子受体。玄武质玻璃还含有其他微生物生长所必需的微量营养元素，如磷和金属等。洋壳中循环的一些流体中含有有机碳、氢气和硫化氢等，这些组分被微生物所利用，代谢所产生的酸，也能够加速洋壳岩石的风化，从而参与洋壳的元素地球化学过程。虽然科学家很早就从暴露于海底玄武岩的蚀变结构中推测微生物在海底洋壳风化过程中起着重要的催化和促进作用（Fisk et al.，1998），然而长期以来一直缺乏对海底洋壳玄武岩岩石中栖息微生物群落的系统调查和微生物学的直接证据。在东太平洋隆起 9°N 和夏威夷附近玄武岩的玻璃表面，栖息的细菌和古菌的丰度远高于周围海水的微生物密度，微生物的丰富度、多样性等也远高于上层海水、深海海水、深海沉积物和海底热液喷口等环境，从玄武岩中检测到的微生物类群并无特殊之处，也是海水以及海洋沉积物中比较常见的类型（Santelli et al.，2008）。虽然东太平洋隆起 9°N 和夏威夷两者在地理上相隔甚远，但它们容纳的微生物群落组成却具有高度的重叠（Santelli et al.，2008）。相比之下，基于 IODP 钻孔取芯的一项针对海底沉积物覆盖之下玄武岩的调查研究却有截然不同的发现。利用 IODP301 航次 U1301 钻孔，人们对胡安·德富卡（Juan de Fuca）洋脊东翼玄武岩栖息的微生物类群进行了调查（Lever，2011），发现在年龄为 3.5 Ma 的玄武岩中栖息着大量与甲烷代谢和硫循环密切相关的微生物类群。随着钻孔深度的增加，出现了自养代谢过程减弱而异养代谢过程增强的现象。这表明暴露于海底的玄武岩与洋壳深部玄武岩中栖息的微生物类群以及它们的代谢活动具有明显的差异。暴露于海底的玄武岩处于冷的环境中，与海水接触发生蚀变反应；相比之下，深部的玄武岩可能处于一个热的环境中，尤其是局部的热液活动也极大地影响微生物的生物地球化学过程以及生理代谢活动。

洋壳年龄是决定微生物分布的重要因素之一。随着年龄的增加，洋壳内的环境逐渐冷却，岩石中的孔隙、裂隙等逐渐被次生矿物充填，洋壳内流体循环的强度和水岩反应的强度均会急剧下降（Suzuki et al.，2020）。在年轻的洋壳玄武岩中，还原态铁和硫的氧化过程能够支撑起一个规模庞大的自养生物量。然而，在更古老的洋壳中，依赖于有氧异养过程的微生物能够驻留在更老的地壳中，并且其丰度可以超过 10^{10} 个细胞 /cm^3，因此这将会大大增加对上洋壳微生物细胞丰度的估计。

下洋壳中同样也发现了微生物栖息的证据。在大西洋亚特兰蒂斯高地下洋壳中，微生物丰度极低，$<10^3$ 个细胞 /cm^3，全部以细菌为主，没有检测到古菌的信号；生物多样性较低，α-变形菌、β-变形菌、γ-变形菌是最主要的组成类群；栖息着一个以降解碳氢化合物这一异养代谢过程为主的细菌生态系统（Mason et al.，2010）。在西南印度洋亚特兰蒂斯岸带（Atlantis Bank）下洋壳辉长岩区域栖息的微生物也有着类似的特征（Li et al.，2020），即微生物的丰度非常低，异养代谢过程的基因远多于自养代谢过程的基因。

二、计量统计结果分析发展态势

截至 2020 年 8 月，海洋极端地质环境微生物这一主题近 20 年（2001～2020 年 8 月）共发表 720 篇科学引文索引（SCI）论文，共被引用 21 468 次，篇均被引 29.82 次，*H* 指数高达 67。年度论文发表量总体呈现波动上升的趋势。发文量最多的年份为 2018 年，达到了 63 篇（图 3-6）。

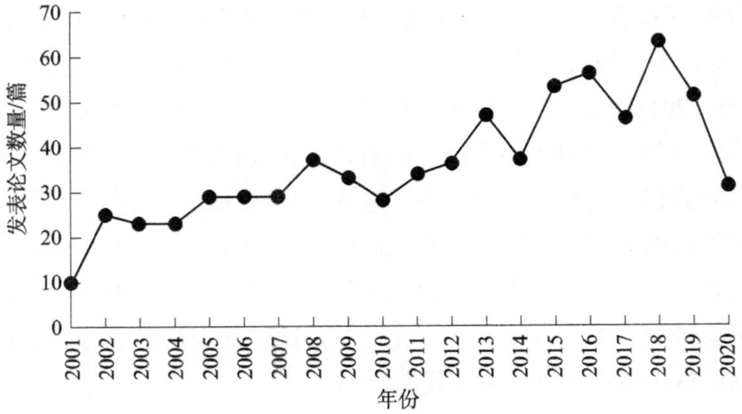

图 3-6　2001～2020 年海洋极端环境微生物年度发表论文数量

就国家 / 地区而言，全世界共有 73 个国家 / 地区参与海洋极端环境微生物的研究。其中，美国的发文量位居第一，占比 26.53%，其次为日本和中国[①]。排前十位的国家 / 地区还有德国、法国、西班牙、意大利、韩国、英格兰和印度（图 3-7）。

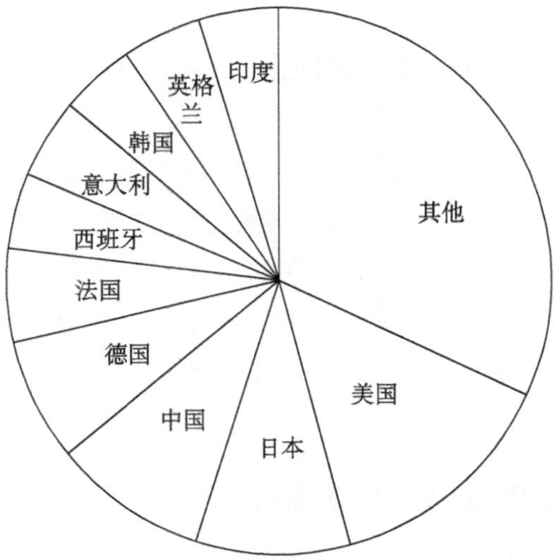

图 3-7　海洋极端环境微生物排前十位的国家 / 地区发文量相对比例

① 不包括台湾地区数据。

就研究机构而言，全世界共有1001个机构参与了海洋极端环境微生物的研究。其中，日本海洋地球科学技术厅的发文量位居第一，占比14.31%。其次为巴斯德研究所、法国国家科学研究中心（CNRS）和马里兰大学。国内机构国家海洋局和中国科学院紧随其后，占比分别5.14%和4.31%。排前十位的国内机构还有中国科学院生态环境研究中心（2.64%）。

在基金资助机构方面，全世界共有777个机构为海洋极端环境微生物研究提供基金资助。其中，中国国家自然科学基金资助的发文量最高，占比10.14%。除中国以外，美国、欧盟、日本、德国、法国也为各国家/地区该主题的研究提供资助。排名前十位的还有中国科学技术部的国家重点研发计划（2.5%）和国家重点基础研究发展计划（973计划，1.94%）。

从论文关键词检索分析，海洋极端环境微生物这一主题的关键词包括嗜极微生物（extremophile）、古菌（archaea）、细菌（bacteria）以及深海热液喷口（deep sea hydrothermal vent）等，这些关键词也体现了当前研究的热点与前沿。近20年来，海洋极端环境微生物这一研究主题的关键词也经历了时间变迁。出现时间相对较晚的关键词有抗微生物活性（antimicrobial activity）、真菌（fungi）、适应（adaptation）、嗜盐古菌（haloarchaea）、生物勘探（bioprospecting）、海洋真菌（marine fungi）、宏基因组学（metagenomics）、生物活性（biological activity）等，说明了这一研究方向的主题变迁。

三、国际上的研究团队与平台

海洋极端环境研究对设备具有较大的依赖，欧美等发达国家起步较早，最早美国的斯克里普斯海洋研究所、伍兹霍尔海洋研究所（WHOI）利用装备优势，组织团队，拓展了海洋极端环境微生物的研究。从2001～2020年的发文量和影响看，目前较为活跃的团队主要来自以色列希伯来大学的Oren Aharon团队，日本海洋研究开发机构（JAMSTEC）的Ken Takai和Fumio Inagaki团队，伊朗德黑兰大学的Amoozegar和Mohammad Ali团队，巴斯德研究所的Mart Krupovic团队，德国比勒费尔德大学的Norbert Sewald团队，中国上海交通大学的肖湘、王风平团队以及自然资源部第三海洋研究所的邵宗泽团队。

第三节　地表极端环境微生物

一、各类极端环境的适应机理

（一）嗜（耐）盐机理

高盐环境对微生物生存的最大威胁在于失水和离子毒性（Oren，2002）。在盐度较高的环境中，微生物胞内的渗透压明显低于胞外，若没有合适的机制来限制这种渗透压差，微生物将迅速脱水死亡。嗜盐微生物主要通过两种机制来适应高盐环境：①通过 Na^+ / 氢离子（H^+）泵跨膜蛋白将胞内多余的 Na^+ 排到胞外，同时利用钾离子（K^+）单向转运系统将胞外的 K^+ 泵入胞内维持渗透压平衡；②绝大多数耐（嗜）盐菌通过基因组编码的蛋白质在胞内合成相容性溶质，如甘油、谷氨酸及脯氨酸等有机物，以适应高盐环境（Oren，2002）。

（二）嗜（耐）酸机理

通过对生长于极端酸性环境中的嗜酸微生物进行生理生化实验以及基因组分析发现，这些微生物在漫长的演化历程中形成了各种 pH 平衡机制来应对强烈的酸胁迫：①通过改变膜脂的结构和组成，以及减小膜通道的孔径和通透性等途径来形成对质子高度不渗的细胞膜，从而抑制质子通过细胞膜流入细胞内（Macalady and Banfield，2003）；②通过主动运输钾离子或者其他阳离子进入细胞内，产生独特的化学渗透梯度，即唐南电势（Donnan potential）来减少质子的内流（Baker-Austin and Dopson，2007）；③通过降解对自身有害的有机酸（乙酸和乳酸），或者使用质子流出系统，例如 H^+-ATP 酶、逆向转运蛋白以及协同转运蛋白等来降低细胞内质子浓度。当嗜酸微生物细胞内多余的质子未能及时转移至外界时，其细胞质缓冲液则可以有效应对小幅度的胞内 pH 变化，从而缓解 pH 降低带来的不利影响（Castanie-Cornet et al.，1999）。此外，相比于不耐酸的微生物，嗜酸微生物细胞内的酶中含有大量的铁蛋白，这有助于在更低的 pH 环境条件下维持酶的结构和功能的稳定性；嗜酸微生物也可以通过合成分子伴侣来修复受损的大分子，从而保证嗜酸微生物重要生命活动的正常进行（Baker-Austin and Dopson，2007）。

（三）嗜（耐）高温机理

高温地质环境微生物为了适应高温环境，从基因组、蛋白质到细胞结构层次，以及生态方面都具有其特异性。①嗜热微生物的基因组较小，基因组结构更紧凑，有利于缩短细胞分裂时间而加快增殖速度，也能减少核酸合成过程中的能量消耗。基因组 DNA 还可通过离子保护作用（如 K^+ 和 DNA 的负电性有助于保持 DNA 的双链结构）、排列和折叠方式（更多的嘧啶-嘧啶或嘌呤-嘌呤组成）、特殊的甲基化方式、更多的 DNA 结合蛋白质等应对机制来增加其热稳定性（Sabath et al.，2013）。水平基因转移可发生在许多不同的生物之间，大到不同的生物界（嗜热细菌和古菌之间），小至相邻的属。②嗜热微生物的蛋白质热稳定性也是保持代谢活性最重要的保护措施。嗜热微生物蛋白质的氨基酸序列较短，所含蛋白质家族种类也较少，由此可以减少易高温变性位点。同时会表达出一系列重要的生物大分子，如热休克蛋白、伴侣蛋白以及一些抗氧化蛋白等，以增加嗜热微生物的蛋白质在高温环境下的稳定性（Wang et al.，2015）。一些嗜热微生物还会利用二硫键或形成复合物来增加蛋白质在高温环境下的稳定性。③嗜热微生物具有特殊链长、不饱和度、环数、极性头基团的膜脂结构，从而增强细胞膜脂在高温条件下的稳定性与通透性。此外嗜热微生物还具有特殊的细胞壁结构。④嗜热微生物之间的共生关系也是增强其应对极端条件的措施之一（Wurch et al.，2016）。

（四）嗜（耐）低温机理

由于长期处于低温环境中，低温微生物进化出了适应低温环境的生理机制，包括特殊的蛋白质结构、冷休克蛋白基因的高表达和细胞膜脂类组成的改变等（Siddiqui et al.，2013；De Maayer et al.，2014）。①在低温条件下降低精氨酸和脯氨酸含量，起到的作用是减少氢键和盐桥，从而增强蛋白质构象柔性，提高酶在低温下的活性。②产生抗冻蛋白或冰结合蛋白，这些蛋白质可以结合并限制冰晶的生长，起到降低细胞质冰点的作用。③在低温条件下调整细胞膜流动性的能力。嗜冷微生物通过吸收甘氨酸、甜菜碱、蔗糖等溶质，降低细胞内的凝固点，从而获得低温下保持活性的能力。④胞外多糖也是一种低温保护剂，嗜冷菌在低温下可以产生大量的胞外多糖。此外，微生物可以通过冷休克反应来应对突然降温的环境，低温下高表达的基因主要是编码冷休克蛋白的基因；低温环境下，嗜冷微生物可以通过提高抗氧化酶合成基因的表达量来应对低温下的氧化损伤；嗜冷微生物还可以通过增加基因组中单糖利用基因的拷贝数或通过获得新的一碳代谢基因应对冰川寡营养环境（Shen et al.，2019）。

（五）嗜（耐）干旱机理

为了抵御干旱生态系统缺水、高辐射等外界压力，干旱生境中的光合细菌（如蓝细菌）及一些产孢细菌（如放线菌）等已经进化出许多适应行为以及一些细胞生理机制。①进入休眠状态，如放线菌和芽孢杆菌类群在缺水条件下会进入代谢极其缓慢的休眠状态（Filippidou et al.，2019），形成厚壁孢子（akinetes），孢子体内部以吡啶二羧酸（DPA）和酸性可溶性小蛋白（SASP）为主，它们可以保护细菌的DNA免受氧化应激损伤。②产生可调节渗透压的物质（Lennon et al.，2012），包括胞外聚合物（extracellular polymeric substance，EPS）或胞内相容性溶质（compatible solute），很多蓝细菌可以产生含脱氧糖苷、多肽和乙酰基团的疏水性EPS，EPS也是形成生物膜（biofilm）的重要组成部分，可显著提高微生物应对逆境等复杂环境的适应性。同时，微生物细胞内部相容性溶质的表达合成量上升，常见的相容性溶质有脯氨酸和海藻糖，相容性溶质可维持细胞内渗透压的平衡。③提高活性氧分子清除酶的表达，如超氧化物歧化酶（SOD），以降低应激压力所产生的过多的ROS的浓度，保护微生物免受ROS的氧化损伤（Katoh et al.，2004）。④产生遮光剂，如黑色素、产孢吸光色素和伪枝藻素（Pointing and Belnap，2012），遮光剂对干旱区土壤表层微生物群落非常重要，因为这种被动保护机制即使是在微生物长时间的休眠期仍能发挥作用，这些环境避难所提供的干燥缓冲区保护了干旱区微生物，使其免受紫外线辐射。

二、计量统计结果分析发展状况与研究队伍

（一）高盐环境微生物

截至2020年5月，在Web of Science科学引文索引扩展版（SCIE）数据库中检索到2000～2019年发表的嗜盐/耐盐微生物文献7307篇。这意味着过去20年中，基本上每天都有一篇与嗜盐/耐盐微生物有关的文章发表。超过10%的研究集中在生物科学（包括环境科学与生态学、微生物学、生物化学与分子生物学、进化生物学、生物技术与应用微生物学、生物多样性保护）、地球科学（地质学、地球化学）、盐湖农业、水科学等领域。

其中，涉及高盐湖泊环境的文献有526篇。从世界范围来看，中国参与的研究数量明显领先于其他国家，涉及的研究机构主要是中国科学院、中国地质大学、云南大学、清华大学。参与相关研究较多的国家还有美国、法国、以色列等。通过对这526篇文献的具体主题进行分析发现，盐湖研究采用的研究手段有培养、组学、同位素地球化学与矿物学，研究方向主要集中在地质、微生物资源、微生物生态、适盐机制、应用微生物与元素循环等方面，研究热点地区集中在美国、中国、德国等国家。

涉及半干旱/干旱区盐碱土壤环境的文献锐减为86篇，这意味着对比其他高盐环境（如深海、盐湖、高盐滩涂等），半干旱/干旱区盐碱土壤环境中的耐盐/嗜盐微生物研究是十分匮乏的。从论文发表现状来看，中国学者参与的研究数量明显领先于其他国家，参与相关研究较多的国家还有印度、美国、伊朗、西班牙等[图3-8（a）]。通过对这86篇文献的具体主题进行分析发现，围绕农业生产（以利用耐盐/嗜盐微生物提高植物抗逆性、改良盐碱地为主）的研究数量远超其他方向，说明改良盐碱地、增加作物产量一直是半干旱/干旱区盐碱土壤环境耐盐/嗜盐微生物研究的热点。此外，极端嗜盐/耐盐微生物的分离及应用和盐碱土壤微生物生态学这两个极端微生物研究的传统方向也占据了相当大的比例。类比地外环境，寻找地外生命方向的3篇文献全部发表于2017～2019年，代表了半干旱/干旱区盐碱土壤环境耐盐/嗜盐微生物研究的全新领域。还有少量研究将这一领域的研究与古气候及生物标志物进行了结合[图3-8（b）]。

(a) 主要参与国家

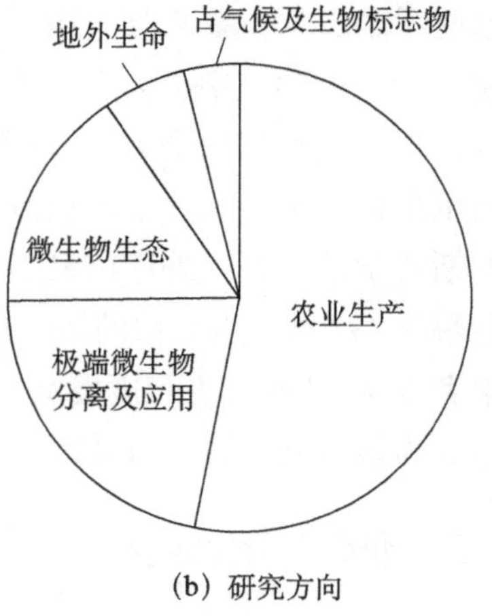

(b) 研究方向

图3-8　半干旱/干旱区盐碱土壤环境耐盐/嗜盐微生物论文的主要参与国家与研究方向

农业生产、极端微生物分离及应用以及微生物生态三个方面的研究较为传统，且往往具有重要的经济、战略价值（如增加粮食作物、经济作物的产量，获取具有各种潜在功能的耐盐 / 嗜盐微生物资源等），因此世界范围内多个国家的研究团体（如西班牙格拉纳达大学、伊朗德黑兰大学、中国科学院等）对这三个方面的研究均较为关注。

（二）热泉环境微生物

Web of Science 数据库 2000～2019 年涉及“热泉”（hot spring）的文献记录有 5381 条。这些文献以研究论文和专利为主，共计 4906 条（占总数的 91.2%）。2017～2019 年发表研究论文共 1205 篇，超过总数的 20%，表明热泉研究尚在起步阶段，同时也表明热泉是热点研究领域之一。这些文献的研究领域主要集中在生物科学（包括环境科学与生态学、微生物学、遗传学、生物化学与分子生物学、生命科学及生物医学、进化生物学、生物技术与应用微生物学、生物多样性保护）、地球科学（地质学、地球化学、地球物理学）以及其他学科（物理学、能源燃料科学、材料科学、高分子科学、水科学）上。其中，生物科学是研究最为集中的领域。主要采取的研究手段有培养、指纹基因（如 16S rRNA）或脂类、组学、同位素地球化学、矿物学等技术。研究主题包括种质资源、微生物生态、适热机理、代谢途径等。研究热点地区集中在美国黄石国家公园、大盆地国家公园，芬兰，冰岛，中国云南腾冲、西藏，东南亚等地。

这些学术贡献主要来自美国、日本、中国。涉及的研究机构主要是中国科学院、云南大学、中山大学、中国地质大学，美国蒙大拿州立大学、佐治亚大学，加拿大阿尔伯塔大学，俄罗斯科学院。其中，中国科学院是国内主要的贡献机构，也是国际上产出最多的机构。Bonch-Osmolovskaya Ea（就职于俄罗斯科学院）是国外产出最多的学者。2015～2019 年来自中国的研究文章的总数为 378 篇，占总数的 20.2%，呈明显的增加态势。尤其是在 2019 年，中国学者发文 88 篇，占年度发文总数的 22.8%，表明中国学者在热泉研究领域已经成为最活跃的科研力量之一。

（三）低温环境微生物

对地表低温环境微生物的研究主要集中在对冰雪和冻土环境中微生物的研究。在 Web of Science 数据库，近 2000～2019 年涉及这一领域的有 5661

篇文献。其中，选择包含微生物学主题的文章，并去除病理学等方向文章后共有 783 篇相关文献，这些文献的研究领域主要集中在环境科学与生态学、微生物学等方面，主要采取的研究手段有指纹基因（如 16S rRNA）、功能基因、纯培养和宏基因组等技术。2010～2019 年，有关冰雪中微生物的功能基因、纯菌分离及宏基因组方面的研究趋势如图 3-9 所示。研究主题包括微生物的种质资源、微生物生态、微生物的适冷机理和代谢途径等。研究热点地区集中在北极、南极、泛第三极地区。

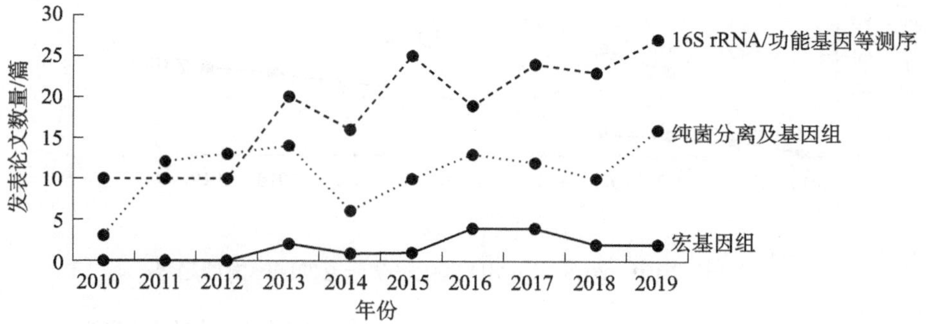

图 3-9　2010～2019 年冰雪中微生物研究的国际发展趋势

这些学术贡献主要来自美国、中国、德国和英国。涉及的研究机构主要有中国科学院、德国亥姆霍兹联合会（HGF）、法国国家科学研究中心、美国加州大学、英国布里斯托尔大学和丹麦奥胡斯大学。其中，国际上产出最多的机构是中国科学院，它也是我国国内主要的贡献机构。

检索到冻土微生物研究文献 706 篇，从 2000 年的 9 篇增长至 2019 年的 74 篇，2020 年（截至 5 月）发表 44 篇。发表文献数量总量排名前五的国家分别是俄罗斯、美国、中国、加拿大、德国。

（四）酸性环境微生物

Web of Science 数据库 2000～2019 年关于酸性环境微生物的文献共 213 篇，其中 16S rRNA 测序分析 163 篇、宏基因组 43 篇、宏转录组 6 篇、宏蛋白质组 7 篇、单细胞基因组 3 篇。过去 20 年，分子生物学方法，尤其是基于高通量测序技术以及相关生物信息学和统计学方法的发展和应用，以前所未有的广度和深度推动了以 AMD 作为模式系统的微生物生态学研究，极大地拓展了科学界对 AMD 微生物多样性、功能与进化的认识。目前，AMD 微生物的研究还是以 16S rRNA 测序分析微生物多样性为主，文章数量从 2010 年的 13 篇增至 2019 年的 158 篇；宏组学研究方法的应用仍处在上升阶段，从 2010 年的 1 篇增加到 2019 年的 51 篇，该领域内相关文章占比迅速上升并维持在 33% 左右。以 2010 年后发表的文章数作为依据，AMD 微生物学研究 2010～2019 年国际发展趋势如图 3-10 所示。

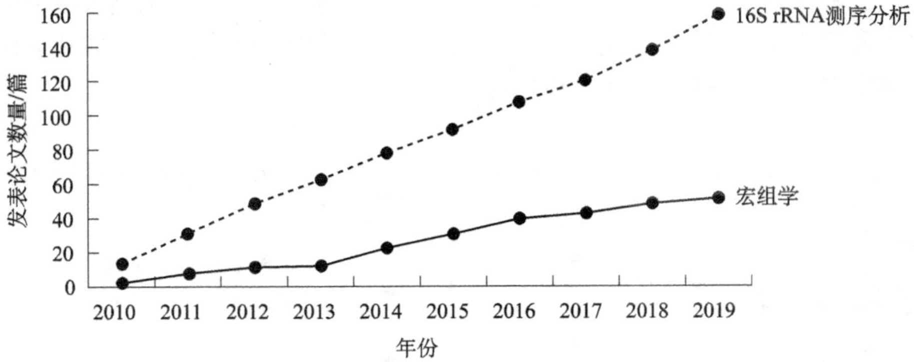

图 3-10　2010～2019 年 AMD 微生物学研究的国际发展趋势

这些文献的研究方向主要涉及 AMD 微生物群落在不同时空尺度上的分布模式和生态驱动力、代谢功能以及群落功能的时空动态变化等。这些学术贡献主要来自美国、西班牙、德国、澳大利亚、法国和中国。涉及的研究机构主要有美国加州大学伯克利分校、西班牙国家研究委员会（Consejo Superior de Investigaciones Científicas，CSCI）、德国联邦地球科学和自然资源研究所、澳大利亚联邦科学与工业研究组织、法国国家科学研究中心以及中国中山大学。国际上产出最多的机构是美国加州大学伯克利分校，中国中山大学生命科学学院是国内主要的贡献机构。

（五）干旱环境微生物

在 Web of Science 科学引文索引扩展版检索到 2010～2019 年关于干旱环境微生物的文献共 111 篇。这些文献主要涉及如下几个方向：①探索各种极端干旱生境中微生物群落结构特征和功能分布；②基于基因组代谢特征，直接分析预测微生物参与 C、N、S 等地球化学循环的功能机制；③通过微生物宏基因组序列拼接的方式，重构获取微生物基因组草图，开展微生物在极端干旱环境的适应性进化研究，揭示演化过程和规律。涉及的研究环境主要为荒漠、峡谷和极地干旱河谷。

这些学术贡献主要来自美国、墨西哥、南非、中国等。涉及的研究机构主要有墨西哥国立自治大学（Universidad Nacional Autónoma de México）、美国能源部、南非比勒陀利亚大学（University of Pretoria）、西班牙国家研究委员会等。其中，国际上产出最多的机构是墨西哥国立自治大学，国内的主要贡献机构是中国科学院。

第四节　深地环境微生物

一、微生物多样性

（一）地下水环境

地下水微生物多样性极高，大量微生物类群亟待被发现。例如，美国科学家通过分析来自美国科罗拉多州含水土层沉积物和地下水样本的超过2500种微生物基因组，发现了地下水系统中惊人的微生物多样性以及47个新种群（Anantharaman et al.，2016）。

由于地下水流速缓慢，微生物长期累积并作用于碳、氮、铁、硫等元素的循环和迁移转化，从而改变地下水的化学组分。目前已发现可能参与氮循环的微生物主要有厌氧氨氧化菌（*Candidatus* Scalindua 和 *Candidatus* Brocadia）；亚硝化单胞菌属（*Nitrosomonas*）以及亚硝化短小杆菌属（*Nitrosospumilus*）和亚硝化球菌属（*Nitrososphaera*）；脱氮硫杆菌（*Thiobacillus denitrificans*）和脱氮黄杆菌（*Flavobacterium denitrificans*）（Kumar et al.，2018）。参与碳循环的主要有盐硫杆状菌属（*Halothiobacillus*）、硫杆菌属（*Thiobacillus*）、氢弧菌属（*Hydrogenovibrio*）和硫磺单胞菌属（*Sulfurospirillum*）等碳固定同化微生物（Stegen et al.，2016），地下水中产甲烷古菌都属于广古菌门，同时还存在大量未被发现的新种群（Wang et al.，2015）。地下水中硫酸盐还原菌以变形菌门为主，主要有脱硫肠状菌属（*Desulfotomaculum*）、脱硫球茎菌属（*Desulfobulbus*）、脱硫叠球菌属（*Desulfosarcina*）、脱硫橄榄状菌属（*Desulfobacca*）等（Li et al.，2017）。地下水中参与铁循环的微生物也有较高的多样性，包括假单胞菌属（*Pseudomonas*）、梭菌属（*Clostridium*）、地杆菌属（*Geobacter*）、希瓦氏菌属（*Shewanella*）和克雷伯氏菌属（*Klebsiella*）等（Liu et al.，2020）。地下水中参与砷循环的微生物主要来自厚壁菌门、变形菌门、放线菌门等。

（二）洞穴环境

测序技术的发展使得人们对陆地深部生物的多样性及空间分布特征的认识有了很大的提高，变形菌门、放线菌门、厚壁菌门和拟杆菌门在细菌群落中占主导（Yun et al.，2016；Ortiz et al.，2013；Wu et al.，2015）。洞穴真菌主要为子囊菌门（Ascomycota）、担子菌门（Basidiomycota）及接合菌门（Zygomycota），且以子囊菌门占优势（Man et al.，2018）。

深成岩溶洞穴含有高浓度的还原性气体（CH_4、H_2S、NH_3 等），是一个类似于海底热液喷口的化能自养驱动的生态系统，化能自养的微生物利用这些还原性的气体获得能量生长，从而养育了鞭毛虫、线虫、钩虾等其他生物（Reboul et al.，2019）。

目前有关陆地深部生物圈病毒的报道还十分有限，主要局限于对洞穴水体样品的研究。例如对波多黎各洞穴公园（Camuy River Cave Park）天坑水样宏基因组的研究表明，病毒在水体微生物组中占 2.87%（Rodriguez-Ramos and Rios-Velazquez，2018）。亚得里亚海海蚀洞穴水体中病毒与细菌的数量之比在 1.3～32，在营养匮乏时期，病毒对细菌自下而上的裂解作用则成为细菌数量的主控因素。此时病毒的数量增多，引起细菌的裂解，导致环境中可溶性有机碳的增加，重新成为细菌生长的物质基础。这一过程将有机碳从捕食链中移除，而将其转移至微生物循环（Krstulović et al.，2013）。但有关病毒在洞穴生态系统中的认识还十分薄弱，它们与宿主的关系以及对洞穴元素循环的作用目前还难以评估。

洞穴蝙蝠携带的病毒是另一个被关注的热点。据不完全统计，已从蝙蝠体内分离到 80 多种病毒，其中一些病毒是人畜共患重大疾病的传染源，严重威胁人类公共健康。已有研究证明，蝙蝠是澳大利亚 1994 年暴发的亨德拉病毒（Hendra virus）以及 1997 年暴发的梅南哥病毒（Menangle virus）、1998 年马来西亚暴发的尼帕病毒（Nipah virus），以及慢病毒（Tioman virus）、埃博拉病毒（Ebola virus）、狂犬病毒（lyssavirus）、严重急性呼吸性综合征相关的冠状病毒（SARS-related coronavirus）、线状病毒（filovirus）、亨尼巴病毒（henipavirus）、副黏病毒（paramyxovirus）等的自然宿主。蝙蝠之所以能携带多种病毒，与其寿命长、迁徙范围大、种群密集、社会交流紧密等多种因素有关（Han et al.，2017；Schountz，2014）。

（三）大陆岩石圈环境

原核生物对温度和压力具有较高的耐受能力，因此是岩石圈中主要的功能群。此外，也有少量真核生物的报道。随着深度的加大，温度与压力升高，岩石孔隙度降低，地表带入的有机质减少，生活在岩石孔隙水的生物量、活性、功能多样性也随之降低。但是，在地下深部由地质过程产生的气体（CH_4、H_2、CO_2 和 He 等）增多，以及长时间的水岩反应导致盐分增高，可以孕育一些深地特有的微生物。已报道的深地微生物群落主要分为两大类：一类以厚壁菌门占优势，另一类则以变形菌门占优势。由于地质环境条件和营养程度的差异，岩石圈生物组成在空间分布上有很大的差异。岩石圈微生物的多样性、丰度与深度之间呈现显著的负相关关系（Magnabosco et al.，2016）。以变形菌为主的群落分布在较浅的地层，地层流体的年龄较小，且盐度相对较低，并且流体与地表流体交换比较频繁；地下深部流体则分布于较古老的地层中，地层流体具有较高的盐度，其中的微生物以厚壁菌门和广古菌门为主。在较浅的年轻岩石与流体中，难以区分时间和空间对地下微生物分布的影响（Davidson et al.，2011）。岩石圈微生物群落在大尺度空间分布的控制因素比较复杂，有时群落的差异与环境条件的关系并不显著（Magnabosco et al.，2018），并且在不同地点微生物类群的差异不能用常见的生态学解释，如地理隔离、物理化学特征、深度以及水的滞留时间等（Lau et al.，2014）。

（四）煤层和油藏环境

煤层和油藏是陆地深部环境中最具资源效应的极端环境，对这些环境中微生物的研究也日渐深入。煤层中占主导的古菌主要为氢营养型和甲基营养型的产甲烷古菌，而细菌种类差异明显，主要为变形菌门、放线菌门和厚壁菌门等（Strapoć et al.，2011）。

油藏环境细菌主要有变形菌门、厚壁菌门、拟杆菌门、放线菌门、绿弯菌门、热袍菌门（Thermotogae）和热脱硫杆菌门（Thermodesulfobacteria）等 15 个门，同时还有厌氧氨氧化细菌等。古菌主要为甲烷囊菌属（*Methanoculleus*）、甲烷砾菌属（*Methanocalculus*）、甲烷绳菌属（*Methanolinea*）、甲烷鬃菌属（*Methanosaeta*）、甲烷食甲基菌属（*Methanomethylovorans*）、甲烷热杆菌属（*Methanothermobacter*）、甲烷杆菌属（*Methanobacterium*）、甲烷胞菌属（*Methanocella*）、盐几何菌属（*Halogeometricum*）、热裸单胞菌属（*Thermogymnomonas*）和热球菌属（*Thermococcus*）等，以产甲烷菌为主（Li et al.，2017）。温度是影响油藏微生物多样性及分布最重要的因素，同时，注水开发会对油藏原位微生物的多样性和分布产生直接影响（Frank et al.，2016）。

二、微生物功能

对陆地深部生物圈功能的认知，少量基于可培养微生物，更多的是基于非培养微生物功能上的预测。像其他极端环境生物研究一样，陆地深部微生物只有少于 1% 的微生物可在室内培养。总体来讲，深地生境的生物量普遍低于地表生境，但是不同生境的生物量差异较大（Magnabosco et al.，2018）。技术的创新以及生物分子提取方法的改良有效地提高了低生物量样品中生物信息的提取率，使得检测方法更加灵敏，如用于团簇同位素分析的超高分辨率的稳定同位素质谱仪（Panorama 质谱仪），二代、三代测序仪等，让科学家能够更深入地研究深部甲烷的来源，并辨别生物成因和非生物成因甲烷（Young et al.，2017），构建陆地生物圈中首个原核生物（Chivian et al.，2008）和真核生物（Weinstein et al.，2019）的基因组，发现更多新谱系、新物种（Becraft et al.，2017；Momper et al.，2017），揭示首个陆地生物圈活跃群落的代谢网络（Lau et al.，2016）等。

（一）地下水环境

从地表补给区到排泄区，地下水系统中的生态位及微生物功能群常呈现规律性分区。地下水微生物从功能上大致可分为有机物分解菌、硝酸盐还原菌（NRB）、铁还原菌、硫酸盐还原菌和产甲烷菌等。地下水微生物介导的地球化学元素循环研究主要集中在 C、N、S、Fe 以及各元素循环相互影响等方面。例如，微生物参与的 C、N、S 和 Fe 等地球化学过程会导致高砷地下水的形成（Li et al.，2017），并且地下水中砷的迁移转化与铁还原菌地杆菌属和铁杆菌属（*Ferribacterium*）、有机物分解菌以及反硝化菌假单胞菌属密切相关（Jiang et al.，2019）。地下水微生物数值模拟的引入，极大地提高了人们对地下水微生物功能的了解。例如，Shapiro 等（2018）采用一种基因组代谢模型，对地下含水层中典型产甲烷菌的代谢过程进行了模拟，并预测了细胞整体代谢过程中单个生化反应的速率。

由于人类活动的不断增加，地下水水量急剧减少、水质不断恶化，地下水中的有害成分如氮、砷、氟、碘、有机物、病原菌等含量显著增高，对人体健康和工农业产生巨大危害。有研究表明，微生物在污染物的迁移转化过程中扮演了重要角色。地下水中 Nitrosoarchaeum、地杆菌属、螺旋体属（*Spirochaeta*）和甲烷杆菌纲（Methanobacteria）等在氧化零价 S 或还原 Fe 的同时可以去除地下水中的污染物钒（Zhang et al.，2018）。地下水中高浓度硝酸盐可通过反硝化细菌去除，而渗透性反应墙技术的应用则可以保障该过程中碳源与电子供体的供应（Gibert et al.，2019）。因此，地下水微生物对地下水供水安全至关重要，将微生物技术应用于地下水污染物的修复被高度关注。

（二）洞穴环境

洞穴奇古菌的硝化作用对沉积物氨氧化的贡献可高达 40% 以上（Zhao et al.，2018）。在地中海扎金索斯岛海蚀洞穴水下岩壁上密集分布着卵圆形和丝状的生物构造，并且具有很强的硫化氢气味，经过 280 天的培养，明显观察到了产甲烷的现象。分子生物学证据表明，这些培养物以硫酸盐还原菌和产甲烷古菌为主，包括甲烷微菌科（Methanomicrobiaceae）、脱硫球茎菌科（Desulfobulbaceae）、脱硫杆菌科（Desulfobacteraceae）、脱硫单胞菌科（Desulfuromonaceae）、弯曲杆菌科（Campylobacteraceae）、梭菌科（Clostridiaceae）、分类未定的科Ⅰ和Ⅱ（Incertae sedis - Family Ⅰ&Ⅱ）（Polymenakou et al.，2018）。

深成岩溶作用形成的富含 H_2S 和 CH_4 的、以化能自养为主的黑暗洞穴生态系统与深海热液喷口的生态系统极为相似，除了硫氧化菌、甲烷氧化菌外，还有多样化的固氮微生物。例如在意大利福拉萨西（Frasassi）洞穴中，通过 $^{15}N_2$ 的标记技术发现，固氮微生物具有显著的生境分化。在沉积物中固氮微生物以异养的地杆菌属、固氮弧菌属（*Azoarcus*）以及脱硫弧菌属（*Desulfovibrio*）（Desai et al.，2013）为主。2018 年，研究者从 Movile 洞穴生物席中分离得到第一株甲烷氧化菌甲基单胞菌属暂定种（*Candidatus* Methylomonas sp. LWB）。宏基因组数据显示，以厌氧环境为主的沉积物中，依然存在好氧的甲烷氧化菌的基因，这可能预示着微氧环境的存在。在沉积物中还检测到铁氧化细菌属（*Sideroxydans*）和披毛菌属（*Gallionella*），同样显示微氧的铁氧化菌在营养物质循环中可能起着重要作用（Kumaresan et al.，2018）。

（三）大陆岩石圈环境

岩石圈中具有特殊的微生物类群。例如，暂定金矿菌（*Candidatus* Desulforudis audaxviator）——一株硫酸盐还原型自养微生物，可以将硫酸盐还原和氢气氧化相耦合，具有完整的碳、氮代谢途径，在地下环境中完全可以自给自足（Chivian et al.，2008）。盐单胞菌（*Halomonas* spp.）对地下高盐环境具有较强的耐受性，且在高盐环境下可能会利用多种营养物质和有机质（Dong et al.，2014）。除此之外，岩石圈环境中还存在多种嗜热的铁氧化还原细菌、产甲烷菌和厌氧甲烷氧化古菌、自养型的硫氧化还原菌等，甚至是真核生物。

此外，岩石圈深部微生物介导的氢气氧化可以和多种电子受体的还原过程相耦合，其中典型过程包括硫酸盐还原菌在还原硫酸根的同时将 H_2 和 CO_2 转化为乙酸，以及氢营养型产甲烷菌将 H_2 和 CO_2 转化为 CH_4（Stevens and McKinley，1995）。上述过程产生的乙酸可以支撑乙酸营养型的产甲烷菌和乙酸发酵，产生的 CH_4 可以用来支撑嗜甲烷（甲烷氧化）微生物的生存（Lau et al.，2016；Simkus et al.，2016；Magnabosco et al.，2018）。另外，多种岩石圈生物可以在消耗 H_2 的同时耦合 NO_3^- 或 NO_2^- 还原，并从中获得能量（Dong et al.，2014；Baker et al.，2016）。岩石圈生物利用氮元素作为能量是最近才受到关注的（Lau et al.，2016），因为氮循环过程是陆地深部隐秘的生物地球化学过程。能耗最低、最原始的代谢途径 Wood-Ljungdahl 是地下化能自养微生物生态系统（subsurface lithoautotrophic microbial ecosystem，SLiME）中常见的固碳途径（Baker et al.，2016），但也有报道指出，有一些优势菌是利用能耗最高的卡尔文循环来获取碳的（Lau et al.，2016）。

（四）煤层环境

煤炭的生物成气通过一系列的功能微生物遵循经典的厌氧发酵“四阶段”理论（Park and Liang，2016），将煤炭中复杂的有机化合物分解为有机酸等碳水化合物、氢气和二氧化碳，最后，经由严格厌氧的产甲烷菌群合成甲烷。此外，有一类产甲烷菌可以直接利用煤中苯甲氧基来生成甲烷（Mayumi et al.，2016）。在厌氧处理系统中，非产甲烷菌和产甲烷菌相互依赖，互为对方创造良好的环境和条件，构成互生关系，同时，双方又互为制约，在厌氧生物处理系统中保持平衡状态。

（五）油藏环境

油藏是一个独特的生态系统，在不同时空尺度上交织着C、N、S等元素循环和能量循环，以及石油烃的厌氧生物降解等物理变化和化学反应，微生物特定的功能基因所编码的酶在这些反应中发挥着不可替代的作用。同时，油藏是一个天然的地质生物反应器，在研究深部地下微生物代谢活动中承担着重要的功能（Hallmann et al.，2008）。氮转化是油藏环境重要的元素循环过程之一。当硝酸盐作为电子受体存在时，NRB能够优先获得能量将NO_3^-转化为NH_3，抑制SRB活动及H_2S的产生，在油藏环境SRB腐蚀生物控制方面具有应用价值。油藏环境石油烃厌氧生物降解是重要的生物地球化学过程之一。从油藏微生物应用及工程意义上来说，目前的关注点是烃降解产甲烷的动力学问题，也就是反应速率、反应限制步骤及其调控机制，动力学问题的解决则依赖于对其反应途径、功能基因及相关的生物标志物的认识（Liu et al.，2019）。

三、计量统计分析资助态势及研究团队

欧美国家以及澳大利亚在国际深地生物圈的研究中处于领先地位。例如，欧洲各发达国家从20世纪80年代中期开始对地下水污染物的微生物修复开展研究，美国从1991年开始实施包括地下水污染生物修复在内的“超基金项目”。美国、德国、英国等国家已开展了土壤、地下水烃类等有机污染及氮污染的原位生物处理实验研究，并成功地完成了多项地下水污染的微生物修复项目，微生物修复技术也被各国家制定政策规定为地下水污染修复治理的首选方法之一。此外，在微生物促产煤层气方面已有企业［如美国卢卡（Luca）科技公司、美国未来能源公司（Next Fuel，Inc.）、美国Ciris能源（Ciris Energy）公司］将生物成气技术进行商业化运营。

按发文量排名，洞穴微生物研究居前八位的国家依次为美国、英国、意大利、德国、西班牙、中国、法国和波兰，发表论文数量居前八位的主要研究机构为西班牙国家研究委员会、新墨西哥大学、弗罗茨瓦夫大学、中国科学院、阿克伦大学、博洛尼亚大学、米兰大学和美国地质调查局。国际上有关洞穴微生物研究的文章总体呈现明显增长趋势（图3-11）。2008～2020年，美国在与洞穴密切相关的喀斯特领域资助的基金项目共计780项，是我国同期资助项目总量的8.5倍，资助金额达23 226.20万美元，其中以地球科学和大气与空间地球科学两个领域资助项目最多，分别为193项和172项。

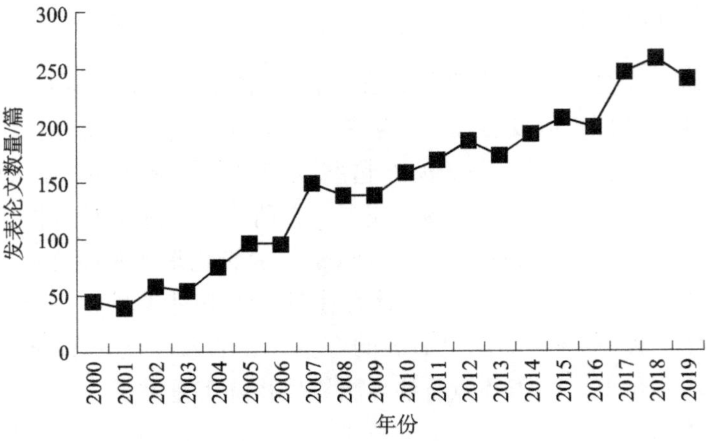

图 3-11　国际上有关洞穴微生物研究的文章数量

自 2000 年以来，地下水微生物学方向的文章数量呈线性增长趋势，论文的作者主要来自美国、中国、德国、加拿大、英国等国家。按发表与地下水微生物相关研究论文的数量排名，位居前十位的机构为美国地质调查局、中国科学院、美国加州大学伯克利分校、德国亥姆霍兹环境研究中心、中国地质大学、美国太平洋西北国家实验室、美国斯坦福大学、加拿大滑铁卢大学、美国橡树岭国家实验室和美国国家环境保护局。

近年来，国际上对岩石圈环境的研究兴趣持续增加［图 3-12（a）］，其中与生物相关的 SCI 论文数量约占 1/10［图 3-12（b）］。美国、德国、瑞典、日本是研究岩石圈微生物较早的国家，其中美国普林斯顿大学、美国新墨西哥矿业技术学院、加拿大多伦多大学、中国地质大学（北京）是岩石圈生物研究具影响力的单位。此外，美国太平洋西北国家实验室、美国橡树岭国家实验室、德国马克斯·普朗克学院、芬兰国家技术研究中心在该领域也具有实力。

早在 20 世纪 90 年代，美国和澳大利亚的科学家基于煤层气碳氢同位素组成和重烃含量提出，微生物在煤层气的形成中发挥着重要作用。目前，国际上煤层微生物的研究主要集中在六个方面：煤层生物甲烷的起源、煤层微生物多样性和群落功能，煤层微生物生长代谢的物质基础，煤层微生物生长代谢活性影响研究，煤层气资源的生物增采，煤层地下水污染的生物治理，煤层微生物参与的碳和硫等元素循环。

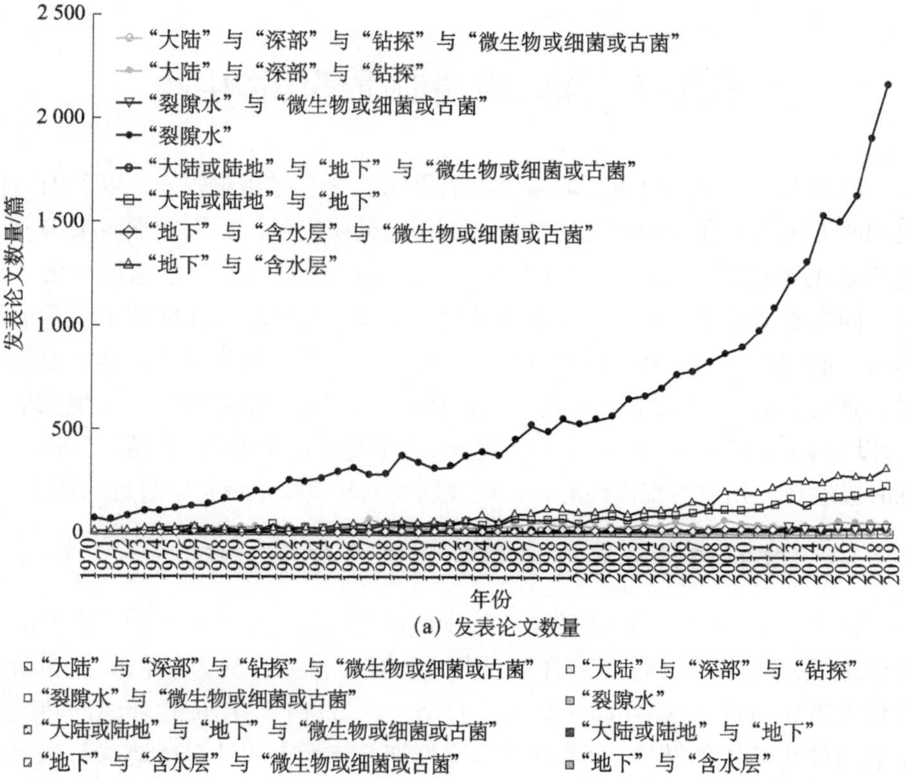

(a) 发表论文数量

(b) 分布领域统计图

图 3-12　与岩石圈环境相关研究论文数量及分布领域统计图（文后附彩图）

资料来源：斯高帕斯（Scopus）文献数据库。

注：搜索用的词组以引号标记，而带"微生物或细菌或古菌"（microorganism or bacteria or archaea）的与生物圈有关。

第五节　深时极端地质环境微生物

国际上深时极端地质环境微生物的研究发展态势相对较好。以美国为代表的西方国家的高校和研究所都成立了相应的研究队伍，围绕深时极端环境地质微生物过程，开展了多方面的工作：①以地质记录为主题，从化石记录、同位素地球化学记录、元素地球化学记录、有机地球化学和沉积学记录等多方面研究生物与环境的相互作用；②利用分子生物学手段，结合地质记录，研究重要微生物功能群的起源和演化；③从实验室模拟角度，开展微生物与环境相互作用机理，以及微生物地质过程的地质记录方面的研究；④选择与深时极端环境相似的现代环境，开展现场观测和生物地球化学过程的研究。

一、深时极端地质环境与生物地球化学循环

目前，深时极端地质环境微生物学的研究已经揭示了一系列重大的深时极端地质环境事件，包括氧含量显著升高期（古元古代大氧化事件和新元古代大氧化事件）、硫化海洋显著发育期（缺氧富 Fe^{2+} 向硫化环境的转变和富氧向硫化环境的转变）、温度显著降低期（新元古代“雪球地球”冰期事件）、温度显著升高期（早三叠世和古新世—始新世之交）、海洋 pH 显著降低期（古新世—始新世之交）和后生动物大灭绝期（奥陶纪末期和二叠纪—三叠纪之交）。

（一）氧含量显著升高期

地球早期主要处于一种无氧的状态，从地球的形成一直持续到古元古代早期（4600 Ma～2330 Ma），大气氧气含量显著低于现代大气氧含量的 10^{-5}（Zahnle et al.，2006）。地球表层氧含量的显著升高主要有两个阶段：第一阶段发生在古元古代早期（2500 Ma～2000 Ma），称为大氧化事件，主要表现为大气开始含有永久性氧气以及表层海洋广泛氧化；第二阶段发生在新元古代晚期（720 Ma～541 Ma），也称为新元古代大氧化事件（Neoproterozoic oxidation event，NOE），主要表现为海洋氧化程度的显著升高。

2011～2020 年来有关 GOE 和 NOE 的研究有明显的升高趋势，其中 NOE 方面发表的成果远高于 GOE（图 3-13）。就 GOE 来说，主要的研究团队集中在美国或与美国有密切合作，发表的文章数量超过了总数的一半，排

在前四位的国家分别是美国、加拿大、南非和澳大利亚，均在 GOE 方面有良好的地质记录。就 NOE 来说，我国的研究表现优秀，与美国处于旗鼓相当的位置。虽然学界对 GOE 和 NOE 都开展了大量的研究，但从极端环境微生物角度开展的工作相对有限（图 3-13），亟须加强。

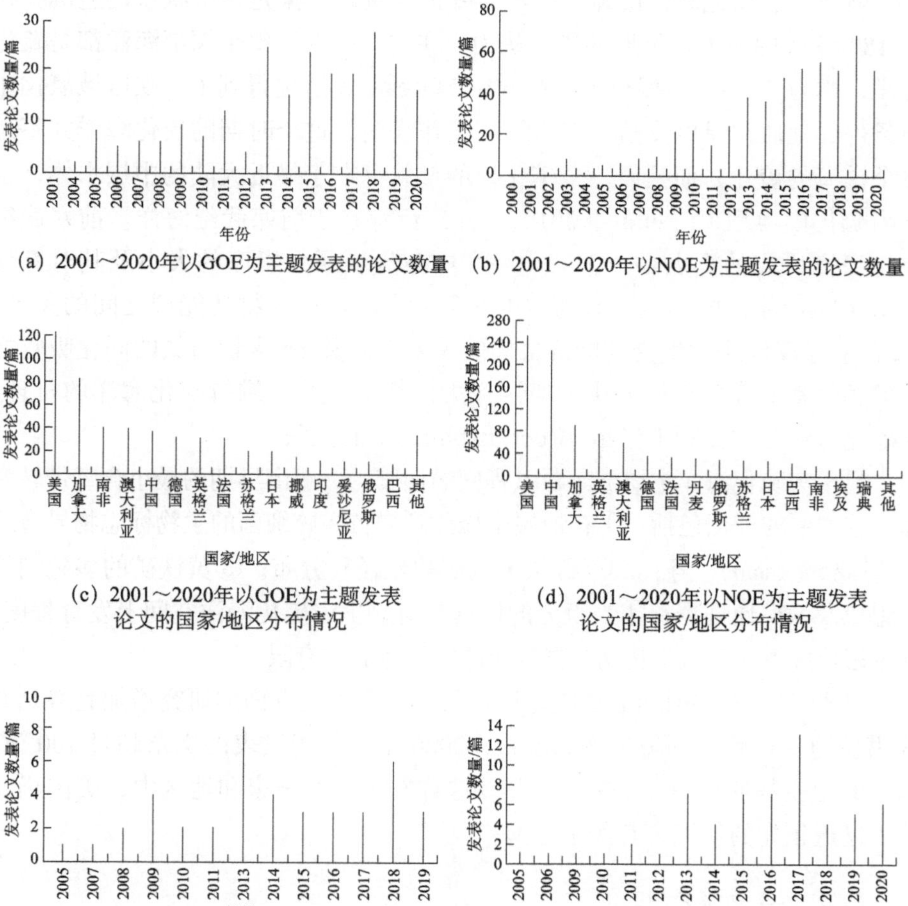

(a) 2001～2020年以GOE为主题发表的论文数量　(b) 2001～2020年以NOE为主题发表的论文数量

(c) 2001～2020年以GOE为主题发表论文的国家/地区分布情况　(d) 2001～2020年以NOE为主题发表论文的国家/地区分布情况

(e) 与GOE有关的微生物方面的论文数量　(f) 与NOE有关的微生物方面的论文数量

图 3-13　GOE 和 NOE 的研究现状

注：因在部分年份中未检索到符合关键词的文献，故年份不连续，下同。

（二）硫化海洋显著发育期

硫化海洋是指缺氧且富集游离 H_2S 的海洋水体。根据背景环境的特点，硫化海洋显著发育期可分为两种。其一是由早期缺氧富 Fe^{2+} 的海洋（又称铁化海洋）向硫化海洋的转换，主要发生在前寒武纪；其二是由早期富氧的海洋向硫化海洋的转换，主要发生在显生宙。这些硫化海洋也都偏离了早期的海洋环境背景，因而也是典型的深时极端地质环境。

前寒武纪铁化海洋向硫化海洋转变最直接的证据是条带状铁建造沉积在约 1800 Ma 的消失。早期的模型认为，在这个时期，整个深部海洋都是硫化海洋，称为“坎菲尔德海洋”（Canfield Ocean），这也得到了反映区域氧化还原条件的地质记录的支持。但后续的工作表明，这个时期的硫化海洋并不是整个深海的硫化，而是以一个硫化楔的形状存在于靠近大陆的中层水体，也称为硫化楔模型（Li et al.，2010），并广泛存在于前寒武纪海洋。前寒武纪硫化海洋研究的重点是：①硫化海洋中微量金属元素库的大小及其对氮代谢微生物功能群的影响，如海水 Mo 含量与固氮微生物功能群之间的关系、Cu 含量与反硝化微生物功能群之间的关系，以及 Zn 含量与氮的同化吸收之间的关系等；②硫化海洋中初级生产力的组成成分、维持硫化海洋的机理；③硫化海洋与“无聊十亿年”（boring billion）的关联。

显生宙硫化海洋主要随着缺氧程度的加剧，一般由早期的最小含氧带转变而来，主要的证据包括：①在地层中检测到来自绿硫细菌的生物标志物异海绵烷（isorenieratane）等；②草莓状黄铁矿的粒径和分布；③黄铁矿的多硫同位素组成及 S/C 比证据；④铁组分的证据。对显生宙硫化海洋空间上发育程度，其生态环境效应及微生物功能群分布的认识还非常有限。

21 世纪以来，国际上在硫化环境及相关微生物方面的研究增加趋势是非常明显的。在硫化环境方面，2011～2020 年，每年发表的文章超过 100 篇，相关的微生物研究占 1/4 左右。开展这方面研究的国家和地区中，美国仍然处于遥遥领先的位置，其次是德国。

（三）温度突变期

温度是影响微生物组成的重要环境参数之一，地球演化史上出现了数次地表温度的快速升降。最引人注目的降温事件是以新元古代成冰纪“雪球地球”为代表的冰期事件，整个地球被冰所覆盖（Hoffman et al.，1998）。此外，显生宙也发生过两次主要的冰期事件，一次是晚奥陶世赫南特期的短暂冰期事件，另一次是石炭纪至二叠纪早期的长时间尺度冰期。温度的显著升高期主要发生在早三叠世、白垩纪和古新世—始新世之交和始新世早期，主要证据是来自各种能反映古海水温度的地球化学指标，包括磷酸盐质生物壳体的氧同位素和浮游有孔虫的氧同位素记录等。另外，新元古代“雪球地球”退冰期间可能也会出现大幅度的升温事件，只是持续时间可能较短。

地表温度的变化一般认为是由大气 CO_2 分压（pCO_2）的变化所引起的，后者又与火山作用等地球内部活动以及地表硅酸盐的风化速率密切相关。除此之外，大气中还有其他重要温室气体，如 CH_4 和 N_2O 等。虽然大气中这些气体的含量要远低于 CO_2，但这些气体的温室效应远高于 CO_2，因而其含量变化也可能对气候变化有重要影响。大气中 CH_4 和 N_2O 的含量与微生物主导的 C、N、S 的生物地球化学过程密切相关。因而，深入研究不同气候波动环境下的微生物地质过程对认识全球气候变化具有重要影响。目前，对深时气候的变化开展了较为深入的工作，特别是有关全球变冷的研究，但有关深时气候变化极端地质环境微生物的研究明显偏少（图 3-14）。

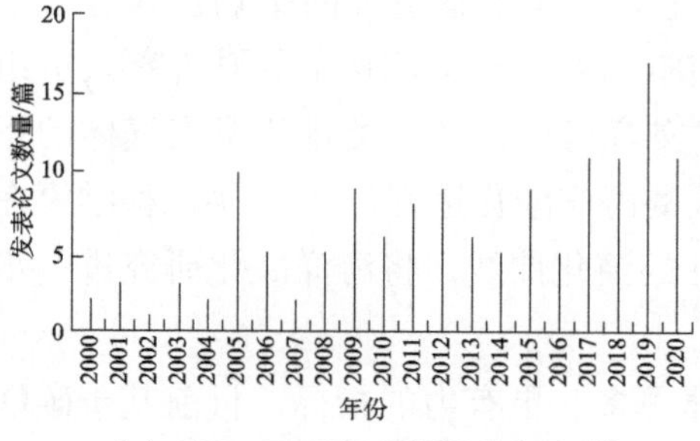

（a）2000～2020年有关深时升温时期极端地质环境的文章数量

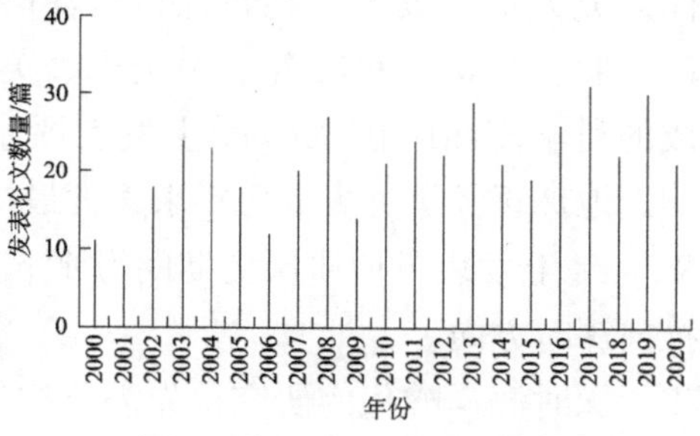

（b）2000～2020年有关深时降温时期极端地质环境的文章数量

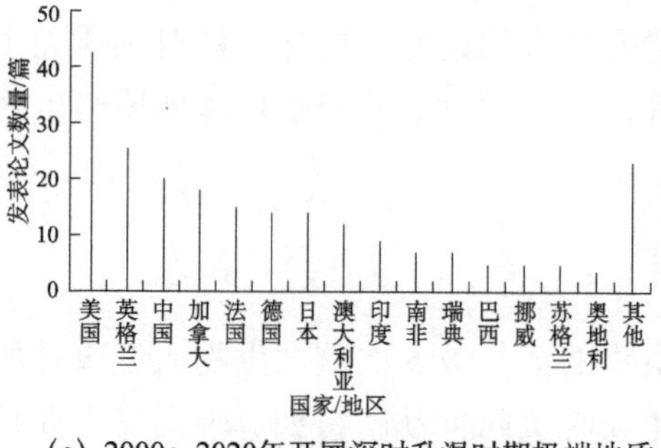

（c）2000～2020年开展深时升温时期极端地质环境研究的主要国家/地区

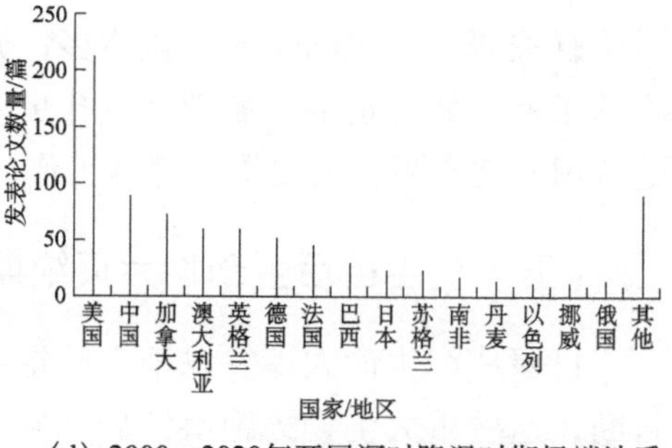

（d）2000～2020年开展深时降温时期极端地质环境研究的主要国家/地区

图 3-14　深时温度变化时期极端地质环境及地质微生物研究现状

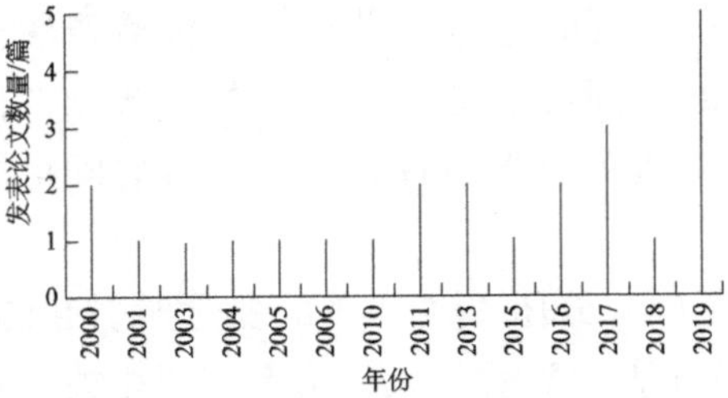

(e) 2000～2020年发表的有关升温时期极端地质环境微生物研究的文章数量

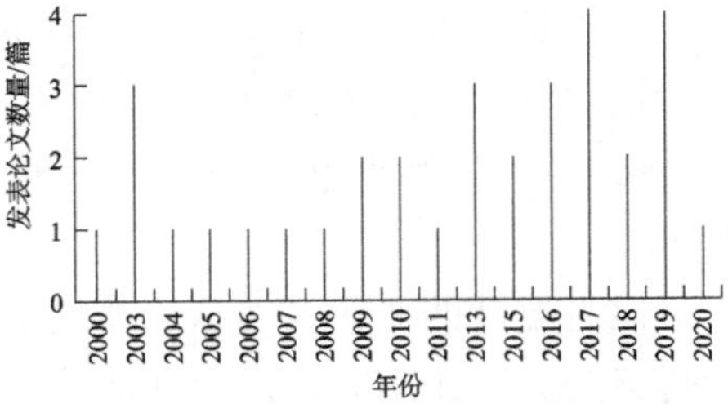

(f) 2000～2020年发表的有关降温时期极端地质环境微生物研究的文章数量

图 3-14　深时温度变化时期极端地质环境及地质微生物研究现状（续）

（四）海洋显著酸化期

自 2003 年科研人员在 *Nature* 上提出“海洋酸化”概念及相关科学问题以来，人类对海洋酸化问题的认识与日俱增。2009 年，全球超过 150 位海洋研究人员齐聚于摩纳哥探讨海洋酸化对海洋生态系统的影响，并签署了《摩纳哥宣言》。2012 年，由美国、英国、西班牙、德国等多国研究人员组成的科学家团队在 *Science* 上发表评述文章指出，受人类排放温室气体的影响，地球正经历过去 3 亿年来速度最快的海洋酸化进程。2017 年，科学家在 *Science* 上发表了地质历史期间海水 pH 的演化曲线，将海洋酸化研究进一步推向深时方向。

深时海洋酸化的研究首先需要构建更多、更有力的指标，目前基于碳酸盐岩硼同位素的分析，已经在一些重大转折期识别出了显著的酸化事件，如埃迪卡拉纪冰期结束期、二叠纪—三叠纪之交等，但还是非常零星。此外，目前有关海洋酸化的研究日益增多，研究较多的是海洋酸化事件对钙质壳生物的影响，如腕足类、珊瑚、有孔虫等，但人们对这些极端环境地质微生物的认识非常有限，需要深入开展工作。

（五）显生宙后生动物大灭绝期

自寒武纪生命大爆发以来，后生生物的演化被 5 次重大生物灭绝事件所打断，这些重大生物灭绝事件分别发生在晚奥陶世赫南特期、晚泥盆世弗拉期—法门期、二叠纪—三叠纪之交、三叠纪—侏罗纪之交和白垩纪—古近纪之交。目前研究认识到，除了白垩纪—古近纪之交的生物大灭绝可能与地外天体的撞击有关，其他几次重大生物灭绝事件主要与地内因素有关，并且可能受多种环境因素的共同影响，如温度升高、海水缺氧、海洋酸化等。

目前，对重大生物灭绝期后生生物的灭绝过程有了较为全面的认识，对当时的古海洋环境背景有了较深入的了解，并发现这些事件一般伴随着显著的C、N、S生物地球化学循环异常。同时研究发现，后生生物的大灭绝伴随着微生物的繁盛，如发现了二叠纪—三叠纪之交大量与微生物有关的岩石类型和生物标志物记录。接下来需要深入揭示微生物功能群的类型，微生物功能群繁盛与后生生物灭绝之间的关系，微生物功能群与C、N、S循环异常之间的关系以及微生物功能群繁盛的环境效应。

生物大灭绝的研究一直是地学和生命科学研究的热点，且每年发表的论文数量一直保持在100篇以上。有关显生宙重大生物危机期微生物的研究相对偏少，其发表的论文数量不足文章总数的1/5。

二、地质环境微生物功能群识别

前文已经论述，微生物的识别和鉴定是深时极端地质环境微生物研究的一个重点和难点问题。一方面，深时地质微生物的研究不可能像动植物一样从化石的角度开展；另一方面，绝大部分现代微生物学的研究手段很难直接应用到深时地质微生物的研究中。地质环境微生物功能群是一个重要的潜在突破口，科研群体从不同层面、不同角度进行了大量的尝试，并取得了较为显著的进展。

（一）稳定同位素组成

稳定同位素地球化学是开展深时极端地质环境微生物功能群研究常用的手段，在C、N、S代谢微生物功能群的起源及重大地质转折期微生物功能群组成的变化方面取得了较好的认识。例如基于碳同位素的记录，认识到产甲烷微生物功能群在约34亿年前就已经起源（Ueno et al.，2006）。

稳定同位素地球化学除了可以解决一些关键微生物功能群的起源问题外，对微生物功能群的空间分布和古海洋水化学亦具有重要意义。无机碳同位素组成的空间差异一般与水体的氧化还原程度密切相关。在分层缺氧的沉积盆地，碳酸盐岩的$\delta^{13}C$存在一个非常大的从浅水到深水不断减小的梯度，如新元古代“雪球地球”冰期结束之后的陡山沱期（Jiang et al.，2007）及早三叠世的极端高温期。同时，有机碳同位素的空间差异与水化学和微生物功能群密切相关，当没有特殊微生物功能群广泛发育时，有机质的$\delta^{13}C$与碳酸盐岩的$\delta^{13}C$有相似的分布，如前寒武纪地层中有机质的$\delta^{13}C$也普遍存在一个从浅水到深水不断减小的梯度（Eigenbrode and Freeman，2006）。当特殊类型的微生物功能群丰度显著升高时，如绿硫细菌，在特定区域（硫化的透光层）会出现有机质 $\delta^{13}C$ 的高值，如二叠纪—三叠纪之交的生物大灭绝期（Luo et al.，2010）。

（二）脂类生物标志物记录

脂类生物标志物是深时极端地质环境微生物研究中常用的手段之一。一方面，脂类生物标志物的研究促进了一些关键地质微生物功能群起源和演化的认识，如澳大利亚北部古元古代末期巴尼溪（Barney Creek）组发现的异海绵烷、绿硫菌烷（chlorobactane）、奥克烷（okenane）和3-甲基藿烷表明绿硫细菌、紫硫细菌和甲烷好氧氧化细菌等至少在16亿年前就已经存在（Brocks et al.，2005）。根据在阿曼南部盐盆地的南侯格夫（Huqf）超群中发现的24-异丙基胆甾烷，推测海绵动物（异养微生物功能群）在新元古代"雪球地球"时期的间冰期就已起源（Love et al.，2009）。脂类生物标志物的研究还解决了一些疑难化石的生物属性问题，如明确指出埃迪卡拉化石狄更逊水母（*Dickinsonia*）是动物。

另一方面，脂类生物标志物的研究揭示了前寒武纪海洋初级生产力（自养微生物功能群）的组成和演化过程。虽然化石记录和分子生物学研究都表明真核生物在中元古代早期就已出现，然而大量的脂类生物标志物的研究在这时期的样品中未发现明显的来自真核生物的甾类化合物。这表明，该时期海洋生态系统还是以原核自养微生物功能群占主导的。C_{27}、C_{28} 和 C_{29} 甾烷的相继出现表明，红藻类和绿藻类等真核光合产氧微生物在生态系统中的比例逐渐升高，并且在新元古代"雪球地球"冰期的间冰期达到非常高的值，反映真核生物与原核生物丰度关系的甾烷与藿烷的比值出现2～3个数量级的升高，约达到1（Brocks et al.，2017）。这与显生宙海洋样品中的比值非常相似，而与这之前前寒武纪海洋样品中的比值截然不同。这表明，这个时期海洋中真核光合生物不仅分异度有明显升高，同时丰度也有显著上升。

此外，脂类生物标志物的研究对认识重大生物灭绝期微生物功能群的变化有重要价值，如在晚泥盆世弗拉期—法门期、二叠纪—三叠纪之交（Grice et al.，2005）和三叠纪—侏罗纪之交的生物大灭绝中，都检测到了来自绿硫细菌的典型生物标志物异海绵烷，说明这些重大生物转折期出现过透光层缺氧现象，也间接说明硫酸盐还原微生物功能群丰度的升高。此外，在多次生物大灭绝期，脂类生物标志物的研究揭示了好氧甲烷氧化微生物功能群的繁盛（3-甲基藿烷，如晚奥陶世赫南特期、二叠纪—三叠纪之交等）、蓝细菌的繁盛（2-甲基藿烷，如二叠纪—三叠纪之交）、藻类自养微生物功能群的变化（绿藻和疑源类的繁盛，如二叠纪—三叠纪之交和三叠纪—侏罗纪之交等）。

（三）矿物学记录

微生物与矿物相互作用是微生物获取能量和电子的重要途径，在这些过程中会产生一些特殊矿物，这些矿物被保存在地质体中。例如，趋磁细菌合成的各种磁小体矿物（磁铁矿和胶黄铁矿）具有粒径均一、晶形独特、化学纯度高等特征，并且在沉积物中仍能基本保持链状排列这一特质。

近些年日益增加的报道显示，微生物代谢活动可以在地表温度和压力条件下催化诱导一些高温高压环境才能形成的矿物，如蒙皂石伊利石化等。这些“低温”矿物可以作为识别地质历史时期微生物活动的潜在标志。例如，Aubineau等（2019）从加蓬弗朗斯维尔（Francevillian）盆地下元古界微生物岩中发现低温成因伊利石，从而提出微生物在古元古代海洋中具有富集海水钾离子行为，进而促进蒙皂石向伊利石转化（蒙皂石 $+Al^{3+}+K^{+} \longrightarrow$ 伊利石 $+Si^{4+}$）。

三、微生物地质演化

有关微生物的地质演化远没有像动物和植物的演化关系被人们弄得那么清楚。上述稳定同位素和脂类生物标志物可以用于探讨部分微生物功能群的起源，但对大部分微生物的起源是无能为力的。随着分子生物学的兴起与基因测序技术的提高，分子生物学在生物系统分类和演化方面发挥了重要作用，如美国著名分子生物学家基于基因的研究最早命名了古菌并提出生命的“三域”（three domains）分类体系（Woese et al.，1990）。利用分子生物学构建的生物系统发育树，结合地质记录可以较好地确定关键地质微生物功能群起源的具体时间，进而研究地质微生物功能群起源的环境背景和生态效应。

近年来，分子生物学在重建重要微生物功能群起源和演化方面取得了很大进展。分子生物学的研究支持了产甲烷微生物功能群最早在太古宙早期或冥古宙末期（3800 Ma～4100 Ma）开始起源的说法，产生的 CH_4 是重要的温室气体之一，解决“黯淡太阳悖论”（Faint Young Sun Paradox）中所面临的问题（Battistuzzi et al.，2004）。这一认识和目前已有的碳同位素记录是比较吻合的（Ueno et al.，2006）。同时，分子生物学的研究为光合微生物功能群的起源提供了许多新的认识，提出了光合微生物功能群最早起源于太古宙早期（>3200 Ma），非产氧光合生物要早于产氧光合细菌，且后者出现的时间可能非常接近古元古代大氧化事件所发生的时间（Xiong et al.，2000；Battistuzzi et al.，2004）。这些演化事件对进一步认识早期地球环境演化和大氧化事件的过程具有重要意义。此外，产氧光合藻类出现的时间要明显晚于大氧化事件，冠群光合藻类的起源时间可能在中元古代晚期（约 1100 Ma）（Yoon et al.，2004）。

分子生物学在氮代谢微生物功能群起源和演化方面也取得了非常重要的进展，这些研究表明，奇古菌中氨氧化古菌（AOA）起源于大氧化事件之后，与大气氧气含量升高密切相关，其后续生态位的扩张可能和海洋的氧化过程有关（Ren et al.，2019）。与 AOA 相似，氨氧化细菌（AOB）是一类好氧氨氧化自养细菌，与 AOA 属于同一功能群。最新的分子生物学表明，AOB 可能起源于新元古代晚期，与 AOA 向深海的生态扩张有时间上的联系。这些成果对深入认识氮循环的特征和演化以及地质历史时期氮同位素的解释具有重要价值。

此外，分子生物学还可以揭示很多微生物演化事件与深时极端地质环境之间的关系。例如，分子生物学的工作揭示了一类新的产甲烷古菌在距今 2.5 亿年前后的二叠纪—三叠纪之交的起源（Rothman et al.，2014）。深入研究这些微生物起源的机制及生态环境效应有助于深入理解深时极端地质环境的地质微生物过程。

四、实验模拟和现代观测

实验室微生物地质过程模拟和类似早期海洋水化学环境中微生物地质过程的观测与研究是深入理解深时极端地质环境微生物地质过程的重要途径。在这些方面，国际上已经开展了较为深入的研究工作。

（一）微生物地质过程的实验模拟研究

微生物地质过程的实验模拟研究涉及不同的方面，包括微生物地质过程在地质记录中的标识、关键微生物代谢过程、环境参数对微生物代谢的影响以及微生物代谢过程中同位素分馏特征。

一般认为，微生物岩包括叠层石和各种具微生物成因的沉积构造的岩石，是由微生物地质过程形成的，然而一直缺乏明确的证据支撑。通过实验室的模拟，建立了一系列比较明确生物成因的沉积构造，如叠层石中的微米级别的孔隙等（Bosak et al.，2004）。

关键微生物代谢最新的进展是在 Mn 的氧化方面，早期的研究一直认为 Mn 的氧化必须有 O_2 及相关氧化剂的参与，因而 Mn 的氧化可以指示环境中 O_2 的存在及产氧光合生物的起源等。最新的实验室模拟研究发现，专性厌氧的非产氧光合微生物可以导致 Mn 的氧化，进而可以形成 GOE 之前的 Mn 的氧化物（Daye et al.，2019）。

环境参数对微生物代谢的影响分别从主量元素和微量元素两个角度开展。对于主量元素，目前主要集中在甲烷厌氧氧化方面，探讨了甲烷厌氧氧化与硫酸盐、硝酸盐和铁-锰氧化物还原之间的耦合关系（Beal et al.，2009），并且开展了一些定量化的研究。例如，研究表明，当水体硫酸盐浓度小于 1 mmol/L 时，甲烷厌氧氧化的速率显著降低（Beal et al.，2011）。在微量元素方面，目前开展最多的工作是探讨微量金属元素浓度变化对微生物代谢的影响，如探讨了微量金属元素对许多氮代谢微生物功能群的影响，包括氮的固定和同化吸收作用（Zerkle et al.，2006）以及 N_2O 产生和消耗的过程（Glass and Orphan，2012）等。

针对微生物代谢过程的同位素分馏，也开展了许多实验室微生物培养或模拟工作，重点集中在 N 和 S 同位素方面。例如，发现以 Fe 作为辅因子的固氮酶固氮过程对氮同位素的分馏效应远大于以 Mo 作为辅因子的固氮酶（8‰ vs. 0‰）（Zhang et al.，2014）；发现单纯的硫酸盐还原微生物功能群可以导致大于 45‰ 的同位素分馏效应，而不需要中间价态硫的歧化代谢微生物功能群的参与（Sim et al.，2011）。这些为从稳定同位素地球化学角度开展深时极端地质环境微生物研究提供了基础。

（二）对与深时极端地质环境类似环境的现场观测和研究

“将今论古”是地质学研究的基本思维方法，深时极端地质环境微生物的研究也类似。开展与深时极端地质环境相似的现代极端地质环境微生物机理研究是认识深时极端地质环境微生物地质过程的重要途径。针对深时极端地质环境的特点，目前开展研究最多的对比环境是低硫酸盐浓度且缺氧硫化的环境和低硫酸盐浓度且缺氧铁化的环境。主要针对的科学问题包括 C、N、S、Fe 代谢的微生物功能群及其分布特征，P 循环特征，Fe 同位素和 S 同位素分馏特征等。这些研究成果极大地促进了对深时极端地质环境中生物地球化学记录的解译和微生物地质过程的探讨。

五、国际上的研究团队与平台

国际上开展深时极端地质环境微生物研究的研究团队非常多。以美国为例，基本上所有的综合性大学都有地球生物学团队，各地球生物学团队都在开展深时极端地质环境微生物方面的研究工作，包括麻省理工学院森蒙斯（Summons）教授等、加州理工学院基尔希文克（Kirschvink）教授和菲希尔（Fischer）教授等、斯坦福大学佩恩（Payne）教授等、哈佛大学诺尔（Knoll）教授和皮尔逊（Pearson）教授等。根据这些研究团队的侧重点，主要可以分为以下几个方向（图 3-15）。

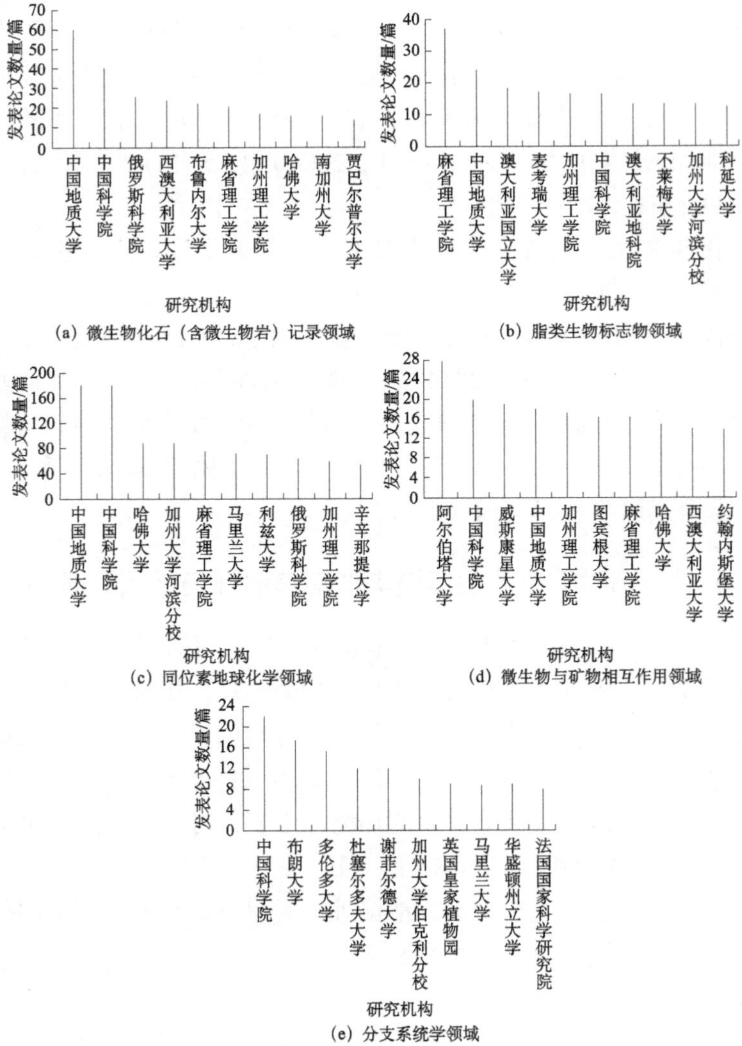

图 3-15 深时极端地质环境微生物相关领域主要的国际研究团队与论文发表情况

（1）从微生物化石（包括微生物岩）记录角度开展工作，揭示早期生命的起源、功能和环境效应等，主要的研究团队有美国哈佛大学的诺尔教授团队、弗吉尼亚理工大学的肖书海教授团队、麻省理工学院的博萨克（Bosak）教授团队、南加州大学的博特杰（Bottjer）教授团队等，英国剑桥大学的巴特菲尔德（Butterfield）教授团队等，比利时列日大学的贾夫（Javaux）教授团队等，德国哥廷根大学的雷特纳（Reitner）教授团队，法国里尔大学法国科学研究中心的塞维斯（Servais）教授团队等。

（2）从脂类生物标志物角度开展工作，探讨深时极端地质环境特征和微生物群落特征，以及两者的相互作用，主要的研究团队有美国麻省理工学院的森蒙斯教授团队、加州大学河滨分校的勒夫（Love）教授团队、宾夕法尼亚州立大学的弗里曼（Freeman）教授团队等，澳大利亚科廷大学的格里斯（Grice）教授团队、国立大学布罗克斯（Brocks）教授团队等，德国马普所霍尔曼（Hallmann）博士团队、基尔大学的施瓦克（Schwark）教授团队等，英国布里斯托大学的潘克斯特（Pancost）教授团队、南安普敦大学的怀特塞德（Whiteside）博士团队等。

（3）从同位素地球化学角度，重点研究微生物驱动的元素循环特征及其对地表环境的影响，主要的研究团队有美国加州大学河滨分校的里昂（Lyons）教授团队、马里兰大学的法夸尔（Farquhar）教授团队等、宾夕法尼亚州立大学的昆普（Kump）教授团队、华盛顿大学的别克（Buick）教授团队等，英国利兹大学的威纳尔（Wignall）教授团队、牛津大学的詹金斯（Jenkyns）教授团队等、伦敦大学学院的希尔兹（Shields）教授团队，德国埃朗根-纽伦堡大学的约阿希姆斯基（Joachimski）教授团队等，法国巴黎地球物理学院的菲利波（Philippot）教授团队等，南丹麦大学的坎菲尔德（Canfield）教授团队等。

（4）从微生物与矿物相互作用角度开展工作，重点探讨微生物的成矿、成岩作用，以及矿物对微生物代谢的影响，主要的研究团队有加拿大阿尔伯塔（Alberta）大学的孔豪瑟（Konhauser）教授团队、德国图宾根大学的卡普勒（Kappler）教授团队、美国爱荷华州立大学的斯旺纳（Swanner）博士团队、澳大利亚西澳大学的梁文希（Wacey）教授团队。

（5）从分子系统学角度开展工作，重点集中在关键微生物功能群的起源和演化方面，主要的研究团队包括美国加州理工学院的菲希尔教授团队、麻省理工学院的福尼尔（Fournier）博士团队等，英国布里斯托大学的多诺霍（Donoghue）教授团队和桑切斯-巴拉卡多（Sanchez-Baracaldo）教授团队等。

国际上开展深时极端地质环境微生物的平台有很多。除每个国家的自然科学基金委员会在地球生物学方面的资助平台之外，其他的主要平台有美国地质学会的地球生物学和地质微生物学分会、NASA 的国家天体生物学研究所（NAI）和戈登研究会议（Gordon Research Conference）的地球生物学会议，以及国际地球生物学学会（Geobiology Society）。此外，还有一些知名的基金会也都为深时极端地质环境微生物的研究提供良好的平台，如美国西蒙斯基金会（Simons Foundation）资助的“生命起源”（Origin of Life）项目。美国阿古伦（Agouron）研究所也资助了大量地球生物学方面的研究，很大一部分涉及深时极端地质环境微生物方面的研究。

第六节　深空环境的生命

一、太阳系内天体宜居环境研究及生命探测

随着航天任务的开展，太阳系中的主要天体都得到了一定程度的探测。生命离不开水，太阳系的许多天体都有水存在或存在过的证据，甚至还拥有能量源，因此具有很大的生命探测潜力。目前除火星外，太阳系内最有可能形成和保存生命的天体还有土卫二、土卫六和木卫二。

（一）火星

火星作为地球的近邻，许多地质和物理特征与地球类似。早期火星表面曾有大量的液态水，现代火星也可能有暂时性地表水（Fischer et al., 2014）、地下水冰（Dundas et al., 2018）或地下湖泊（Orosei et al., 2018）。但现代火星大气稀薄，表面受到的辐射强度高，对生命的生存存在巨大的挑战。

火星表面曾存在大量地表水活动，并在其表面刻画出数量繁多、形态各异的沟谷，保留下了古湖泊甚至古海洋的遗迹。“火星快车”（2003 年）、“机遇号”（2004 年）、“勇气号”（2004 年）和火星侦察轨道器（2005 年）都在火星表面发现了冲沟、河谷地貌等水活动的形貌证据，以及层状硅酸盐、水合硫酸盐、赤铁矿等指示水环境的矿物证据。水冰是重要的火星水资源储存库，目前火星表面的水冰主要赋存在两极冰盖、中纬度地区的地下冰层中。但直接观测到的现今流水作用较为罕见，可能存在小规模的地表水活动，且

最可能以卤水的形态出现。

早期火星被认为处在太阳系的宜居带中，宜居环境最可能出现在湖泊和热液系统中，现代火星的宜居环境则可能在地下或浅表暂时性卤水出现区。“好奇号”火星车的着陆区——盖尔撞击坑在 40 亿年前是一个中性、低盐度、持续了几百年到数万年的古湖泊，湖水中不同氧化还原状态的铁离子和硫离子可能为化能自养型微生物创造了生存条件（Grotzinger，2014）。同时，火星的古热液系统也有生命存在的可能（类似地球热液系统活跃着的嗜热嗜酸微生物），热液活动停止后，温度降低，为保存生物特征提供了条件（Michalski et al.，2018）。

20 世纪 70 年代发射的“海盗号”系列飞船最早将寻找火星生命作为探测目标任务，并且检测到了有机碳和高氯酸盐相互作用产生的氯甲烷和氯苯（Navarro-González et al.，2010）。“好奇号”不仅证实了“海盗号”的发现，还在古老泥岩（3.5 Ga）中检测到了热解产生的噻吩类、芳香族和脂肪族化合物（Ming et al.，2014；Eigenbrode et al.，2018）。尽管目前无法确定这些有机物的成因，但为了解火星有机碳的保存以及碳循环提供了有利信息。

（二）地外海洋世界

对太阳系冰卫星的探测主要来自木星及其卫星探测器“伽利略号”（1989 年）和土星及其卫星探测器“卡西尼-惠更斯号”（1997 年）。“伽利略号”对木卫二进行了长期探测，发现了木卫二的冰下海洋、南半球的羽状喷发现象（Manga and Wang，2007）。“卡西尼-惠更斯号”也观测到了土卫二的羽状喷发现象，喷发量约为 200 kg/s，喷发物富含水蒸气，还包括分子氢、有机碳、其他挥发物和固体颗粒［包括氯化钠（NaCl）晶体和冰粒］（Hurford et al.，2007）。Russell 等（2017）提出了一种假说：土卫二和木卫二的洋底由热对流造成物质交换，为紧邻在冰壳下层的化能自养型微生物圈提供营养物质。微体生命很可能被携带至星球表面或通过喷发羽喷出地表。“伽利略号”还揭示了土卫六浓厚的大气（98.4% 的氮气、1.4% 的甲烷和其他微量气体）和以液态烷烃为主要成分的湖泊（Hayes et al.，2008），使其成为太阳系内生命探测的热点。

二、地球作为天体生物学研究的基石

地球生命及其过程为研究地外天体中可能存在的生命提供了唯一的类比

对象。认识地球生命的起源、演化及其与环境的相互作用是天体生物学的重要研究领域，因此，地球生物学也是天体生物学研究的基石。地球类比研究在天体生物学中的应用包括：①类似行星环境的早期地球环境、生命起源及其化石和生物标志物等研究；②类似行星环境的地球极端环境中现生生物代谢机制、环境适应策略和生物标志物等研究；③将地球样品置于模拟的太空或行星环境中，模拟矿物、生物标志物和微生物体系变化等实验研究（Martins et al.，2017）。

（一）地球生命起源

生命起源分为四个阶段，即从无机小分子物质生成有机小分子物质、从有机小分子物质生成生物大分子物质、从生物大分子物质组成有机多分子体系、从有机多分子体系演变为原始生命。在早期地球环境中，从无机小分子自然生成有机小分子的过程是可以实现的（Miller，1953）。氨基酸等有机小分子合成蛋白质与核酸等重要生物大分子的可能性也得到证实。如果前生命化学过程需要的几个必要条件（固态岩石圈、适宜的温度、具有适当化学成分和物理条件的大气圈等）具备后，早期地球可能在仅仅数千万年甚至500万年时间内就演化出最简单的生命体系（Lazcano and Miller，1996）。

尽管目前对生命演化主要阶段的认识趋于一致，但对非生命到生命这一关键演化节点所发生的环境及其过程还存在极大争议。目前提出的生命产生环境包括“温暖水塘”（Darwin，1871）、海底热液喷口（Martin et al.，2008）、陆地热泉（Djokic et al.，2017）、矿物-水界面（Cairns-Smith，1982）等。

（二）地球极端环境与宜居性

地球最初的宜居带局限于可以促使非生命向生命转化的相对苛刻的环境中。一旦生命演化突破其“达尔文阈值”（Darwinian threshold），宜居性即由环境条件及其生物作用共同控制，并最终取决于生命对环境的改造能力。极端地质环境微生物对环境条件的强大耐受力拓宽了我们对生命极限的认识，以地球极端环境为地外天体的类比对象能为寻找地外生命提供借鉴。

截至2017年，以地球极端环境的地质和生境特点为参照，针对太阳系天体的特征，在全球建立了37个天体生物学类比研究点（Martins et al.，2017）。

其中类比火星极端环境的有21个，类比木卫二和土卫二的共有11个，类比土卫六的有3个，地球早期生命起源的研究区有2个，其中典型区域的现场景观如图3-16所示。

图3-16　全球用于天体生物学研究的极端环境典型区域的现场景观（文后附彩图）

资料来源：分布图改自 Preston and Dartnell（2014）。

火星类比研究点最多。西班牙力拓河是火星古代湿润环境的良好类比之地，中国柴达木盆地被列为火星古湖泊及现代干涸湖泊的类比研究基地，以美国黄石国家公园为代表的地球活跃热液系统是进一步了解火星古代热液活动的重要类比对象，南极洲的干谷作为极端干旱和永冻区的类比。木卫二和土卫二的类比研究点包括死海、大西洋中脊的失落之城、马里亚纳海沟等。地球上的沥青湖是土卫六湖泊的类比研究点，如位于特立尼达和多巴哥（拉丁美洲岛国）的天然沥青湖。

三、系外生命探测

（一）系外行星宜居性

现今已经发现了数千颗太阳系以外的行星，其中相当部分是类似于地球大小的行星。影响系外行星宜居性的主要因素是行星质量、行星轨道和宿主恒星的特性（光度、年龄、紫外强度和星风活动性等）。当行星在这些方面的条件满足目前对行星宜居性的认识时，就可以被称为疑似宜居系外行星，其宜居性的确认最终要靠对行星大气成分或行星表面的观测来确定。

NASA 的开普勒天文卫星发现了十几颗疑似宜居系外行星。但开普勒天文卫星所发现的目标都在距地球 500 光年以外，难以用望远镜来观测以确定其宜居性和生命痕迹。目前国际上正在进行中的地基和空间望远镜搜寻系外行星的计划有近 50 个。基于目前系外行星的统计数据，疑似宜居行星出现在太阳类型恒星周围的概率为 5%～20%，出现在红矮星（质量约为太阳质量的一半以下）周围的概率为 20%～50%。由于红矮星的数目远超太阳类型恒星，目前搜寻宜居系外行星的努力主要针对红矮星。

（二）系外行星生命探测

根据对地球生命活动所产生的生命信号以及探测方法的理解，系外行星生命探测的方法主要包括：①观测位于几光年远的系外行星上夜晚的灯光；②聆听系外智慧生命所发出的无线电信号；③探测系外行星叶绿素的生命信号；④研究行星大气在生命存在与否条件下的可能化学成分。

在观测方面，在恒星-行星光度比悬殊的条件下获得行星反射光谱或发射光谱，或者利用掩星事件获得行星大气透射光谱，并通过分析得到行星大气中生命信号的特征是对国际科学界的一大挑战。就目前国际科学界的动向来看，对太阳系附近比较明亮的恒星周围宜居行星生命信号探测的工作可望在 2035～2040 年通过发射大型空间望远镜取得重要进展。最近在金星大气中探测到了大量的磷化氢（Greaves et al.，2020），这为探测系外生命提供了新的思路。

四、生物入侵及行星保护

生命在不同天体之间的迁移可能会对原居住生命/生态系统造成巨大影响，导致生态系统不可逆转的变化。其中，微生物尤其是一些嗜极微生物是最有可能的“入侵者”。人类在开展深空探测活动中，既有可能把地球微生物带到其他天体表面，也有可能将地外微生物带到地球，从而导致生物灾难。

（一）行星保护的发展历程

行星保护的第一步是在发射太空飞船时尽量少地携带地球微生物。对于现在复杂的空间飞船设计来讲，达到真正意义上的完全灭菌是极大的挑战。如果飞船要降落到地外天体表面的话，就必须遵循严格的规则来对飞船进行处理，以最大限度地减少对目的地环境或生命的威胁。如果探测器从其他天体返回地球的话，则应该尽可能避免地外生命对地球的入侵。

在“阿波罗”时代，从月球返回的宇航员和样品都进行了隔离检查，以确定是否带回完全不同的生命形式。“阿波罗计划”后期，由于已经确信月球上不存在生命而不需要进一步呈报从月球返回地球时所带回的材料。

迄今，人类还没有从火星带回样品，尚不能排除火星上存在生命的可能性。在未来的工作中，需要隔离从火星返回到地球的样品和参与样品处理的宇航员，以切断火星生命与地球生物圈之间的直接接触。此外，在任务发射之前，应建立基于四级生物安全的样品接收实验室。美国空间研究委员会在2002年提出“特殊区域”的概念：地球生物有可能被复制到火星的特殊区域。考虑到生命系统中水的普遍存在和生化反应的温度条件，特殊区域将有液态水的存在，包括冰点更低的卤水等。基于行星保护考虑，迄今在火星上的科学考察都避开了这些特殊区域。

（二）行星保护的微生物学参数

微生物在地球上无所不在，能够适应多种极端环境条件。极端微生物特殊的生理-生态学特征是准确认识行星保护的基础。

（1）极端嗜酸微生物。它们可以在pH低至0的水环境中生存，可能在金星大气的硫酸云中生存和代谢。前往金星大气或地表的任务需要避免此类微生物的污染。

（2）耐辐射微生物。火星地表复杂的辐射条件使得绝大多数地球微生物难以存活。但分离自核废料的耐辐射球菌（*Deinococcus radiodurans*）可以承受超过15 000 Gy的辐射剂量，是地球上其他生物承受能力的1000倍以上，如果被着陆器带到火星表面，很可能会生存并繁殖（Dartnell et al.，2010）。

（3）化能自养型微生物。海底“黑烟囱”附近的生物群落完全由来自行星内部的地球化学能量支撑。该环境中的微生物有可能利用岩石地核在高温下与海洋相互作用产生的碳、金属和能量进行代谢，从而有在火星深部和冰卫星的冰下海洋中存活的可能性。

（4）耐或嗜低温微生物。地球高纬度地区和广阔的海洋深部分布着大量耐低温或嗜低温微生物。它们在极端低温或缺失营养物质的条件下会休眠，直至会随着环境条件改变而调整为生长繁殖状态。一旦被携带到火星和冰卫星上，此类微生物将可能幸存于冰下海洋中。

（三）行星保护的科学技术要求

进行有天体生物学目的的太空探索时，需要进行严格的向前污染控制：能够在飞船出发前就对其整体进行灭菌而携带尽量少的微生物细胞前往，在飞船环绕飞行探测时避免坠毁，在对其他行星环境进行接触研究时避免将地球微生物传给这些星体。对空间飞船结构的清洗程序包含挥发性去油污、液态超声和乙醇擦拭，并且利用超临界 CO_2 流体在低温下对大型构件进行有效灭菌。

五、空间舱内的生物实验

（一）空间生物实验

在人类不断地进行太空探索的同时，微生物也一直伴随着人类的太空飞行。自 2015 年以来，NASA 对国际空间站（ISS）材料表面、空气和水进行了持续的微生物监测和研究，建立了相应的空间微生物数据库，并对某些微生物的生物安全风险进行了评估。ISS 环境表面上可培养的微生物密度为 10^4～10^9 CFU/m^2，主要由细菌（放线菌门、拟杆菌门和变形菌门）和真菌（子囊菌门和担子菌门）组成。

在空间环境下生长的细菌为了增加对环境的耐受性和存活率，易表现出许多不同于地面的生物学特征，如生长速度加快、迟缓期缩短、毒力增加、抗菌药抗性增强、更易形成生物膜等，并且随着空间飞行时间延长，这些生物效应明显加剧。从 2019 年开始，NASA 对 ISS 上细菌和真菌的致病性和抗菌药抗性进行了定量分析，还将开展黑色素和 DNA 修复对深空真菌适应性和存活性的作用研究，以确保宇航员在未来的任务中不受到来自微生物的健康威胁。美国、欧盟国家等还开展了大规模的空间环境舱外暴露研究项目，苏联/俄罗斯也开展了大量的空间环境下舱外微生物暴露研究工作。研究发现，大部分的微生物（如氮循环微生物）在太空暴露后，回到地面还可以重新活化（Lindeboom et al.，2018）。

未来长期太空探索或建立永久性月球、火星基地，需要构建一个类似地球生物圈的小型生态系统——生物再生生命保障系统（BLSS）。植物是BLSS中的核心部件，而植物系统微生物能直接或间接地参与植物的生长、生理等过程，保证植物正常生长。因此，植物系统微生物问题在空间舱室中越来越受到重视。过去的研究主要集中在空间生命保障体系内可控环境植物栽培实验中的微生物群落研究。近两年来，NASA 借助 ISS 开展的相关研究主要集中在有益的植物-微生物相互作用和机制、提高植物生长或抑制病原活性的方法、开发和降解空间复杂有机物等方面。在 ISS 上产生的废物需要进行安全、经济的处理。最近发现，丝状真菌和一些独特的细菌菌株对航天飞行中复杂有机物的分解具有潜在作用（Romsdahl et al.，2020）。

（二）临近空间

临近空间是海平面之上 20～100 km 高度的空间范围，包括平流层的大部分、整个中间层以及少部分热层（林巍，2020）。临近空间是地表环境与近地轨道空间的过渡区，环境非常极端，具有高辐射、低温、干燥、低气压等复杂环境条件，难以在实验室中模拟。临近空间的复杂极端环境可以在一定程度上类比现代火星的表面环境和早期地球的高辐射环境，是开展天体生物学研究的理想区域。

目前已经在临近空间 41 km 高度发现微生物存在的确切证据，在最高 77 km 的高空也可能存在生命（Smith，2013）。20～40 km 高度区域的微生物数量可达 10^4～10^5 个 /m^3。由于临近空间的空域范围甚广，可以推测这一空间中存在着一个生物数量较多但相关认知很少的生物圈（Bryan et al.，2019）。

样品的获得是开展临近空间天体生物学研究的基本前提，这需要能够到达临近空间高度的特殊飞行器和可以顺利开展生物实验的有效载荷。临近空间研究和探测的飞行器主要包括探空火箭、高空飞机和浮空器等。自 20 世纪 30 年代以来，人们已在临近空间开展了多次天体生物学实验，这些实验可以分为两类，一类是生物暴露实验，一类是生物原位收集实验。目前已经在临近空间的不同高度分离培养了多株细菌和真菌，其中以能够形成孢子等休眠体的菌株为主，基于 16S rRNA 基因的分子生态学研究也获得了类似的结果，表明这些微生物主要通过形成休眠体来适应临近空间的特殊环境。此外，一些生物可能通过附着在颗粒上或形成聚集体等方式来抵挡临近空间高辐射对细胞的伤害。

六、计量统计结果分析态势

截至 2020 年 9 月，利用 Web of Science 对国际上的天体生物学研究检索发现，2000～2019 年共发表文章 3069 篇，随时间呈稳定的线性增长趋势。从检索主题看，研究最多的方向是生命起源和极端环境与宜居性（图 3-17）。

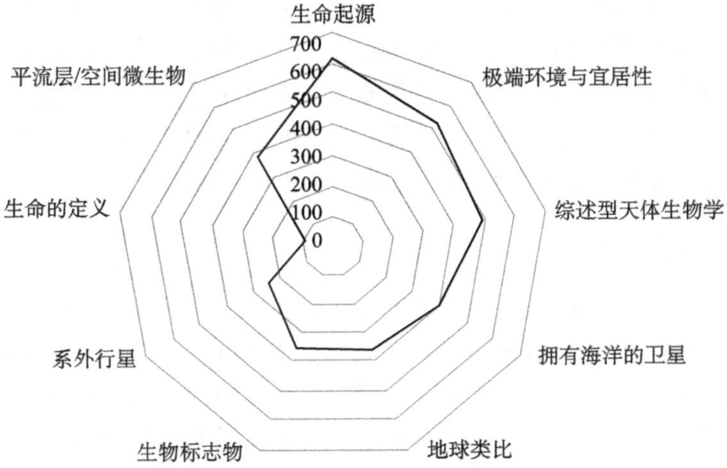

图 3-17　不同研究方向的英文文章数量雷达图（单位：篇）

极端环境与天体生物学的关系密切，因为地外行星和卫星的环境在地球的极端环境中易于寻找到二者的相似性。从地球的极端环境入手，相对于空间探测来说，成本低，样品容易获取，因此研究程度较高。地球类比和生物标志物方向都是以地球极端环境为研究对象，是支撑地外生命寻找的有利判据来源。类比研究中以对火星的类比为主，占 77%。生物标志物研究中，针对寻找火星可能生命痕迹的研究也占了 55%。此外，平流层 / 空间微生物的研究主要分布于医药卫生（24%），微生物（22%），以及分子生物、地质微生物和天体物理（三者共 22%）学科。系外行星的天体生物学研究在 2010 年以前的文章仅占 9%，之后飞速发展，原因可能是 2009 年 3 月 NASA 发射开普勒太空望远镜专门用于寻找太阳系外的行星。

同时，以地球为研究对象的文献（1286 篇）约是以外太空为研究对象的文献（866篇）的1.5倍，对地球极端地质环境下生命痕迹的研究是天体生物学的主要贡献力量。

七、国际研究团队和平台

国际上已经有多个天体生物学组织和研究团队，主要分布在欧美等发达国家和地区，代表性的有以下几个。

（1）美国NAI。1998年成立，主要任务是开发天体生物学领域的创新研究方法，为空间任务提供科学依据。NAI是一个虚拟组织，现有10个团队、约700名研究人员分布在近100所院校。主要工作包括：开展跨学科合作研究，培养下一代天体生物学研究人员，为太空任务提供科学技术指导，探索新方法、新应用，进行公共科普与宣传。

（2）加拿大天体生物学网络（CAN）。由加拿大相关研究机构和研究人员组织、参与天体生物学研究，目标是促进加拿大科学家与NAI之间的合作。“NASA天体生物学路线”大多数由该组织提供。主要工作包括：与天体生物学有关的类比研究、会议组织、人才交流与培养、公共科普与宣传。

（3）欧洲天体生物学网络协会（EANA）。2002年初成立，为NAI的附属合作伙伴，目前由来自17个欧洲国家的成员组成。EANA旨在建立共享数据库，培养年轻科学家，促进经费互利和设施共享，利用极端环境微生物的研究解决欧洲环境问题，进行公共科普与宣传。

（4）西班牙天体生物学研究中心（CAB）。成立于1998年，是美国国家航空航天技术研究所（INTA）和西班牙国家研究委员会的一部分，隶属于NAI。CAB的研究领域主要为极端环境与地球的类比、极端环境微生物以及寻找宜居环境，还参与了包括“好奇号”在内的多次火星探测任务的载荷研发。

（5）东京工业大学地球-生命研究所（ELSI）。2012年成立，主要研究地球的形成和演变、地球生命的起源和演化，该研究所特别强调在早期地球系统的背景下研究生命是如何出现以及如何演化的。

（6）澳大利亚天体生物学中心（ACA）。2001年成立，位于新南威尔士大学，主要研究方向为微生物学、古生物学、有机地球化学、行星科学、天文学和科普。研究内容包括地球早期生命的地质证据、极端地质环境微生物、现代和古代环境的生物标志物、地球化学以及系外行星。

（7）俄罗斯科学院生物医学问题研究所（IMBP）。在国际空间站建立了舱内外研究平台，开展了大量空间舱内和舱外暴露微生物实验研究。

第七节　高压环境微生物

以上的极端地质环境微生物都受到多种环境因子的影响。下面以高压环境为例，阐述单一环境因子对微生物的影响。

一、国际总体发展现状

20世纪中叶之前，人们一直忽视在像深海这样高压环境中的生命，深海微生物学也未被确立为一个可被研究的方向。但不可否认，深海高压环境中存在着各种各样的微生物，这些微生物也对整个海洋环境起着重要的作用。尽管早期人们在高压环境中发现了许多微生物，却很难解决“深海高压环境下的微生物是否处于最适条件”的问题。

随着高静水压技术的发展和应用，对耐压和嗜压微生物的研究显得更加可能且重要，这些微生物不仅仅可以提供很多的耐高压酶，其可以耐受高静水压的特殊结构还具有非常高的研究价值。在对从深海中分离出的耐压和嗜压微生物进行深入研究前，就有对一些常见模式菌株的压力耐受探索。例如，对大肠杆菌的研究发现，静水压从55.304 MPa向上移的过程中极大地扰乱了大肠杆菌细胞分裂和蛋白质合成的速率（Welch et al.，1993），并且随着环境静水压力的增加，大肠杆菌细胞的损伤也不断增强，到200 MPa时细胞就失去活性。

随着深潜技术和采样技术的不断发展以及高压培养技术的不断革新，很多耐压和嗜压微生物从深海中被陆续发现和分离。嗜压菌大多具有嗜冷特性，只能在20 ℃以下培养，这类嗜冷嗜压菌大多属于γ-变形菌。曾经报道过11株纯培养的嗜冷嗜压菌，包括希瓦氏菌属、发光杆菌属（*Photobacterium*）、科尔韦尔氏菌属（*Colwellia*）、摩替亚氏菌属（*Moritella*）和一个未知的属（DeLong et al.，1997）。其中，深发光杆菌（*Photobacterium profundum*）的SS9菌株成为目前一株研究较深入的模式菌株，其压力调控和适应机制已有多篇报道。希瓦氏菌属的一些菌株对压力适应性强并在海洋中广泛分布，因此也成为热点研究模式菌株。其中紫罗兰希瓦氏菌（*Shewanella violacea*）DSS12菌株作为一株操作方便、研究较多的高压适应深海细菌，在基因表达的压力调节和细胞色素d在高压下细胞生长中的作用等研究中被广泛关注。1998 年从日本海沟中首次分离出日本海沟摩替亚氏菌（*Moritella japonica*）DSK1 菌株，该属中的另一菌株海洋摩替亚氏菌（*M. marina*）MP-1 是该属的模式菌株，却没有压力适应的特征，这证实了即使进化地位非常接近，不同种间也存在着是否具有嗜压属性的差异。

对于 γ-变形菌中的深海嗜冷嗜压细菌，生理学和压力适应相关的研究较多并获得了一系列研究成果，例如，在压力控制的基因表达调控方面，通过将在深海希瓦氏菌（*S. benthica*）DB6705 菌株中获得的高压生长激活启动子克隆至大肠杆菌中，验证了该启动子下游两个可读框为压力调节操纵子。在紫罗兰希瓦氏菌 DSS12 菌株中也发现了类似的操纵子，在压力提高时基因的表达会得到增强。在压力对呼吸链的调节方面，通过对 *S. benthica* DB172F 菌株细胞色素 c 的纯化和研究，推测在 0.1 MPa 和 60 MPa 时该菌株中存在着不同的电子传递链。同时，在对不同门类微生物的压力适应研究方面，发现细胞膜、DNA、蛋白质、营养物质都会受到压力的影响，也有研究对其机制进行了探讨。但是，严格嗜压菌作为一种只有在高压环境中才能较好生长的微生物，由于其分离纯化和培养的难度较高，导致对严格嗜压菌的研究并未同兼性嗜压菌一样广泛。目前已知的严格嗜压纯菌中，细菌中仅有 γ-变形菌门中的 10 株菌和 α-变形菌门的 1 株菌得到纯培养。其中，从马里亚纳海沟中分离出的狂热海洋科维尔氏菌（*Colwellia marinimaniae*）MTCD1 菌株的生长压力范围为 80～140 MPa，在 120 MPa 时最适生长温度为 6 ℃，该菌株为目前已知的生长压强最高的纯培养细菌。

除了具有嗜冷特性的耐压 / 嗜压细菌，深海中还存在着另外一类耐压 / 嗜压菌，它们分离自海洋中一种特殊的环境——最高温度可高达 400 ℃的深海热液口。然而，该环境中古菌为优势类群，其最适生长温度常常高于 80 ℃。与嗜冷嗜压细菌不同，嗜热菌尤其是超嗜热菌中的严格嗜压菌被发现且获得纯培养的数量更少。2011 年分离出的亚亚诺西火球菌（*Pyrococcus yayanosii*）CH1 菌株是目前发现的唯一严格嗜压的超嗜热古菌，该菌株分离自大西洋中部 4100 m 深度的热液口，生长压强范围为 20～120 MPa，最适生长压强为 52 MPa。对 *Pyrococcus yayanosii* CH1 压强适应机制的一系列研究证明，其中并不存在特定的严格嗜压控制基因，仅二十几个点突变就可以使其获得常压下生长的能力（李学恭等，2013）。此外，在嗜冷和嗜热嗜压菌研究过程中，均发现当培养的压强提高到一定程度（超过 0.1 MPa 或超过原位压强）时，有些菌株的最适 / 最高生长温度将会进一步提升，这也提醒我们，高压适应可能并非一种独立作用的因子，而是通过细胞内的温度–压强等共适应机制来调控细胞的活动，而研究细胞的高压耐受边界以及通过提高压强的方法来拓宽生命的温度边界也成为新的研究思路和方向。

目前，马里亚纳海沟作为典型的深海高压环境，有着目前已知的海洋最深处——挑战者深渊，该深渊成为人们研究高压下生命的一个天然实验室和关注点。很多关于适应高压的微生物从马里亚纳海沟中被分离并纯培养，进而开展后续详细的研究。利用“挑战者号”在马里亚纳海沟采集到的沉积物样品，分离得到了 38 株放线菌（Pathom-Aree et al.，2006），这些放线菌能够产生一种新的自然化合物，意味着这种深海放线菌存在一些以前未被揭示的催化活动（如腈水合酶的催化活动）。

总的来说，目前高压环境下微生物的探索和研究，一方面，侧重对一些分离自常压环境中微生物耐压性能的探索；另一方面，更多地着力于对来自深海高压环境的耐压 / 嗜压微生物的研究，包括其分离培养（如何高效分离及避免分离培养过程中嗜压性状的丢失）、适应高压条件的微生物在高压适应过程中的生理特性和基因表达方式的研究、高压环境中的蛇纹石化反应对有机物的生成以及对这种环境中生命起源的影响、高压环境下生命边界的探索、高压环境中基因组的进化、高压与其他环境因子的互作和共同适应等研究方向。对高压下的生命，尤其是广泛分布于深海环境中微生物的研究方兴未艾，尚有很多未知领域等待我们去探索和发现。

二、高压与生命起源、生命边界

（一）高压环境与生命起源

生命起源仍存在多种的可能性，并且在历史上未必只发生了一次，也未必是通过某种途径形成单个细胞的。但确切的生命起源过程不得而知，仍有很多研究对生命起源时期的环境特征进行描述，包括早期的海水盐度、金属元素以及其他微量元素的浓度等。但在有些方面还存在争议，比如生命早期环境的酸碱性等。比较明确的是，生命是在地球上水大量聚集之后产生的。从宇宙元素的产生到地球上水的产生并逐渐形成早期海洋的过程都存在一定的偶然性，但水却成为生命必需的前提条件。从无机物产生有机物并且最终组成复杂的生命形式需要在水环境中进行。在此基础上，静水压力为生命的产生提供了辅助条件。高压环境帮助了早期生命稳定初始的细胞结构，并可以保护其免遭强烈的辐射和陨石冲撞。

从米勒（Miller）实验到后续更加完善的实验模拟和色谱分析，给出了从无机物生成有机物的实验证据，而在深海高压环境可能存在区别于米勒实验中通过模拟闪电产生有机物的方式。深海热液口和深渊被认为很可能是生命起源的位点，也是研究的热点。在热液环境中，海水向地壳渗透后形成热液喷发，在这个过程中持续输出还原态的金属离子和氢气、硫化氢等气体。由于热液羽流与周围海水理化性质的差异形成氧化还原梯度，发生化学反应并且形成一些有机物。通过实验室对热液环境的原位观测及实验室模拟都已经证实了这种可能性。

在深渊中，大量底栖生物的发现打破了原先超高压环境没有生命存在的认识。其中，存在的严格嗜压微生物不能在常压下生存。这些微生物既有可能来自上层水面微生物的沉降并且经历了适应高压的过程，又有可能起源于底层超基性岩蛇纹石化过程。

由于高压对生命起源的独特意义，对深海高压环境微生物的研究能够从微生物的基因组和代谢途径等反演生命早期所处的环境特征，从而帮助理解生命是如何从大分子组装而成的等关键科学问题。

（二）高压与生命边界

人们对生命边界进行了一定的探索，如微生物生长的最高温度边界（122 ℃）、微生物生长的最高压力边界（140 MPa）、微生物生长的最低温度边界（预估 −25 ℃～−10 ℃）等。2009 年 *Pyrococcus yayanosii* CH1 的发现突破了目前已知的绝对嗜压古菌的温度和压力的耐受极限（图 3-18），该菌株可以在高达 108 ℃环境下生长，并且可以耐受 120 MPa 的压强。

深海作为一种高压低温或高温（特指热液口）高盐的特殊极端环境，其特定环境下生存的微生物很有可能代表着生命可以存在的边界条件，而且目前已知的最高压强 / 温度条件下可以生长的微生物均来自深海环境。随着培养技术和模拟设备的进一步发展，模拟深海超高压力条件已经成为可能。如何利用现有的样品和设备进行生命存在边界条件的探索是人们下一步需要关注和解决的问题。

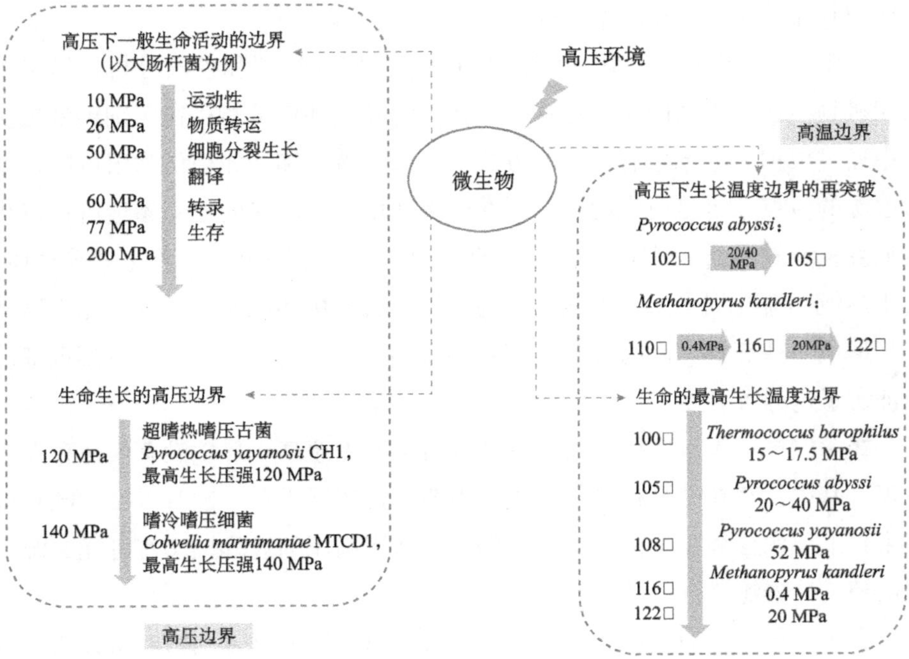

图 3-18　高压下的生命边界

注：深海火球菌（*Pyrococcus abyssi*）、坎德勒甲烷火球菌（*Methanopyrus kandleri*）、嗜压热球菌（*Thermococcus barophilus*）。

三、高压下的共适应机制

微生物功能群在不同的地质历史时期面临一种或多种环境压力，理解共适应机制有助于人们正确认识不同极端环境条件下微生物功能群的作用范围与大小。在深海环境中，除了压力，微生物还需要长期面对多种极端条件，如低温、高温、高盐、氧胁迫等，从而进化出多种代谢途径以对抗这些不利的生存条件。热液口来源的热球菌目（Thermococcales）作为典型的超嗜热微生物，最初的研究主要在于通过不断地寻找嗜热酶及特有的嗜热基因以揭示其嗜热机制。随着不断发现新的超嗜热微生物类型，人们认识到嗜压热球菌同时具有温度、压力、pH 与狭窄的盐度生长跨度，并且其基因组较小，仅为 2 Mb 左右（赵维殳和肖湘，2017）。

AOA 是深海中丰度最高的微生物类群之一，起源于早期厌氧的陆地环境，在演化过程中增加了抗氧化能力而逐步进入深海。除海水表层以外，

AOA 在现代海洋的整个海水垂直范围内都有较多的数量，在向深海迁移的过程中必须同时适应低温、高压、寡营养等多种不利条件。

基于现有研究数据发现，微生物针对不同环境胁迫的响应确有重合。与适应性相关的机制包括能量代谢、相容性溶质、氨基酸代谢、膜脂合成、抗氧化系统等。压力适应与微生物代谢的标识见图 3-19。

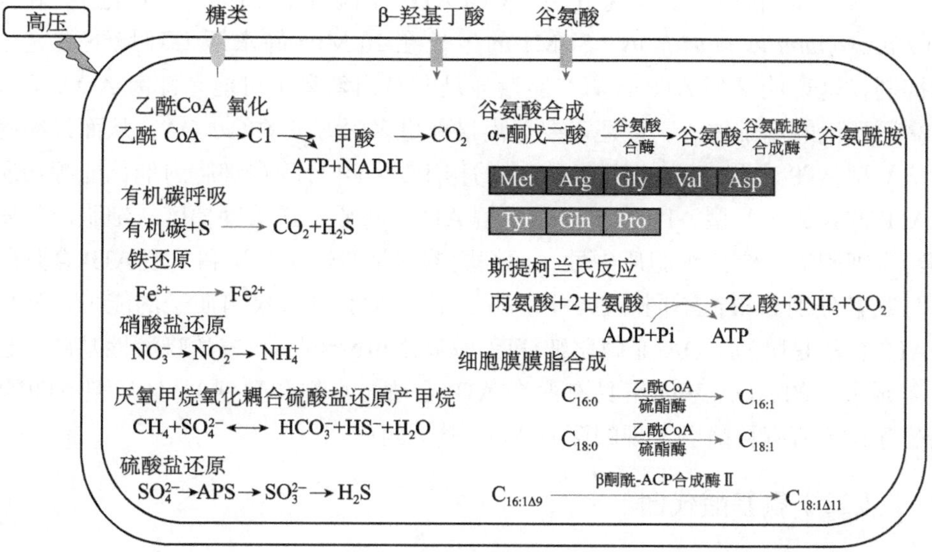

图 3-19 压力适应与微生物代谢的标识图

注：甲硫氨酸（Met）、精氨酸（Arg）、甘氨酸（Gly）、缬氨酸（Val）、天冬氨酸（Asp）、酪氨酸（Tyr）、谷氨酰胺（Gln）、脯氨酸（Pro）。

（一）能量代谢

古菌、细菌和真核生物在细胞膜上都拥有可以产生能量的 ATP 合酶，且这些 ATP 合酶有着共同的祖先。在具有广泛生长范围的超嗜热 / 嗜压古菌广热热球菌（*Thermococcus eurythermalis*）A501 菌株中，能量代谢在酸、碱、高盐、低盐、高压、低温、高温等多种胁迫中具有不同的响应。与常见的细菌和真核生物中质子驱动的 ATP 合酶不同，在热球菌目菌株中，产能是利用钠离子驱动的 A_1A_0 ATP 合酶完成的，并在该目激烈火球菌 DSM 3638 中首次报道了这种钠离子驱动的 A_1A_0ATP 合酶。除了具有特殊的 ATP 合酶，在严格厌氧的热球菌目中，不存在复杂的氧化呼吸链，而是广泛存在着一类膜上复合体——膜上结合氢酶（membrane bound hydrogenase，MBH）和膜结合氧化还原酶（membrane bound oxidoreductase，MBX），利用这些膜上复合体中的 Na^+/H^+ 反向转运蛋白在胞内外形成 Na^+ 梯度，以此驱动 A_1A_0 ATP 合酶。除了 MBH 和 MBX 构成简单的电子传递链，以诺利纳斯热球菌（*T. onnurineus*）为代表的部分热球菌目菌株还可以利用甲酸代谢途径产能，该途径是目前已发现的最简单且最古老的能量代谢途径。

共适应机制也体现在 AOA 的 ATP 合酶中，海洋亚硝化侏儒菌（*Nitrosopumilus maritimus*）SCM1 的生长在 20 MPa 静水压力环境下被完全抑制，这说明Ⅴ型 ATP 合酶在能量充足时泵出氢离子可能是深海 AOA 适应高压环境的基础之一，但还需要分离培养更多深海来源的 AOA，从而最终确认Ⅴ型 ATP 合酶在适应高压环境中的作用。ATP 合酶在细胞内催化能源物质 ATP 的合成，Ⅴ型 ATP 合酶可以水解 ATP，将质子逆浓度梯度从细胞质基质转入细胞器，维持细胞质中性，而细胞器内呈酸性，A 型和Ⅴ型 ATP 合酶同时或者分别分布在不同 AOA 中。由于消耗 ATP 具有泵出质子的能力，Ⅴ型 ATP 合酶是嗜酸 AOA 亚硝化细杆菌（*Ca.* Nitrosotalea）属的嗜酸性基础。超深渊来源的 AOA 基因组具有两套 ATP 合酶（A 型和Ⅴ型），且Ⅴ型 ATP 合酶存在于所有目前已发现的深海 AOA 基因组中。

（二）氨基酸代谢

对热球菌目中的嗜压菌压力适应研究发现，氨基酸合成与压力适应相关：对严格嗜压菌 *Pyrococcus yayanosii* CH1 的研究发现，在高压和低压的胁迫条件下，氨基酸合成途径都有明显的响应；在嗜压热球菌 MP 中发现组氨酸（His）、Asp、苏氨酸（Thr）和 Pro 的合成与高压和低压下的胁迫响应密切相关，而后不同压力条件下必需氨基酸利用的差异实验表明，在高静水压下需要额外的 3 种氨基酸——天冬酰胺（Asn）、半胱氨酸（Cys）和 Tyr。生长必需氨基酸与环境胁迫响应之间的关系目前仍在研究中，推测可能与胁迫条件下能量供应的限制有关。现有的热球菌目在不同环境胁迫下的转录组及蛋白质组显示，氨基酸的合成与利用在热球菌目的环境适应中起到了至关重要的作用。在柯达卡兰热球菌（*T. kodakaraensis*）的研究中，通过蛋白质组的研究方法，比较了高温、氧化和高盐等不同的条件下菌株在蛋白质表达上的差异，发现氨基酸合成是 3 种胁迫响应共有的上调表达途径。

（三）相容性溶质

在所有已研究的嗜压微生物中都发现了相容性溶质（compatible solute）

的积累。相容性溶质是微生物应对极端环境胁迫时细胞内积累的一类分子量较小的有机化合物，在极端环境胁迫下起到维持细胞渗透压稳定性、保护胞内生物大分子稳定性，从而维持细胞活性的作用。

相容性溶质主要包括以下三类：糖类及其衍生物、氨基酸及其衍生物以及含有磷酸基团的多元醇衍生物。嗜热微生物主要利用海藻糖（trehalose）和甘露糖甘油酸酯（mannosylglycerate，MG）或葡萄糖甘油酸酯（glucosylglycerate，GG）作为相容性溶质。其中海藻糖被认为是一种应对多种环境胁迫的广谱共同保护剂，在高盐、高温、酸性和氧化胁迫下均有报道。在中温菌中常用于应对渗透压胁迫的相容性溶质有谷氨酸（Glu）、脯氨酸、甘氨酸、甜菜碱（glycine betaine）和四氢嘧啶（ectoine），而在超嗜热微生物中常用谷氨酸作为相容性溶质。在超嗜热古菌热球菌目中，除谷氨酸之外，还可以利用天冬氨酸作为相容性溶质。含有磷酸基团的多元醇衍生物也可作为相容性溶质。在超嗜热古菌和嗜热细菌中常发现含有带负电荷的磷酸基团的多元醇衍生物作为相容性溶质，在热球菌目的研究中被报道与高温、高渗透压及高压等胁迫均有关，磷酸二肌醇磷酸酯（di-myoinositol-phosphate，DIP）是其中分布较广的成员。

在热球菌目中被报道主要积累和利用的相容性溶质为DIP、Glu、Asp、甘露糖甘油酸酯和海藻糖。现有的实验数据表明，热球菌目中的相容性溶质主要在高温、高渗透压及压力胁迫条件下积累，不同菌株之间略有差异，且不同的相容性溶质一般主要用于应对不同的胁迫条件，但不同相容性溶质之间也有一定的协同作用。陆地来源的氨氧化古菌 Nitrososphaerales 目中存在潜在的甘露糖基-3-磷酸多糖合酶，表明它们有能力从甘露糖-6-磷酸中合成甘露糖甘油酸酯，其合成酶也在维也纳亚硝化暖菌（*N. viennensis*）蛋白质组中被检测到了。基于基因组分析，几乎所有的深海 AOA 都具有产生 DIP 的潜力，而 DIP 可作为渗透压调节剂来应对深海中的高静水压力和低温环境。此外，暂定亚硝化广栖菌属（*Ca.* Nitrosocosmicus）含有预测为海藻糖转运功能的蛋白质，表明这类具有多种功能的相容性溶质也可能在 AOA 中发挥作用。

（四）膜脂组分

在嗜压细菌细胞膜中，高压能够诱导不饱和脂肪酸的积累从而降低饱和脂肪酸的比例。膜脂的流动性对维持细胞生命具有重要的作用。细菌可以合成饱和或不饱和脂肪酸，古菌可以合成含不同数量环戊 / 己烷的甘油二烷基甘油四醚（glycerol dibiphytanyl glycerol tetraether，GDGT），这些特殊脂类的变化能够增加或降低膜流动性，使其能在高温下维持完整膜结构，或避免在高压和低温条件下固化。

超嗜热古菌热球菌目的膜脂相对其他古菌较为简单，目前的研究表明，其膜脂组分为由 20 个碳原子（C_{20}）组成的磷酸甘油二醚阿克醇（sn-2, 3-diphytanylglycerol diether，archaeol）和由 40 个碳原子（C_{40}）组成的不包含环戊烷部分的甘油四醚 GDGT-0。

所有的 AOA 都含有高丰度的泉古菌醇核心的 GDGT，并且该核心膜脂是奇古菌所特有的。GDGT-泉古菌醇在海洋中普遍存在，包含一个环已烷和四个环戊烷环，它的结构类似于由（超）嗜热泉古菌门（Crenarchaeota）合成的 GDGT-4 的结构，但其双植烷烃基链内部环化形成了一个环已烷，故流动性大大提高，这可能也是 AOA 能适应深海极端环境的基础之一。

（五）抗氧化途径

压力能够诱导细胞内的过氧化反应。在极端环境条件下，氧化还原平衡的破坏是对微生物体最主要的生理伤害之一。热球菌目最主要的抗氧化酶是超氧化物还原酶（superoxide reductase，SOR）。此外，还存在于好氧微生物氧化清除系统中类似的二铁黄素氧化还原蛋白（flavodiiron 蛋白，FDP）。不仅针对氧化胁迫条件，针对热球菌目其他环境胁迫的研究还表明，厌氧氧化清除机制与多种环境胁迫的适应相关：在 *T. kodakarensis* 中，在高温胁迫下与氧化还原相关的过氧化物氧化还原酶（peroxiredoxin）和氧化还原酶（oxidoreductase）的浓度均上调，在高盐胁迫下与氧化还原相关酶类，如 peroxiredoxin、铁氧还蛋白（ferredoxin）和硫氧化还原蛋白还原酶（thioredoxin reductase）的浓度也均上调。

AOA 虽然是好氧微生物，但目前分离的浅海 AOA 由于抗氧化系统不完善［含有 SOD，但缺乏过氧化氢酶（catalase）、过氧化物酶（peroxidase）等，清除活性氧能力不足］，导致其对 ROS 较为敏感，因而 AOA 是否存在其他机制的补偿或共适应还需进一步的探究。

（六）基因组演化

高压诱导基因组的演化已经在原核与真核生物中被发现（Wang et al., 2015），但未开展深入细致的研究。在自然界中，微生物通常生活在持续变化的环境条件中，很难出现微生物的最适生长环境，因此微生物时刻受到来自环境的胁迫作用。然而微生物可以通过多种方式和策略对抗这样的环境胁迫，其中的一种方式就是通过其自身基因组的演化来适应环境。基因组是微生物的遗传物质，是复制和转录的模板，对微生物自身的生活方式、代谢途径以及能量平衡等起到了最基本的支撑作用。

从早期的原始生命到如今拥有巨大多样性的微生物功能群，微生物的基因组伴随着漫长的地质环境的变化而不断地发生变化，使微生物始终能够适应环境并有利于自身的生长和繁衍。因此，其演化的过程就是微生物从生命原始的起源到不断扩张，最终占据地球上包括各种极端环境在内不同生态位的过程。生命复杂度的提升也是微生物不断演化的结果，如古菌和细菌的融合产生了真核生命。

微生物受到的演化驱动力主要包括突变、自然选择、基因漂变以及基因重组。虽然基因组演化方面的研究已经积累了相当多的数据和理论，但是演化按照哪种理论进行仍存在争议，原因是对于不同的微生物类群、不同的生态环境，微生物的演化可能通过不同的方式进行，因此难以使用同一种理论来解释。

此外，微生物自身的突变速率与微生物的基因组大小和有效种群大小有关。有效种群较大的微生物主要受到自然选择的影响，反之，则主要受到基因漂变的影响。例如，花中间原体（*Mesoplasma florum*）与大肠杆菌在自发突变上相差 1.5 个数量级，是因为 *Mesoplasma florum* 以内共生的形式生存，有极小的有效种群大小。

对于海水寡营养环境，微生物基因组有缩小的倾向以应对能量限制。一部分因突变而丢失功能的基因会被抛弃以节约代谢消耗的能量。在氮限制的环境中，微生物倾向于利用 A—T 碱基对代替 G—C 碱基对。对于高压环境而言，深海环境十分复杂，拥有多种不同的生态环境，如热液、冷泉、深渊等，且不同区域理化参数差异巨大，使得其中的微生物可能存在多种演化方向策略以适应不同生境。在高温高压的热液环境中，微生物的基因组较小，而在超高压的深渊环境中，由于能量输入较少，微生物代谢缓慢，代时较长，故一般有较低的基因突变速率。

截至 2021 年，国内对海洋环境微生物基因组演化的研究主要集中在常见的表层海水微生物上，例如远洋杆菌（SAR 11）和玫瑰杆菌（Roseobacters）等，研究了菌株的谱系、演化历史、驱动力和方式特征等。虽然通过构建系统发育树和比较基因组学等手段进行研究，取得了一系列的成果，积累了较为成熟的理论体系、研究方法和技术路线，但是对高压深海环境来源的微生物的研究还有待深入开展。

四、国际上的研究团队与平台

国际上研究高压环境微生物的著名团队及其优势概述如下。

（1）JAMSTEC。作为日本海洋科学技术研究与发展的核心机构，JAMSTEC 在海底观测、大洋钻探、海洋微生物等领域都有深入的研究。拥有“地球号”科学钻探船、载人深潜器“深海 6500”和超级计算机“地球模拟器”等一批先进的科考船和大科学装置，本部拥有保压转移与深海微生物培养系统等一系列设施。

（2）CNRS。该机构是欧洲最大的基础科学研究机构，依托多所法国大学的科研平台。CNRS 下属的国家地球科学和天文学研究所（INSU）长期致力于海洋微生物、原生生物生态学的研究与应用。CNRS 与欧洲的各个基础研究机构合作，包括海洋观测站、板块观测站系统、欧洲海洋钻探研究联盟等。研究人员可以使用各类最新的研究设备，比如高压微生物模拟培养、显微平台和海洋科考船等开展广泛的研究。

（3）加州大学系统（University of California System）。美国加州大学圣迭戈分校下属的斯克里普斯海洋研究所是一所综合性海洋科学研究机构。其中，海洋生物技术与生物医学中心（CMBB）的重点研究方向包括海洋生物医学、海洋药物研发、深海海洋微生物特性、海洋动物基因工程等。斯克里普斯海洋研究所拥有 5 艘科考船、水力学实验室和海洋科学发展中心。针对高压微生物的研究，其静置式高压培养系统可以模拟深海环境进行微生物高压培养。

（4）德国亥姆霍兹联合会。它是德国最大的科研机构，包括 19 个研究中心。其中亥姆霍兹基尔海洋研究中心（GEOMAR）开展从海底地质学到海洋生物化学的一系列研究。该中心拥有 4 条科考船、德国唯一载人潜水器 JAGO、深海无人遥控机器人（ROV）KIEL6000 等。此外，GEOMAR 的多个实验室可以进行分子生物学、海底可视化、稳定同位素等研究。

（5）美国伍兹霍尔海洋研究所。它是世界著名的综合性海洋科学研究机构，有 1000 多名教职人员和六大跨学科中心。海洋化学与地球化学系（MC&G）的研究重点包括生命与生物地球化学循环的共同进化、深海热液喷口和冻土等不同环境中的极端微生物代谢途径。伍兹霍尔海洋研究所拥有 2 艘科考船，分别是“亚特兰蒂斯号”（Atlantis）和“阿姆斯特朗号”（Armstrong）。该所还提供各种研究设施，包括深海数字成像实验室、共振质谱仪、营养分析仪、静水压力测试系统等，且最早研发了高压微生物培养系统。

第四章
国内发展现状与发展态势

第一节　学科整体发展情况

一、在国际上逐渐产生影响

我国科学家对各类极端地质环境的微生物都开展了研究，涉及的极端地质环境包括海洋极端环境（热泉和冷泉区、深渊、洋底沉积物、缺氧区等），陆地深部地下环境（油气藏、洞穴、地下水、超深钻等），当代地表极端环境（陆地热泉、酸性矿坑水、盐碱湖、雪冰等），深时极端地质环境，深空环境等。

在国际上，涉及极端地质环境研究如此广泛的国家还是比较少的，从这个层面来说，我国在国际上处于不可替代的地位，在一些领域处于国际领先地位，如有关海洋深古菌门、二叠纪—三叠纪之交的地质微生物研究等；在一些领域处于“并跑”地位，如热泉微生物、雪冰和洞穴微生物、油气藏微生物、酸性矿坑水微生物研究等；在一些领域的研究具有鲜明的特色，如盐碱湖微生物、超深钻微生物、地下水微生物、陆地热泉微生物研究等。

二、不同分学科领域的发展存在较大的差异

在各类不同的极端地质环境中，微生物研究存在很大的不平衡。有些领域有很大的优势，有些领域则需要大力加强。

（一）优势领域

在我国，相对比较有优势的研究领域是海洋极端环境中的热泉、深渊和洋底沉积物，陆地深部地下环境的油气藏和洞穴，地表极端环境的酸性矿坑水和雪冰等，二叠纪—三叠纪之交等若干关键地质时期的极端地质环境。在这些优势领域，我国不仅在研究成果方面比较突出，也有规模比较大的研究队伍，后续还会进一步快速发展。

海洋极端环境的研究主要得益于国家的海洋战略以及国际上的大洋航次，在国际上发表的代表性研究成果包括提出了海洋深古菌门及其相关的生态功能研究、热泉微生物的系统研究、有关压力对微生物的影响研究等，在部分领域处于国际领导地位。特别是有关深古菌门的研究，我国分别在《自然-微生物学》（*Nature Microbiology*）、《美国国家科学院院刊》（*PNAS*）、《国际微生物生态学会会刊》（*ISME J*）上发表了具有重要国际影响的成果，并得到了国际学者的正面评价。

油气藏和酸性矿坑水环境的研究主要得益于我国大量的资源能源开发及其面临的一系列环境问题，由此带动了这些环境的微生物研究，并已经部分在工业上得到了应用，如微生物驱油等，一些工作在国际上具有重要影响。洞穴和雪冰微生物的发展主要得益于国家对全球变化领域的重视以及微生物资源的重要性，代表性成果是青藏高原雪冰和冰川微生物以及长江中游洞穴微生物的工作，特色鲜明。

深时极端地质环境微生物的研究主要得益于我国比较具有优势的地质学以及典型剖面，代表性成果主要是有关二叠纪—三叠纪之交生物大灭绝期间的各类微生物功能群及其生物地球化学的研究，我国学者在该领域具有领导地位。深空环境微生物的研究主要集中在柴达木盆地与火星环境的对比性研究。

（二）薄弱领域

在海洋极端环境方面，我国学者开始研究深渊微生物、冷泉微生物。海底沉积物微生物的研究是一个广阔领域，亟待加强。

在陆地极端环境方面，雪冰、洞穴等与全球变化相关的微生物研究需要进一步加强，提取各种微生物种质资源。充分利用各类超深钻技术，加强深地微生物的现场原位监测与研究。

深时极端地质环境微生物由于受到地层剖面和研究技术的限制，也是国际上的薄弱领域。我国可以利用一些地层优势，把一些薄弱环节加强以形成系统性的国际影响，包括新元古代“雪球地球”时期的微生物研究、显生宙其他生物大灭绝时期的微生物研究。特别是有关微生物的地质演化研究，在国际上都很薄弱。

在深空环境微生物研究方面，国际上的工作都极其薄弱，我国可以抓住这一难得的机会迎难而上，加强以柴达木盆地为核心的对比研究，积极参与国际竞争，与国际上其他地区的对比研究结合起来。

在微生物种类方面，各类极端地质环境的病毒正引起全球科学家的关注。我国对各类环境的病毒研究还比较薄弱，对各类极端地质环境的病毒研究则更少。

三、技术平台有待拓宽

极端地质环境微生物的突破，首先要在技术上进行革新。有三方面的技术平台，包括野外钻探与保真取样技术、野外原位观测与模拟实验技术、实验室模拟培养与单细胞分析技术。虽然这三类平台技术都相当重要、缺一不可，但与常规环境微生物研究相比，前两类技术平台对极端地质环境微生物研究尤其关键，是典型的“卡脖子”技术。然而，在我国，恰恰是前两类技术平台相对比较薄弱，需要进行自主创新。实验室模拟培养与单细胞分析技术平台相对要好些，许多是直接从国外引进的，部分是自主设计的。

（一）野外钻探与保真取样技术

钻探技术对海洋极端环境和陆地深部地下环境微生物的研究尤其重要。因钻探过程会对其中可能存在的微生物造成致命损伤，采用低速钻探工艺或大口径钻头可以很好地保护位于中心部位的微生物，同时又能有效避免外源污染。与钻井技术相比，掘进技术可以提供深部直接取样的机会。在我国，钻爆法仍然是最主要的岩石巷道掘进方法。大洋钻探需要钻探船，我国要发展深海科学，就必须在现有国力的基础上建造具有中国特色的第三代大洋钻探船。这种船既要避免像日本船那样规模过大，又要吸收美国、日本、欧盟国家的先进技术而成为高效率、高水平的新设施。

地层系统经钻探后需要马上取样，在采样过程中维持原有的极端和无菌环境是进行后续研究的基础，采用先进的保真采样法是关键。目前市面上尚无商品化的全封闭式保真采样系统，主要借鉴类似的改进型钻探取芯工具，也可以采用无菌、封闭、低压的泥浆循环系统进行岩心采样。部分生物学家将石油钻探等取芯技术进行改进，应用于岩层微生物研究中。针对流体取样，目前比较常用的是井孔渗透压流体采样器（osmo sampler）。深海热液保真采样系统适用于科考船或水下运载器。

（二）野外原位观测与模拟实验技术

绝大部分深地极端环境微生物不能在实验条件下培养，原位环境模拟研究技术对深地极端环境微生物的后续研究具有重大意义。对于深海极端环境，建立海底观测站至关重要。循环观测回返系统（circulation obviation retrofit kit，CORK）是海底观测站的一个典型例子。原位检测微生物有直接方法和间接方法。间接方法主要是通过检测流体物理化学性质的变化进而了解微生物活动。地下原位直接检测微生物可以用一种深紫外（＜250 nm）-激光诱发的荧光光谱。另外，循环式渗透压取样器可以改装成培养系统，安装在带有 CORK 装置的钻孔里，从而能在钻孔所处地层流体内进行实验。

极端地质环境原位模拟研究是指在一个较小空间内模拟微生物原有生存环境条件下，保持微生物原有生理及生长状态，即原位状态，并在此状态下动态探索微生物生长过程中物质、能量代谢等一系列变化的研究。极端环境原位模拟研究将打破现阶段无法培养极端微生物的局限，为极端微生物的研究带来更多的可能性。四川大学徐恒等（2017）设计了一种针对深地极端环境微生物的原位模拟实验舱和原位研究系统。原位保真采样设备采集样品后，也可以通过生命舱来模拟深地原位的极端环境。

（三）实验室模拟培养与单细胞分析技术

实验室的模拟工作主要是针对不同极端环境条件开展的。热液金刚石压腔（hydrothermal diamond-anvil cell，HDAC）是一种可用于高温高压微生物模拟培养的实验室设备。这方面我国有很好的基础。

在分析测试技术方面，随着新技术的进步和新仪器的开发，微生物学逐渐从基于培养的、群体细胞的研究向不基于培养的、单细胞亚微米-纳米尺度的研究发展。以纳米二次离子质谱、原子力显微镜、电子显微镜等为代表的显微分析仪器极大地拓展了微生物学研究的广度和深度，我国在这方面已经具备很好的研究条件。

四、学术组织、研究队伍和资助力量比较分散

我国已经有许多与微生物有关的学术组织，包括中国微生物学会地质微

生物分会、中国古生物学会地球生物学分会、中国海洋湖沼学会微生物海洋学分会，以及中国矿物岩石地球化学学会沉积学分会等。这些学会分别涉及地球科学与生命科学两大自然科学。尽管我国已经对各类极端地质环境微生物开展了全方位的研究，但目前还没有专门针对极端地质环境微生物的学术组织。2018 年 10 月，国际深碳观测组织在上海交通大学成立了国际深部生命研究中心，这是国内唯一专门针对极端地质环境微生物的研究中心。

在各类极端地质环境微生物方面，我国都没有形成系统性研究成果，在海洋极端环境、陆地深部地下环境、当代地表极端环境中都没有形成整体优势。深时地质环境微生物也仅在某一两个关键地质时期形成了一些重要成果，但尚未形成整体影响。这与研究队伍分散有很大关系，没有形成极端地质环境微生物的学术共同体。

在国家级平台方面，缺乏专门针对极端地质环境微生物的国家重点实验室。中国科学院微生物研究所微生物资源前期开发国家重点实验室、中国地质大学（武汉）生物地质与环境地质国家重点实验室、上海交通大学微生物代谢国家重点实验室、厦门大学近海海洋环境国家重点实验室、同济大学海洋地质国家重点实验室等涉及极端地质环境微生物的研究主题，但研究力量相对分散，没有形成合力。

在资助力量方面，我国尚未设立专门针对极端地质环境微生物的单独学科体系或资助计划，相关学科体系或资助计划都是依托其他项目或学科进行的。科技部也未设立单独的专项计划或其他大型项目来资助这方面的研究，中国科学院也缺乏这方面的专项计划。这些资助机构都在其他大型项目中列有一些分散性的课题来开展极端地质环境微生物的研究。

第二节　海洋极端环境微生物

一、优势和薄弱领域

（一）优势领域

近年来，我国对海洋权益日益重视，将海洋战略提到了前所未有的高度。十八大报告指出，“提高海洋资源开发能力，发展海洋经济，保护海洋生态环境，坚决维护国家海洋权益，建设海洋强国”。与此相适应，我国已完成载人潜水器“蛟龙号”7000 m 海试，一批海洋科考船已经下水或正在建造；电视抓斗、海斗深渊着陆器、可视多管取样器、深海生物组合取样器、深海水体过滤器、深海浅钻、深海深钻、6000 m可视生物拖网、遥控潜水器（ROV）以及海洋观测网等不断完善；深海环境模拟培养平台基本建设完成，包括覆盖全水深的静置高压培养系统（1～1600大气压）和迄今唯一国际上正在运行的气液混合高压流动培养系统。该系统能够模拟深海冷泉、热液等不依赖于阳光的化能合成生态系统（1～350大气压、0～150℃）。此外，同位素示踪技术、组学技术、单细胞分离/单基因组测序技术、微生物高通量培养技术、微群落技术、生物信息学和复杂体系模型构建等不断发展完善。我国已经具备了开展深海海底生命地质地球化学功能研究的装备和技术支撑。在科学领域，我国已经在深海古菌、海洋微生物碳泵领域形成了研究优势。

（二）薄弱领域

我国缺乏分子和细胞生物学、进化系统学、生物化学、信息学、数学、计算机和工程学等学科与地质科学在海洋极端微生物研究方向的深度交叉；缺乏深海海底生态系统生态群落结构及功能群的大范围系统调查及深入解析；缺乏微生物参与的生物地球化学过程代用指标的开发及向地质时间尺度的拓展；缺乏不同时间尺度海水、沉积物中生物地球化学过程的重建、示踪与解析；缺乏人工智能、大数据分析在深海极端环境微生物研究领域的探索与拓展。

二、研究队伍和平台

近年来，我国科学家在地球生物学研究领域已经产生重要的影响力，形成了一批以中青年科学家为主的研究队伍，在深海微生物研究、古环境古气候研究、环境磁学、生物磁学等方面取得了一系列重要的原创性研究成果，分别发表在《科学》《自然》《自然-地球科学》《自然-微生物学》《美国国家科学院院刊》等国际顶尖学术期刊上。厦门大学焦念志院士领衔的海洋微生物研究团队，在国际上提出了“微型生物碳泵”假说，开拓和引领了海洋微生物研究的新方向；同时，焦院士团队的张锐教授在深海病毒方向开展了大量工作，提出深海病毒很可能在深部微生物演化和物质循环过程发挥了重要功能。上海交通大学王风平教授领衔的微生物研究团队发现并命名了一个新的古菌门——深古菌门，并对其代谢机理进行了系统性研究，通过系统研究深海生态系统的动态演化和丰富的代谢活动，提出了深部微生物在N、S等重要元素生物地球化学循环中的驱动机制。上海交通大学肖湘教授领衔的团队搭建了世界上唯一长时间周期运转的高压气体流动培养系统（模拟深海热液、冷泉系统），成功富集培养了甲烷厌氧氧化微生物体系，提出了高压微生物共适应理论。中国科学院海洋研究所张鑫研究员带领课题组已自主研发了多套应用于深海极端环境的原位探测和采样设备，这些深海探测综合取样设备已经基本覆盖了深海探测和取样的各个方面，为深海环境的探测提供了重要的技术保障。中国科学院微生物研究所黄力研究员、东秀珠研究员团队长期致力于古菌及其环境适应机制的系统研究，在极端嗜热古菌染色体结构及热稳定性、DNA 复制及其调控、热泉古菌病毒、产甲烷古菌等方面取得了多项重要成果。上海海洋大学陈多福教授研究团队自 2000 年起在国内率先围绕冷泉流体活动的沉积记录以及产生这些沉积记录的关键生物地球化学过程等科学问题开展了一系列工作，得到了国际同行的广泛关注和认可。

自 2010 年"蛟龙号"载人深潜器成功海试以来，我国大力发展全水深深潜取样技术并取得了突破性进展，多项技术在国际领先。我国已拥有一批深海和深部科考装备，包括"蛟龙号"载人潜水器、"海龙号""海马号"无人遥控潜水器、"天涯号"深渊着陆器、海洋观测网以及不断完善的生物组合采样器等。"奋斗者号"万米载人深潜器于 2020 年 10 月 27 日在马里亚纳海沟成功下潜突破万米，并于 11 月 10 日在马里亚纳海沟成功坐底，深度为 10 909m。这些硬件设施的建设以及相关人才队伍的培养为深部生物圈的研究提供了坚实的支撑和保障。2016～2018 年，"探索一号"科考船等在马里亚纳海沟挑战者深渊连续开展了系列综合性万米深渊科考活动，获得了大批的样品，质量与数量达到甚至超越了国际同行水平。但在深渊科学研究方面，中国还处于起步阶段，目前仅在上海交通大学、同济大学、中国地质大学、中国海洋大学、国家海洋局第二海洋研究所、国家海洋局第三海洋研究所、中国科学院深海科学与工程研究所、中国科学院南海研究所、北京大学、厦门大学等科研机构中有一批从事深部生物圈和深部碳循环研究，并活跃在国际科学舞台的科学家，他们承担了国家重点基础研究发展计划、国家高技术研究发展计划（863 计划）、重大科技专项、国家自然科学基金等一批重大重点项目，取得了一些初步的研究成果。例如，分离和保藏了大批深海微生物资源，获得了一批具有新颖结构的海洋天然产物，提出了微型生物碳泵理论，提出了沉积物中古菌为核心介导的深部碳循环假说等。

2018 年 10 月在上海交通大学成立的深部生命国际研究中心（The International Center for Deep Life Investigation，ICDLI），由上海交通大学生命科学技术学院、海洋学院、船舶海洋与建筑工程学院牵头，联合德国不来梅大学、日本海洋与地球科学学院、美国斯克里普斯海洋研究所、美国南加州大学、美国比奇洛（Bigelow）海洋研究所以及美国伍兹霍尔海洋研究所等国际顶尖科研学术机构共同建设。中心由肖湘教授担任主任、王风平教授担任副主任，现有来自上海交通大学生命科学技术学院、海洋学院、物理与天文学院、环境科学与工程学院的 11 位专职教师，以及来自德国、日本、美国的 6 位国外专家参与，另外，中心现有博士后研究人员 3 人，硕士 / 博士研究生 39 人。深部生命国际研究中心作为全球共同的深海与深部生物圈模拟研究平台，得到了国际深部生命团体（Deep Life Community）的全力支持，与 Deep Life Community 旗下另一中心——美国暗能量生物圈调查中心（C-DEBI）(含公共信息库）互为补充支撑。深部生命国际研究中心围绕深部元素循环，深部生命极限和深部生命的适应、演化三大主题开展多学科（如地球化学、地球物理、地质构造、分子生物学、生物信息学等）高度交叉融合的系统性研究。

三、主要领域的进展

（一）寡营养开放大洋

在寡营养开放大洋方面，戴民汉提出了寡营养盐海区真光层的双层结构概念。双层结构概念认为寡营养盐海区的真光层可分为两个区域，即营养盐匮乏层和营养盐充足层。位于营养盐跃层之上的营养盐匮乏层的新生产力主要由大气沉降、生物固氮等新营养盐支撑，而营养盐充足层的主要生产力来自下层的新营养盐的补充。这一概念的提出也获得了国家自然科学基金重大项目的资助，该项目针对北太平洋副热带环流区这一典型的寡营养盐海区正在开展相应的研究。寡营养开放大洋洋底微生物在国内的相关研究进展将在后续分别叙述。

（二）深海、深渊与海沟

随着国内深海装备的发展，如 7000 m 水深载人深潜器“蛟龙号”、4500 m 载人深潜器“深海勇士号”、万米全海深载人深潜器“奋斗者号”等，我国逐步具备了深海、深渊及海沟研究的能力。近年来相关成果呈井喷式出现，其中，中国科学院深海科学与工程研究所王勇团队利用着陆器，对挑战者深渊微生物基因组进行了最全面解析，对其原位代谢活性进行了首次报道，揭示了深渊特殊极端环境对微生物种群分化的驱动作用。该研究团队还对挑战者深渊水体中最大类群海微菌门（Marinimicrobia）的基因组学特征和进化过程进行了专门报道。中国海洋大学田纪伟教授团队利用自主研发的大体积深海采水器，采集了马里亚纳海沟的多层次大体积海水样品（最深达 10 500 m），并获得了多个万米水深沉积物样品。研究发现，深渊水体中的微生物呈现明显的垂直分层现象，在万米深水体中快速转变为以烃类降解菌为优势类群的微生物群落。同时，在深渊底部沉积物中鉴定出了大量生物来源的烷烃，这些烷烃可能支持了烷烃降解菌在马里亚纳海沟底层水的勃发。该研究结果揭开了世界最深处水体生命的神秘面纱，指出烷烃可能是万米水体微生物的重要“燃料”，为深入理解深渊海沟的生命过程与机制奠定了重要基础（Liu et al.，2019）。

（三）热液区

我国从 1988 年开始进行海底热液区的调查研究，起步较晚，基础也比较薄弱，前期以国际合作为主。中国第一次参与海底热液调查是在 1988 年 7～8 月中德合作执行的 SO57 航次，该航次对马里亚纳海底热液硫化物、西菲律宾海盆锰结壳（核）矿产资源及其海洋地质环境进行了调查研究。1988 年 9 月～1989 年 1 月，我国派人参加了苏联“维诺格拉多夫号”的太平洋综合调查，在东太平洋海隆附近取得热液沉积物样品。1992 年，中国科学院海洋研究所对东海陆架和冲绳海槽的热液活动进行了调查。1998 年 11 月，经国家海洋局和中国大洋矿产资源研究开发协会安排，我国“大洋一号”科考船于 DY958 航次第 5 航段在马里亚纳海槽开展了大洋热液矿点试验调查，为我国海底热液活动及硫化物资源的调查研究积累了经验。2003 年至今，我国以独立组织形式先后对东太平洋海隆、南太平洋斐济海盆、西南印度洋中脊、跨大西洋地质断面和南大西洋中脊等海区进行了海底热液调查。2003 年 11 月～2004 年 1 月，我国“大洋一号”在东太平洋海隆首次获得了少量宝贵的热液硫化物样品。2005 年 4 月～2006 年 1 月，升级改造的中国“大洋一号”科考船圆满地完成了我国首次横跨三大洋的环球科考任务，并创造了在印度洋探测到新的热液异常区，获得完整的热液硫化物烟囱体样本、热液活动区生物样本、热液沉积物样本等多个国内首次纪录。

2007 年执行我国 DY10519 航次大洋科考任务的“大洋一号”科考船在水深 2800 m 左右的西南印度洋中脊成功发现了新的海底热液活动区，并获得了约 120 kg 的热液硫化物样品，标志着我国已跃入世界上发现大洋中脊海底热液活动区的少数国家行列。2008～2009 年，我国 DY11520 航次大洋科考任务取得历史性突破，在东太平洋海隆发现5处海底热液活动区，在西南印度洋中脊发现5处热液区和1处非活动热液区。2009～2010年我国大洋第21航次科考又分别在东太平洋海隆发现了2处海底热液活动区，在南大西洋中脊发现了2处热液活动区，在西南印度洋中脊发现了1处非活动热液区。通过“十一五”时期几乎从零开始的大洋硫化物调查，到现在我国已在印度洋、太平洋和大西洋三大洋发现了17处海底热液区，发现总量已居世界前列。这些成果的获取极大地提升了我国在国际海底区域事务中的发言权，同时，也为我国深海科技的发展提供了重要平台。2010年12月8日，“大洋一号”科考船再次起航，开始了我国第22航次的大洋科考任务，这也是我国第3次执行大洋环球科考任务，其调查重点仍以深海热液活动和热液区环境生态为主。可以说，如今我国已拥有了一支能独立进行国际深海热液调查研究的人才队伍。随着研究的深入，国内逐渐由原先的资源调查为主转向以科学研究为主，近年来也取得了众多有显示度的成果，例如，2020年上海交通大学研究团队在对来自东太平洋9°N～10°N热液区域的活跃喷发烟囱体和近期熄灭烟囱体中微生物群落分析的基础上，对比综合全球已报道不同状态烟囱体的微生物群落结构，发现并提出了热液区微生物群落的演替规律和模式：海底热液区微生物生态系统的演替主要受到能量来源的转换控制，在典型的热液硫化物烟囱体从初始形成到完全熄灭的过程中，栖息在其中的微生物群落由热液主控型演变成矿物主控型，而且这种转换的发生非常迅速，可以在烟囱熄灭后的数年内完成，并可以长期（千年以上）保持稳定。

（四）冷泉区与海洋沉积物

通过地质勘探，我国在周边海域发现20多个冷泉，其中绝大部分是已经消亡的冷泉，活跃的冷泉主要有两处，分别是台西南冷泉和海马冷泉。中国地质调查局所属的广州海洋地质调查局“海洋六号”船于2019年历时36天，航程3000多n mile，完成了来自中国国内不同研究机构的多项科考任务，取得了一系列联合创新成果，完成了对多类型潜水器协同开展冷泉调查，并获得了新发现。科研人员运用潜水器在南海北部陆坡西北部海域开展联合调查，发现了新的海底大型活动性冷泉，基本查明了其分布范围、地形地貌、生物群落、自生碳酸盐岩及流体活动等。系统全面地对南海北部活动冷泉区进行了系统调查，运用多种手段对海底活动冷泉进行取样和观测，获取了一大批与冷泉系统相关的生物、水、气体、沉积物等样品及数据，为冷泉系统演变过程及机制等科学研究提供了基础支撑。冷泉作为海底的绿洲，滋养了包括从海底菌席等微生物到海底贝类、蟹类、海虾类等高等生物的一个完整生态系统，是由来自海底地层中的含气流体（主要为甲烷气体）以喷涌或渗漏方式注入海水中而形成的。海底冷泉的调查研究对天然气水合物战略资源，地球物质运动、能量交换（传输）与环境效应，生命起源与演化等前沿科学领域均具有极其重要的意义，引起世界各国政府和学术界的高度重视。

（五）海洋低/缺氧区

国内对缺氧区的研究起步较晚，主要涉及长江口外、珠江口、渤海等海域低氧区的发生发展机制、我国近海与国外缺氧区的比较研究、微生物厌氧氨氧化和反硝化作用过程等。2006 年发表了第一篇关于长江口缺氧区较小型底栖生物分布的文章，截至 2020 年，发表相关文章约 48 篇。国内对海洋脱氧机制的研究主要从两大方面着手：一是从地质历史时期研究大洋缺氧事件，二是对目前我国近海缺氧区进行研究。后者主要包括东海及其长江口、黄海及其胶州湾、渤海海区，以及南海的部分沿海海区等水体和沉积物，其中长江口及东海区域研究较多，共有 25 篇相关文献，其次是珠江口、黄海、渤海。国内关于缺氧的研究主要从物理与化学的角度进行，包括对河流、风力、水团、沉积物等物理化学因素的调查研究，通过数据模拟等方法探究缺氧形成的原因。但从生物角度进行的研究较少，约有 8 篇文献对缺氧海区的生物变化进行讨论，主要是研究浮游植物、浮游动物的变化。国内对缺氧区域微生物的研究很少，仅有 3 篇文献有所涉及，关注了缺氧区中厌氧氨氧化作用。

长江口水域近底部水体颗粒有机碳、溶解无机磷、溶解无机氮和表观耗氧量的耦合变化表明，缺氧是由浮游植物大量暴发生长期间产生的有机物降解所驱动的。珠江口发现除有氧呼吸外，微生物的硝化作用等过程对耗氧量也有很大贡献，并可能对沿海水域的碳酸盐平衡和 pH 产生重大影响。渤海夏季低氧区与初级生产力导致的微型生物呼吸增强密切相关，它的低氧面积（700 多 km^2）和达到缺氧水平的水体积比长江口和珠江口外的要少，但脱氧伴随着海水酸化会对渤海钙化底栖生物构成严重威胁。

厦门大学、华东师范大学、中国海洋大学、山东大学、自然资源部第二海洋研究所等单位的研究人员参与了海洋缺氧的相关国际合作。华东师范大学研究人员作为共同作者在 *Science* 上发表了全球脱氧方面的论文。全球海洋氧网络（The Global Ocean Oxygen Network，GO2NE）工作组于 2019 年在厦门大学举办了首届全球海洋氧气网络国际研究生暑期学校，旨在给学生教授海洋氧循环研究的理论、实验技术和数值模拟等。国际海洋研究科学委员会工作组134（SCOR-WG134）“海洋微型生物碳泵”工作组成立于2008年10月，旨在研究海洋惰性溶解有机碳的微生物生产机制及环境效应，其中也涉及微生物呼吸内容。在2020年SCOR工作组申请中，有以“海洋中层呼吸：整合生态、生物地球化学与模型评估”为主题的项目，旨在研究海洋中层水体（200～1000 m）的微生物呼吸对脱氧的贡献、标准化及发展微生物呼吸测量方法、比较区域全球物理-生物地球化学模型等，其成员包括来自华东师范大学、厦门大学的两位研究者。

（六）海山与多金属结核区

除冷泉外，我国科研人员还发现了一系列天然气水合物藏矿区、海底金属成矿和热液活动区，这应该是我们开展海底深部生物圈微生物研究的重点海域，关系到我国未来海底能源和矿产资源的开发利用，并有助于揭示微生物对全球气候环境变化的影响以及与海底矿藏形成的相互关系，促进对地球生命起源和进化的探索，推动海洋生物资源的开发利用。国内对海山的研究虽然起步较晚，但是通过大洋专项、海山先导专项等，国内科学工作者开展了海山微生物资源评价、群落结构及多样性分析等一系列研究，对海山生态系统及其参与的硫、氮、金属元素循环等重要生物地球化学过程取得了初步认识。但总体而言，我国对海山微生物的研究仍未成体系，深水海山调查不足，在微生物多样性、资源评价和应用研究方面与国外发达国家仍有不少差距，并且具有重要应用价值的海山微生物资源相对缺乏。

多年来，海山与多金属结核区的研究主要由中国大洋矿产资源研究开发协会主导，其中2011年该协会的项目“多金属结核和富钴结壳资源调查与评价研究”从两方面开展了研究：①对富钴结壳资源调查评价展开了系统深入的研究。该项目编制了12个海山1∶25万地质图；以特定海山为示范区，对海山地形地貌、结壳类型、厚度、地球化学、生长结构、分布规律和控制因素进行了系统研究；综合运用地质、地球物理、海底视像、海流测量等资料，探讨了结壳的成矿分布规律；进行了样品加工的K值试验；提出了海山富钴结壳资源量的估算方法和指标，估算了调查海山结壳的资源量。在此基础上，提出了富钴结壳勘探区的选区原则及评价指标，研究提出了矿区组合申请方案和优选建议，为我国向国际海底管理局申请富钴结壳勘探区提供了重要依据。②对我国多金属结核合同区的资源勘探评价展开了系统深入的研究：圈定了矿块和试采区，使用了各种先进的方法进行资源评价，提出了首采矿段建议，确定了勘探工程间距，划分了资源量类型；进行了包括 Mn、Cu、Co、Ni 四种传统金属元素和 Mo 元素在内的五种元素综合利用的矿石试验研究，为我国履行多金属结核勘探合同的工作计划提供了技术支撑。

（七）洋壳岩石圈

目前我国海底深部生物圈微生物的研究才刚刚起步，与国际研究水平还存在一定的差距。但我国越来越重视对海洋的研究，制定了一系列海洋开发战略，对周边海域微生物资源能源的开发研究投入了大量的人力物力。在国家的大力投入下，取得了一系列的成绩。为了进一步提高研究成效，需要有侧重点，更紧密地结合国家的海洋和经济发展战略，有突破性地着手解决一批重大重点基础前沿科学问题，并进行海底深部生物圈特殊微生物资源的开发利用。

2016 年，上海交通大学王风平教授团队对来自大西洋中脊西侧翼北部的玄武岩基底进行了微生物多样性调查。该研究建立了一系列营养源添加实验，包括添加碳源（碳酸氢钠、乙酸钠、甲烷）和氮源（硝酸钠、氯化铵）的培养。培养 6 个月后发现，只有添加氮源的富集物中微生物大量增殖，微生物的活性也明显增强。这些数据表明，氮源是深部生物圈洋壳微生物生长繁殖的重要限制因子之一。同时该团队对岩石和富集物样品进行了 16S rRNA 高通量测序，发现 γ-变形菌中的海杆菌占主导地位，并通过梯度试管的方法成功分离到一株好氧中性铁氧化菌海杆菌，说明洋壳微生物很可能利用玄武岩中丰富的还原态的铁，通过铁氧化作用获取能源并造成岩石的氧化蚀变。

2020 年，同济大学李江涛博士与美国伍兹霍尔海洋研究所合作，在 *Nature* 上发表文章，该研究在洋壳近 800 m 的深度剖面上自上而下选取了 11 个岩石样品用于分析测试。从细胞染色计数的结果来看，下洋壳辉长岩中的生物量极低，细胞丰度为 131～1660 个细胞 /cm^3，远低于正常海洋沉积物及上洋壳表层玄武岩中的生物量（为 10^5～10^9 个细胞 /cm^3）。此外，一系列证据表明，岩石中这些微生物的细胞是完整且能够进一步生长发育的，证据包括：对岩石进行长期的富集培养，检测到了甲烷浓度的增加，表明产甲烷过程的发生（甲烷产率为 2.0×10^{-4}～5.66×10^{-2} nmol /g 岩石）；从大部分岩石中检测到了一定的 ATP 浓度，证明了活跃微生物的存在；在部分样品中检测到了碱性磷酸酶的活性［0.04～2.3 pmol/（$cm^3\cdot h$）］等。该项研究结果证实了下洋壳深部生物圈的存在，拓展了生物圈在地球圈层内部分布的下限。鉴于下洋壳在全球海底的广泛分布及其巨大体积，栖息于此的微生物群落即使具有极低的生物量和较缓慢的生长速率，也仍然可能对全球物质循环产生重要的影响。

第三节　地表极端环境微生物

一、高盐环境微生物研究进展与研究团队

（一）高盐湖泊环境嗜盐/耐盐微生物

截至2020年5月，在中国知网（CNKI）中，检索到1999～2019年“高盐湖泊环境嗜盐/耐盐微生物”相关的研究文献490篇。就文章数量来看，CNKI与Web of Science的检索结果类似。环境微生物多样性是群落生态组成的主要特征，也是环境优劣评估的重要指标。目前，环境微生物多样性的研究方法有选择分离培养、碳源利用生理指纹、分子标记、磷脂或脂肪酸生物标记、16S rRNA基因指纹图谱、16S rRNA基因高通量测序、生物标志物以及宏基因组学等。环境中可培养的微生物仅占自然界微生物的极少部分（<10%），高通量测序技术可以从不同角度研究微生物多样性，一方面可对多样品的多个可变区同时测序，另一方面在测序速度和测序通量上也都有进一步的提升。国外利用该技术研究盐湖微生物多样性的报道颇多，包括以色列大学、俄罗斯科学研究院、澳大利亚墨尔本大学等。国内研究起步稍晚但更为深入，蒋宏忱团队先后用高通量测序技术研究了青藏高原盐湖中微生物的多样性，系统揭示了我国青藏高原地区细菌和古菌的多样性及其对环境的响应与生态反馈。李文均团队用纯培养方法和宏基因组手段研究了新疆地区盐湖的微生物多样性。

（二）盐碱湖泊环境嗜盐/耐盐微生物

截至2020年5月，在CNKI中，检索到1999～2019年“盐碱湖泊环境嗜盐/耐盐微生物”相关的研究文献245篇。就文章数量来看，CNKI与Web of Science的检索结果类似，印证了这一领域的研究数量在国内外均相对较少。通过对比不同年份的发文数量可以发现，国内中文相关研究在1999～2019年热度变化的趋势和世界范围基本一致，最近几年，无论是国内

还是国际范围内，“盐碱湖泊环境嗜盐/耐盐微生物”这一领域的研究热度正处在迅速上升的起步阶段。目前国内相关领域的研究方向主要集中于以下几个方面。

1. 盐湖资源利用

盐湖作为重要的生物资源库，含有大量的嗜盐生物、耐盐植物等资源。盐湖资源是我国西部地区的优势资源。开展盐湖重要生物资源的研究，对拓宽研究领域、利用盐湖资源、改善荒漠和生态环境有着重要的推动作用。

中国科学院微生物研究所向华团队对盐湖嗜盐古菌的研究显示了盐湖微生物丰富的资源多样性和广阔的应用前景，中国地质科学院矿产资源研究所孔凡晶团队展示了我国内陆盐湖杜氏藻、内蒙古盐湖菌藻及青海尕海湖卤虫资源与环境变化研究进展，中国科学院青海盐湖研究所、兰州大学、中国石油大学、青海大学等单位在盐湖资源开发利用、油气开发、沉积环境方面有大量研究。已知最具有实用意义的盐湖生物有盐藻、嗜盐菌和嗜碱菌等，某些嗜碱菌已开始用于工业生产，嗜盐菌（视）紫膜是理想的光电转化材料，具有广泛的应用前景；盐生植物和嗜盐菌及嗜碱菌中的抗盐基因可以作为特殊基因库，可以培育抗盐或抗盐碱的新生物品种发展盐湖农业，可以保护盐湖生物资源，改善和恢复恶化的盐湖生态环境。

2. 极端微生物的分离与应用

从统计结果来看，关注盐碱湖泊微生物丰度、多样性等内容的微生物生态方向论文共计 152 篇，主要集中在青海大学沈国平团队、云南大学徐丽华团队、浙江大学吴敏团队、中国地质大学（武汉）蒋宏忱团队、江苏大学崔恒林团队等，他们在青藏高原盐湖、新疆盐湖、内蒙古碱湖及运城盐湖方面都做了大量微生物多样性研究。

在耐盐/嗜盐微生物分离及应用方面的研究中集中于利用耐盐/嗜盐微生物进行有机物的降解和耐盐菌株的培育。国内开展嗜盐/耐盐菌株分离研究的有深圳大学郑易之团队、中山大学李文均团队、云南大学唐蜀昆团队、浙江大学吴敏团队、新疆师范大学刘会强团队、青海大学朱德瑞团队，研究方向集中于嗜盐菌的分离及降解某些嗜盐古菌产生的一类胞内聚合物，这些微生物形成了独特的代谢途径，能够产生普通生物没有的生物活性物质。盐藻富含胡萝卜素，可以直接作为食品农作物抗盐性改良的微生物基因资源。嗜盐/碱菌能够在 5% 以上的盐浓度至饱和盐度的环境中生活，因此极端微生物可能利于提高耐盐碱农作物品质和改造传统工业。

国内团队在盐湖矿产开发利用上已形成初步规模，但在微生物应用研究上力量较薄弱。目前国内团队有浙江大学吴敏团队和内蒙古大学许继飞从盐水中分离出的中度嗜盐的交替单胞菌属（*Alteromonas*）菌种能够降解有机磷化合物，以及一些耐受重金属污染物的嗜盐菌，用于处理重金属污染盐水。虽然，嗜盐微生物在生物可降解塑料、污水处理等领域具有广阔的应用前景，但相关研究仍处于试验阶段，需要投入更多的科技力量加强该领域的研究及开发利用。

（三）干旱 / 半干旱地区盐碱土壤环境嗜盐 / 耐盐微生物

截至 2020 年 5 月，在 CNKI 中，检索到 1999～2019 年“干旱 / 半干旱地区盐碱土壤环境嗜盐 / 耐盐微生物”相关的研究文献 98 篇。就文章数量来看，CNKI 与 Web of Science 的检索结果类似（国际发文总数为 86 篇），印证了国内外在这一领域的研究相对较少。通过对比不同年份的发文数量可以发现，国内中文相关研究在 1999～2019 年的热度变化趋势和世界范围基本一致，1999～2010 年，国际及国内的发文量均在 5 篇以下，2010～2016 年，该领域的发文量处于相对稳定的状态，文章数量的大幅增长出现在 2017 年以后（图 4-1）。这样的趋势表明，最近几年，无论是国内还是国际，“干旱 / 半干旱地区盐碱土壤环境嗜盐 / 耐盐微生物”这一领域的研究热度已经进入迅速上升阶段。目前国内相关领域的研究方向主要集中于以下几个方面。

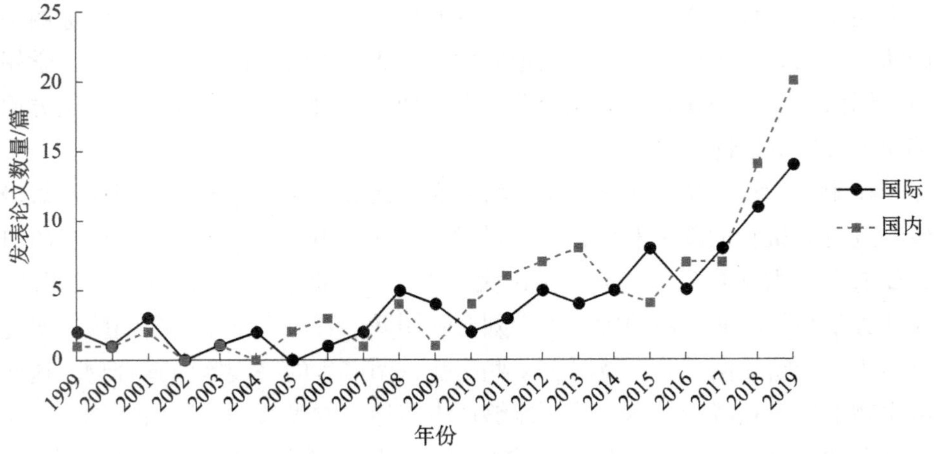

图 4-1　1999～2019 年“干旱 / 半干旱地区盐碱土壤环境嗜盐 / 耐盐微生物”研究年度发文数量

1. 农业生产

以提高作物抗逆性、改良盐碱耕地为主要目的的农业生产方向占据了全部研究的38%，这与我国耕地面积紧张、人口众多、盐碱地面积大的国情相符。针对中国不同盐碱土壤分布区域（如松嫩平原、河套平原等），来自浙江农林大学及东北林业大学盐碱地生物资源环境研究中心的柳参奎团队、黑龙江大学的宋福强团队、哈尔滨工业大学的祁佩时团队、吉林大学的张世宏团队、江苏大学生物质能源研究所的孙建中团队以及中国科学院烟台海岸带研究所的解志红团队等在这一领域均有系统的研究。

2. 微生物生态（侧重多样性研究）

从统计结果来看，关注盐碱土壤微生物丰度、多样性等内容的微生物生态方向研究占比偏低，约为14%。在这方面开展研究的团队包括中国科学院东北地理与农业生态研究所的田春杰团队、华中农业大学农业微生物学国家重点实验室的彭楠团队、宁夏大学的代金霞团队、西北农林科技大学的来航线团队、华中农业大学农业微生物学国家重点实验室的王革娇团队等。

3. 极端微生物分离与应用

国内中文研究中，高盐土壤环境耐盐 / 嗜盐微生物的分离与应用占据了相当大的比例（48%），这与国际英文研究的主题构成形成了鲜明的对比，这可能与我国盐碱土壤环境广泛分布有关。这部分研究的来源较为分散，国内多所机构的研究团队都有涉及，如东北林业大学的张杰团队、哈尔滨工业大学的祁佩时团队、四川大学的田永强团队等。在耐盐 / 嗜盐微生物分离及应用这部分研究中，有17%关注了利用耐盐 / 嗜盐微生物进行土壤有机污染物的降解及环境恢复的研究。这一比例也远高于国际研究，南京农业大学的陈亚华团队、沈阳农业大学的依艳丽团队、滨州学院的谢文军团队、南京大学的姚志刚和王君团队等在这一方向都有丰富的研究经验。

整体来看，中国科学院（如东北地理与农业生态研究所、干旱区生物地理与生物资源重点实验室等）、东北林业大学盐碱地生物资源环境研究中心及华中农业大学农业微生物学国家重点实验室等研究机构（平台）在国内的“干旱 / 半干旱地区盐碱土壤环境嗜盐 / 耐盐微生物”研究中发挥着重要的带头作用。结合我国国情，围绕农业生产和微生物资源开发进行大量的研究是实际且必要的，但对比国际上该领域的研究分布可以看出，国内中文相关领域的研究主题较为单调，在利用盐碱土壤进行行星生物学研究等新兴方向上仍有待进一步提高和发展（图 4-2）。

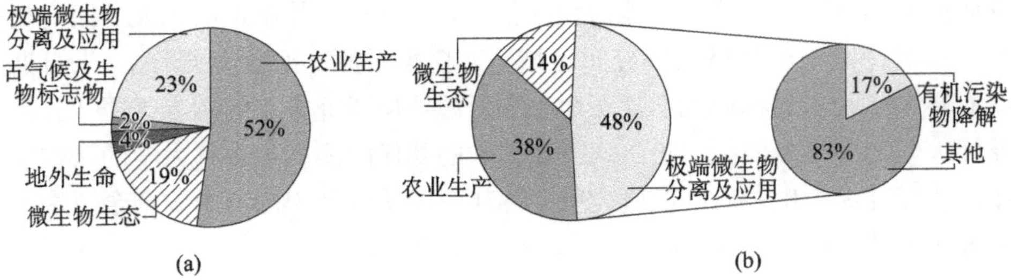

图 4-2　1999～2019 年国际（a）和国内（b）干旱 / 半干旱地区盐碱土壤环境嗜盐 / 耐盐微生物研究中不同方向的发文比例

二、高温环境微生物研究进展与研究团队

（一）微生物多样性

大多数微生物不可培养历来是微生物多样性研究的主要障碍。随着分子生态学的发展，基于功能基因的免培养技术加快了热泉微生物多样性的研究速度。国内相关研究起步稍晚但较为深入，李文均和蒋宏忱等团队先后用高通量测序技术研究了西藏和云南大部分热泉中的微生物多样性，系统揭示了我国西部热泉中细菌和古菌的多样性，同时结合地球化学参数，解析了其中热泉微生物群落的构建机制。

宏基因组学的发展为研究未培养微生物的多样性及进化提供了新的机遇，李文均等中美科学家合作发现了一个广泛存在于包括云南热泉在内的全球热泉的细菌门类 *Candidatus* Kryptonia，其核糖体基因不能与通用引物匹配，故一直以来未被发现，这也表明热泉中还有类似的大量未知微生物。通过宏基因组技术从云南热泉中获得了 6 个曙古菌门（Aigarchaeota）基因组，并对其代谢和进化进行了深入研究，随后又获得 15 个奇古菌的基因组，基于进化及比较基因组学研究，揭示了甲烷代谢的进化规律并提出甲烷氧化微生物的热起源。

（二）热泉环境特定生物学过程及特殊功能微生物

在热泉菌席中，有大量微生物参与了碳、氮、硫、磷等重要生命元素的生物地球化学循环。其中碳固定由绿弯菌门的丝状不产氧光合菌（FAPs）和蓝细菌门的一些类群参与完成。FAPs 利用 3-羟基丙酸双循环进行厌氧不产氧的光合作用，在弱光下高效地固定 CO_2。蓝细菌的聚球藻（*Synechococcus*）则利用卡尔文循环途径进行高效的产氧光合作用。氮循环的生物学动力有两个：一是作为电子受体驱动能量循环，二是作为重要的生源元素进入物质循环过程。固氮微生物广泛分布于各种缺氮生境。由于缺少可利用氮源，这些生境微生物种群的初级生产往往受到限制。生物固氮通常与碳固定及自养代谢偶联。*nifH* 基因编码固氮酶 H 亚基，是生物固氮途径的关键基因。*nifH* 通常被用作靶标基因来检测环境中的固氮微生物。目前已在多种化能自养生态系统中确认了固氮菌的存在。在 76～89℃的陆地热泉系统中也检测到固氮酶的转录活性，但是相关固氮微生物的生物学特征仍不明晰。在热泉菌席中，目前已识别出的固氮微生物主要为蓝细菌。其中，聚球藻是一类单细胞蓝细菌，在非硫热泉中分布广泛。它们在菌席的厌氧层中进行固氮，为菌席其他微生物提供氮源。例如，Alcamán 等（2015）发现，在整个 38～69℃的菌席中，鞭枝藻属（*Mastigocladus*）蓝细菌的固氮能力足以满足整个群落的氮需求。在 50℃以下的热泉菌席中，丝状蓝细菌（可形成异形胞，以制造固氮酶所需的厌氧微环境）或异养细菌参与氮固定；53.5～63.4℃的菌席中聚球藻的 *nifH* 等基因在厌氧条件下大量表达，同时固氮酶的活性加强，证明它可用发酵反应获取能量进行固氮作用。

还原型硫化合物和 H_2 是重要的能源物质。热泉中有大量的厌氧或微氧环境，高价态的硫酸盐作为一种丰富的无机物，广泛存在于热泉环境中，其中的厌氧微生物以之为电子受体氧化有机物获取能量。在热泉菌席中分离到的闪烁杆菌属（*Fervidobacterium*）细菌通过发酵有机物产生氢气。另外，聚球藻固氮过程中 H_2 作为副产物生成（$N_2+8H^++8e^- \rightarrow 2NH_3+H_2$），这在一定程度上消耗了低氧条件件微生物发酵产生的 H^+，维持了群落的中性 pH。H_2 浓度存在昼夜变化规律，H_2 在有光照时可被 FAPs 利用，作为硫还原的电子供体，橙色绿弯菌（*Chloroflexus aurantiacus*）在厌氧光自养和厌氧光异养条件下都能将硫化物氧化成单质硫并积累到胞外，表明还原态的硫为 FAPs 提供了能源。事实上，微生物以高价态含硫化合物呼吸来获取能量是一个古老的代谢方式，有证据表明这种代谢方式可以追溯到 34.7 亿年前。以 *dsrAB* 为靶标基因的研究推测热泉菌席存在 4 种硫酸盐还原菌（SRM），即嗜热脱硫弧菌属（*Thermodesulfovibrio*）及其他三个近源基因型菌。它们分布在热泉菌席的近表层，其最强的硫酸盐呼吸发生在无光合作用的夜间。嗜热脱硫弧菌属是一种严格厌氧的硫还原微生物，隶属于深度分支的硝化螺旋菌门，已发现的成员都分布在热水生态系统中，在原位环境中，它们以氢气或其他小分子有机物作为电子供体获取能量。在缺乏硫酸盐的条件下，该类群的某些基因型菌种可以执行互营降解有机物来传递电子获取能量。嗜硫栖热厌氧杆菌（*Candidatus Thermonerobacter thiotrophicus*）来源于热泉菌席，通过宏基因组拼接获得的一个分箱（bin），执行异化硫酸盐还原过程，隶属于拟杆菌/绿菌（*Chlorobi*）超门。通过转录组分析发现，其末端氧化酶在氧气丰富的时段仍然能够大量表达，很可能兼备有氧呼吸的特性。其进化关系接近硫氧化菌（SOM），是进一步研究微生物异化硫代谢进化的好材料。

（三）高温环境膜脂化合物及与环境关系

高温环境中的膜脂化合物研究主要以古菌合成的类异戊二烯甘油二烷基甘油四醚脂类化合物（iGDGTs）为主，而在多种iGDGTs中，GDGT-0和泉古菌醇往往占最大比重。GDGT-0在实验室培养的古菌中是最常见的。除了嗜盐古菌，它几乎在所有主要的古菌种属中都有发现。除了GDGT-0，嗜热泉古菌和广古菌还合成含有1～4个五元环的GDGT化合物。在美国黄石国家公园热泉样品中还发现了具有5～8个五元环结构的iGDGTs。生活于海底热液喷口的古菌布氏地栖嗜酸菌（*Aciduliprofundum boonei*）不仅可以合成含0～4个五元环的GDGT化合物，还能合成少见的烷基链之间存在共价键的T型GDGT。泉古菌醇是具有5个环的GDGT化合物，该化合物含有1个六元环和4个五元环，其最初发现于海相沉积物中，后来在热泉中也发现了，暗示着一些嗜高温古菌也可以产生泉古菌醇。之后人们在黄石亚硝化暖菌（*Candidatus* Nitrosocaldus yellowstonii）和加尔加亚硝化暖菌（*Ca.* N. gargensis）等嗜高温古菌中发现泉古菌醇占据着主导地位，验证了泉古菌醇来源于嗜高温古菌的可能性。张传伦等发现，泉古菌醇与泉古菌醇和GDGT-0的比值在温度为40～45℃的热泉中存在着最大值，以此推测泉古菌醇合成的最适温度为40～45℃（Zhang et al.，2006）。对膜脂化合物和环境关系的进一步研究表明，GDGT五元环的数量随着温度的升高而增加，因此，四醚指数（TEX_{86}）古温度指标被建立，并广泛用于恢复古环境温度。随后，人们发现各种GDGT与pH具有强相关性。其他的环境参数也可能影响GDGT脂类化合物的分布，如GDGT五元环的环化指数（RI）与重碳酸盐有正相关关系。在国内，张传伦的团队发现，云南热泉中极性、非极性古菌脂类化合物的分布特征不仅与温度和pH有关，而且与其他环境因素（如溶解氧）和群落结构有关（Xie et al.，2014）。近年来，一些国外科学家采用整合基因组学和脂类组学的方法，对微生物膜脂化合物的合成机制进行了研究，国内科学家尤其是研究热泉的科学家在这一方面的研究还不够，该合成机制是今后值得研究的一个发展方向。

其他研究还包括比较基因组学、宏基因组学、宏转录组学、新培养菌株的生态、生理学、生态网络建模、噬菌体-宿主相互作用以及进化理论模型。

三、低温环境微生物研究进展与研究团队

国内对地表冰雪环境微生物的研究工作主要包括以下几方面。一个是中国科学院青藏高原研究所刘勇勤团队的研究，主要包括：①从低温环境中，特别是从青藏高原地区的冰川、冰雪及冰湖中发现并分离未知微生物，开发建立了青藏高原极端环境微生物菌种库，揭示了微生物对冰川极端环境广适温性和多碳源利用的适应机制；②通过高通量测序等方法揭示、比较不同空间、不同时间的低温环境中微生物的数量、多样性和群落组成特征，并将这些数据与环境数据进行关联分析，探究影响微生物群落结构的因素；③揭示了冰川微生物群落特征受气候环境影响的现代环境过程，首次建立了高分辨率冰芯细菌的年际变化序列，发现了近 50 年来冰芯中细菌丰度显著增加、细菌群落组成日益趋同的时间演化特征，拓展了青藏高原冰芯气候环境变化的新指标。另一个是中国科学院微生物研究所、中国普通微生物菌种保藏管理中心的周宇光、刘洪灿、刘庆等的团队对低温环境微生物的菌种分离和适冷机制的研究，包括从低温环境中发现并分离未知微生物，结合高通量测序技术完善微生物基因组信息，并进行了试验以测试分离微生物对一系列理化条件的耐受性，研究低温环境中新型微生物形态、生长温度、耐盐度和耐酸碱等生理特性，预测其在低温生态系统中的潜在功能和适应机制。

国内在低温冻土环境微生物领域的研究团队主要包括武汉大学的彭方等的团队和河南师范大学的常绪路团队，主要是对冻土环境中的微生物进行菌株分离，分析其序列、形态学特征和脂肪酸特征，并进行物种鉴定和命名。清华大学的周集中团队主要研究冻土微生物对增温及其他环境变化的响应。另外，中国地质调查局油气资源调查中心的祝有海团队研究青藏高原的微生物，如祁连山冻土的微生物，包括甲烷氧化菌和产甲烷菌的群落结构、丰度、多样性，研究了环境因子对微生物群落结构的影响，还探究了冻土消融对微生物的群落结构、代谢过程以及土壤某些性质产生的影响。吉林大学的陈霞团队主要研究长白山冻土中植物对微生物群落结构的影响，还探究植

物物种入侵对土壤性质、微生物群落、土壤养分循环以及植被生态系统的影响。

相对于国外的研究团队，国内研究团队也已经发展壮大，尤其在低温环境中未知微生物的纯菌分离方面有更强大的队伍，获得了更多的研究成果。国内研究团队利用高通量测序技术继续大范围地研究了极端低温环境中微生物的组成和特征，同时也开展功能微生物的研究，如产甲烷菌和甲烷氧化菌在低温冻土环境中的研究。国内研究团队大部分的研究地点位于青藏高原地区，还包括长白山高海拔冻土地区，相比国外的研究而言，在南北极的研究相对较少，但在中国本土尤其是青藏高原地区的冰川雪和冻土等低温环境的微生物研究中，拥有很强的团队和喜人的研究成果。

四、酸性环境微生物研究进展与研究团队

目前，国内AMD微生物研究方法还是以16S rRNA测序为主，宏组学尤其是宏转录组和宏蛋白质组的相关应用仍比较少。以2010年后发表的文章数作为依据，近10年AMD微生物研究国内发展趋势如图4-3所示。56篇相关文献一共发表在37种SCIE期刊上，其中有4篇发表在*ISME Journal*上，1篇发表在《微生物学趋势》(*Trends in Microbiology*)上，共占总发表数的8.9%。

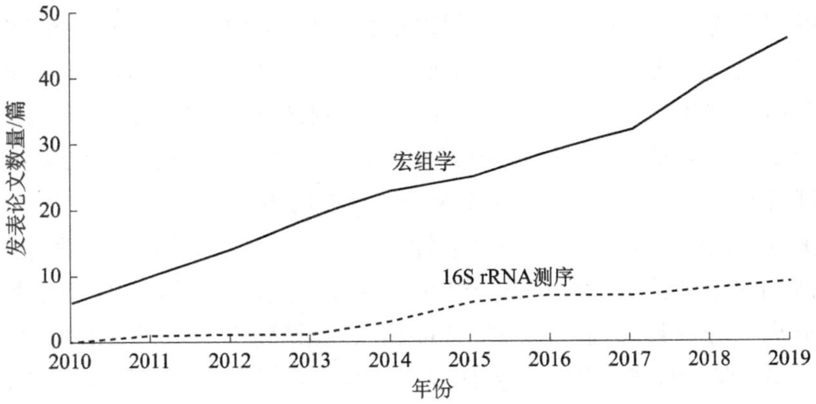

图4-3 2010～2019年国内AMD微生物学研究发表的论文数量

国内开展酸性环境微生物学研究的团队主要集中在南京大学、中南大学、中国科学院微生物研究所、中山大学与华南师范大学。其中，南京大学陆现彩课题组主要聚焦矿山极端环境微生物分解矿物获得能量的微观机制，

建立了 AMD 和尾矿库中微生物分解矿物的动力学模型以及相关元素氧化还原路径，在微生物-矿物相互作用机制研究方面取得了创新性成果。代表性成果包括：①揭示了金属硫化物表面微生物附着的分子机制，识别出以 Fe^{3+} 化学氧化为主的间接作用机制和附着细胞直接侵蚀矿物的接触作用机制，确认了化学氧化和微生物氧化协同浸出机制的主导作用。阐明了微生物氧化毒砂过程中的电子传递路径和次生矿物形成的次序，构建了微生物氧化作用引发矿物表面多种元素参与的复杂的链式氧化还原反应路径。②通过厌氧环境微生物-分解矿物模拟实验、尾矿精细矿物学分析和化学动力学模型研究，查明了铁异化还原菌分解黄钾铁矾的微观机制，评价了微生物-矿物体系中制约电子转移的影响因素，揭示了微生物驱动的黄钾铁矾→绿锈→针铁矿→磁铁矿、菱铁矿逐级转化过程，定量评价了微生物还原分解尾矿矿物的环境效应。

中南大学团队刘学端课题组长期致力于 AMD 微生物群落结构和功能、微生物的适应性进化机制以及生物冶金等方面的研究，代表性工作主要包括：通过从全国 20 多个典型矿区采样，建立了有 1125 个菌株的 AMD 微生物菌种资源库；在国际上首次构建了极端嗜酸微生物功能基因芯片，实现了硫化矿物氧化溶解过程中 AMD 微生物群落与功能的同步检测；通过比较基因组和功能分析，探明了硫化矿物溶解关键菌种的铁、硫氧化功能基因及其代谢网络；确定了水平基因转移和基因丢失是 AMD 微生物进化的主要驱动力之一；推广了完善的生物冶金技术。夏金兰课题组长期综合利用传统谱学、同步辐射谱学、显微分析技术、微生物组学和密度泛函理论（DFT）计算及分子动力学模拟等技术和方法，得到了许多关于微生物-矿物相互作用的新见解。同时，对硫化矿物与微生物作用的微界面的生物化学反应过程、微生物代谢及矿物与溶液间的物质能量传递过程的研究不断深入。

中国科学院微生物研究所姜成英课题组长期从事酸性矿山排水形成机制和生物冶金的研究，在酸性矿山排水菌种分离、微生物组解析、微生物多样性、新基因挖掘、硫代谢途径等方面开展了一系列的研究工作。代表性工作主要包括：在微生物菌种分离方面，已建立了高通量微流控菌种培养筛选平台，获得了大量与 AMD 形成相关的嗜酸微生物菌种；在微生物组研究方面，对云南蒙自、大宝山矿山、西藏热泉等地的酸性矿山排水及酸性热泉的微生物组特性进行了分析；在 AMD 微生物群落研究方面，确定了决定微生物结构与功能的关键因素；在新基因的挖掘方面，发现了一些硫氧化、铁氧化及代谢的关键功能基因；在微生物硫代谢途径发掘方面，解析了AMD中关键微生物的基因组信息，构建了代谢网络模型，揭示了AMD微生物的代谢多样性。

中山大学与华南师范大学的束文圣和黄立南课题组长期聚焦于AMD微生物生态学研究，在AMD微生物多样性和生物地理、群落功能和动态、病毒生态学和矿业废弃地生态修复等方面开展了系统且具有创新性的工作。主要包括：①利用16S rRNA基因焦磷酸测序技术，确定了不同生态尺度下的环境差异（特别是pH）是极端酸性环境中微生物多样性分布格局的主要决定因子，同时结合基因芯片技术，揭示AMD微生物群落动态在功能基因水平上能得到更好的预测；②联合利用宏基因组和宏转录组方法，基于“分而治之”的组装策略，重建了AMD稀有/未培养物种的基因组，并确定了它们的生态位角色，从群落和物种水平确定了AMD微生物对不同环境条件的适应机制；③聚焦AMD系统中新的微生物类群如里奇蒙德（Richmond）矿嗜酸纳米古菌（ARMAN）和嗜酸脱硫杆菌目（Acidulodesulfobacterales），阐明了这些微生物类群的分布、多样性和潜在的生态角色；④发现病毒群落在尾矿生态系统中随其深度变化而变化，病毒多样性和功能与其宿主和不断变化的环境因子之间存在紧密的联系，揭示了病毒在深层尾矿硫循环中扮演着重要角色。相关研究获国内外同行的广泛关注，应邀为*Trends in Microbiology*和*Current Opinion in Biotechnology*撰写综述文章。此外，团队积极开展矿山生态修复工作，利用生态学手段从源头控制AMD污染，相关技术体系获得2019年广东省科学技术进步奖一等奖。

第四节　深地环境微生物

一、优势领域与研究队伍

我国陆地深部生物圈的起步相对较晚、发展较慢，但在某些领域逐渐显示出优势。20世纪90年代至今，由于人们对地质作用过程中微生物研究以及地下水系统功能恢复生物技术的迫切需求，尤其是现代微生物学研究方法与手段的不断发展，我国研究人员在国家自然科学基金、国家高技术研究发展计划、国家“八五”科技攻关等项目的资助下，大大推动了在污染地下水系统微生物修复、微生物降解有机污染物如石油烃类的降解转化、地下水微生物地球化学作用，如N元素的循环，以及原生劣质地下水微生物成因，如高砷地下水的微生物作用等方面的研究，并取得重要进展。中国地质大学（武汉）在山西大同深入开展了原生劣质高砷地下水成因、迁移机制的研究，并对其进行了原位修复，取得了良好的效果。此外在湖北仙桃建立的地下水监测基地已经并入国际关键带监测（critical zone observatories，CZO）网络，为开展地下水微生物长期演替以及与地下水的交互作用提供了宝贵的场所，是我国深部地下生物圈研究中的重要标志性成果。特别是自2000年以来，相关文献数量呈现指数级增长态势，相关的研究机构主要包括中国科学院、中国地质大学、清华大学、吉林大学和同济大学等。

我国喀斯特地貌具有面积大、分布广、生态脆弱性突出等问题，使得国家对喀斯特地区的生态、地质、环境、生物等格外关注。对喀斯特地区近年来的研究集中在岩溶系统的碳循环及生态效应上，并得到国家自然科学基金委员会、科技部重大研发计划项目的资助。2000～2019年，国家自然科学基金委员会资助喀斯特领域的基金项目共计122项，其中洞穴直接相关基金有30项。从研究内容来看，主要集中在高等生物多样性与元素循环的研究上（图4-4）。从承担项目的单位来看，主要有中国科学院亚热带农业生态研究所、中国地质大学（武汉）、贵州师范大学、广西师范大学、中国科学院广西植物研究所等。值得注意的是，近年来，中国地质大学（武汉）洞穴研究团队已经先后在湖北清江和尚洞获得4个重点基金的资助以及8项与石笋有关的面上项目的资助，使得和尚洞成为洞穴研究中的明星。该研究团队分别从洞穴现代过程监测、地球化学、微量元素、同位素地球化学、洞穴脂类化合物与古气候重建、洞穴地质微生物功能群的时空分布及环境意义、洞穴微生物与全球变化等不同方向全方位开展联合攻关，是目前研究最系统的洞穴之一。2007年之后，我国洞穴微生物的研究论文中文发表量基本维持在每年10篇左右，研究内容主要集中在微生物的多样性以及微生物资源的开发与利用上。2011年后，受国家自然科学基金委员会重点基金的资助，中国地质大学（武汉）王红梅教授团队在洞穴微生物领域的国际影响力增加，在洞穴细菌、古菌以及真菌的多样性、空间分布以及与环境因子之间的关系等方面取得了系列成果，她本人于2013～2018年担任国际洞穴联合会微生物与地质微生物分会副主席。

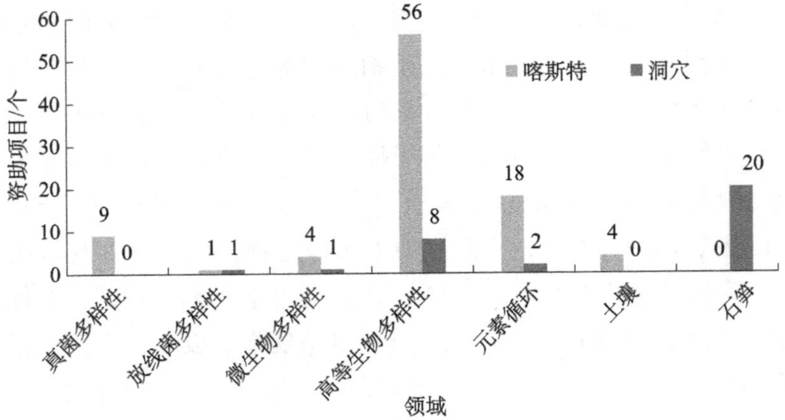

图 4-4　2009～2019 年国家自然科学基金委员会资助的喀斯特与洞穴领域的基金项目

资料来源：http://fund.sciencenet.cn。

我国在洞穴微生物的研究上虽然起步晚，但具有鲜明的特色。近年来，洞穴微生物与全球气候变化的研究备受关注，尤其是岩溶作用在全球碳汇中的作用更是研究中的热点。表生岩溶作用形成的洞穴中甲烷的浓度通常较外界大气甲烷浓度低，这意味着洞穴中的甲烷被消耗了，而同位素及分子生物学证据均显示，微生物在消耗洞穴甲烷中起着关键作用，因此洞穴可能是陆地生态系统中除土壤以外另一个被忽视的大气甲烷汇。由于我国喀斯特地貌面积大、分布广、洞穴数目多，阐明洞穴微生物氧化大气甲烷的速率、评估洞穴作为大气甲烷汇的大小及其环境驱动机制等，不仅可以回答洞穴这一极端寡营养系统中代谢甲烷的微生物类群、空间分布及其与环境和其他生物类群之间的互作关系等科学问题，更为重要的是，该研究结果可以为我国在国际碳减排的谈判博弈中提供新的科学素材和实验数据，增加谈判的筹码。此外，由于洞穴也是蝙蝠的重要栖息地，而蝙蝠往往携带大量的病毒，可能会危及人体安全，因此洞穴病毒的多样性、分布及其与宿主之间的关系、潜在的生物安全等也将成为今后洞穴微生物关注的热点。目前的研究仅局限在洞穴水体病毒的研究中，对其他载体病毒的研究还十分局限。

二、快速发展领域与研究队伍

国内过去和现在对深地生物圈的研究，主要集中在和 ICDP 合作的深部钻探上。最早开展的中国大陆科学钻探（CCSD）项目，钻探地点在大别—苏鲁超高压变质（UHPM）造山带的东部，钻探深度为 5118 m，为深部地下微生物学研究提供了一个独特的机会。尽管CCSD项目的钻孔选在超高压变质带，但是这些超高压岩石有足够大的孔隙使微生物能够生活其中。局部岩心存在气体和流体异常，为寻找独特的微生物提供了前景。16S rRNA基因分析显示出岩石细菌类群主要以类变形菌门为主导，大多数主要参与硝酸盐还原，栖息于偏冷的盐碱性环境中。流体中的细菌群落更加多样，主要包括变形菌门、拟杆菌门、浮霉菌门等；流体中的微生物具有喜高温、厌氧、碱性、喜盐等特性，以硫酸盐还原菌、三价铁还原菌和产甲烷菌为主，在无氧条件下，依靠金属元素和硫酸盐摄取地下化学能进行生命活动。超高压岩石中的微生物主要栖息于液体/气体包裹体中，但是钻探流体中的微生物可能主要来源于地下流体。此外，CCSD项目结果似乎都表明，大约4500 m是本地生物圈的下界，相应位置地温约121 ℃，这一深度接近于已知微生物的最高承受温度。

由中国地质科学院勘探技术研究所承担的ICDP钻探项目“松科1井”，完钻井深为1811.18 m。通过16S rRNA初步多样性分析发现，在深绿-黑色样品中扩增得到的克隆序列较接近于金属硫酸盐还原菌，指示一种缺氧的还原环境；红色样品中扩增得到的克隆序列更接近于亚硫酸盐和硫的氧化菌，指示一种富氧的氧化环境。这一结果与国外对白垩纪海相地层的研究结果可对比。从2014年4月13日开钻至2018年5月26日完井的“松科2井”，完钻井深为7018 m，是亚洲国家实施的最深大陆科学钻井和ICDP成立22年以来实施的最深钻井，是全球第一口钻穿白垩纪陆相地层的大陆科学钻探井，这为研究更深的深地极端微生物提供了新的机遇。

尽管我国对煤层微生物的研究相对较晚，但是最近几年进展较快。Tang等（2012）对鄂尔多斯盆地两个不同煤化程度煤层中微生物群落的研究发现，样品中微生物群落整体多样性较低，其中产甲烷菌群丰度也比预期少，并且两个不同煤化程度的煤中微生物群落也存在差异。Guo等（2012）发现，在鄂尔多斯盆地，煤层水和煤岩中的细菌组成较接近，但是和煤层气差异却很大。He等（2016）从山东赵楼一处不产煤层气的煤矿中取回煤层水样品，经研究发现，该样品中同样存在种类丰富的产甲烷功能菌，并且该功能菌群可以利用褐煤生产甲烷。目前有关煤层中真菌的报道不多，我国沁水盆地煤层水中的真菌主要有子囊菌门、担子菌门，而从属水平上主要有担子菌门的隐球酵母属（*Cryptococcus*）、马拉色氏霉菌属（*Malassezia*）、红酵母属（*Rhodotorula*）和子囊菌门的曲霉属（*Aspergillus*）、青霉菌（*Penicillium*）、瘤霉属（*Phoma*）、被孢霉属（*Mortierella*）、丛赤壳属（*Nectria*）（Guo et al.,

2017）。丁芳芳（2013）和汤浩（2014）从煤层水中分离了产甲烷菌和硫酸盐还原菌，并对其相互作用进行了分析。王爱宽等（2010）研究发现，云南褐煤族组分中的饱和烃是微生物优先降解的主要成分，到后期长链烷烃才被明显降解。陶明信等（2014）对淮南煤田的研究发现，该地区煤层中的可溶有机质遭受过强烈的微生物降解，其中正构烷烃、类异戊二烯烷烃等组分被优先降解，可溶有机质、CO_2、H_2、气态重烃和水都可以成为产气的母源物质。苏现波等（2011）在生物产气条件优化方面开展了较多的研究工作，他们研究了 pH、盐度等对产气的影响。林海等（2012）通过对生物产气微生物进行诱变处理来提高产气效率。

国内开展深层煤炭微生物相关研究的单位主要有中国矿业大学、河南理工大学、北京科技大学、太原理工大学、中国地质大学、中国石油大学、中国科学院大学、农业农村部沼气科学研究所、合肥工业大学、内蒙古工业大学和山西晋城无烟煤矿业集团有限责任公司（煤与煤层气共采技术国家重点实验室）等。国内开展油藏微生物研究的高校主要有华东理工大学、北京大学、南开大学等，中国石油化工集团有限公司、中国石油天然气集团有限公司也各自有小部分从事微生物驱油的研究队伍。

总之，我国陆地深部生物圈的研究目前呈现出良好的发展势头，尤其是地下水微生物、洞穴微生物及岩石圈微生物研究，中国科学家和有关单位已经在国际上呈现出领先的趋势，并且在地下水微生物研究的野外监测基地建设上也取得了令人瞩目的成果。但深地微生物的代谢机制、生物地球化学循环过程等很多科学问题有待解决。由于缺乏大科学计划的支持，深地生物圈的研究目前依然比较分散，学科交叉依然有待深入，科研院所、高校与企业和地质调查单位之间的深入合作亟须加强。深地生物圈的从业人员目前依然偏少，应加强宣传，加大人才培养力度，亟须建设一支稳定的深地生物圈研究队伍，建立平台和技术，与国际一流团队进行实质性的合作，尽快提高我国深地生物圈的研究地位。

第五节　深时极端地质环境微生物

总体来说，除少数研究领域外，我国深时极端地质环境微生物的研究相对于国际上来说显得更加薄弱，还有更大的发展空间。本节首先总结了我国在深时极端地质环境微生物学研究方面的优势领域和薄弱领域，然后分别从研究队伍、学术组织和资助力量几个方面介绍了我国深时极端地质环境微生物学研究的发展态势。

一、优势领域

首先，在我国，深时极端地质环境微生物研究的优势是我国有非常好的地质资源，涵盖各类深时极端地质环境，如新元古代以“雪球地球”为代表的极端寒冷环境，以及退冰期间的极端高温环境和显生宙不同环境背景下的生物大灭绝事件，如寒冷冰期背景的奥陶纪末生物大灭绝、温度显著降低的晚泥盆世弗拉期—法门期生物灭绝事件、二叠纪—三叠纪之交和三叠纪—侏罗纪之交升温背景下的生物大灭绝事件等。其次，针对这些重大地质突变期，我国已经开展了非常好的地质学基础研究，建立了精细的生物地层和年代地层框架、沉积环境背景和事件序列。这些都为深时极端地质环境微生物的研究打下了很好的基础，并在一些领域的地质微生物研究中取得了重要进展，代表性成果主要是有关二叠纪—三叠纪之交生物大灭绝期间的各类微生物功能群及其生物地球化学的研究，我国学者在这方面贡献巨大，成绩显著。

二、薄弱领域

相对于国际上深时极端地质环境微生物学的研究来说，我国的薄弱环节主要体现在几个方面：①研究思路和手段相对单一，一般局限于用常规的地质学、同位素地球化学和有机地球化学等手段解决传统的地学问题，很少考虑地质微生物相关的科学问题；②与现代极端环境微生物学的联合交叉偏弱，主要体现在深时极端地质环境微生物的研究很少借用现代微生物学研究的方法，如从系统发育的角度研究重要微生物功能群的起源和演化研究，也很少开展与深时极端环境相似的现代环境的地球生物学研究，后者主要由现代微生物学家从生物学层面开展工作；③分析技术方法较为落后，包括无污染岩心样品的钻取、微量有机分子的分析技术以及单体化合物的同位素分析技术等，都还有待深入发展与提高。

三、研究队伍、学术组织和资助力量

我国现代极端地质环境微生物研究的队伍是非常庞大的，与之相对的深时极端地质环境微生物研究队伍是非常弱小的，分散在一些大学和科研院所的古生物学或沉积学的研究队伍中，主要的团队有：①从与微生物作用密切相关的沉积学角度开展研究的团队，如北京大学的刘建波团队、中国科学院地质与地球物理研究所的吴亚生团队、中国地质大学（北京）的史晓颖和梅冥相团队、中国地质大学（武汉）的王永标团队和河南理工大学的齐永安团队；②从脂类生物标志物角度开展研究的团队，如中国科学院广州地球化学研究所的彭平安团队、中国地质大学（武汉）的谢树成团队、中国科学院南京古生物地质研究所的曹长群团队、中国石油大学（北京）的王春江团队、同济大学的贾国栋团队和浙江大学的孙永革团队；③从C、N、S等与微生物代谢密切相关的同位素角度开展研究的团队，如北京大学的沈冰团队、中国地质大学（武汉）的李超团队；④从微生物矿物学角度开展工作的研究团队，如中国科学院地质与地球物理研究所的潘永信团队。

2018年是地质微生物学发展的重要一年，成立了两个与地质微生物密切相关的学术组织，一个是中国古生物学会成立的地球生物学分会，另一个是中国微生物学会成立的地质微生物专业委员会。相对于地质微生物专业委员会来说，地球生物学分会成员中从事深时极端地质环境微生物研究的比重较少，缺乏专门的极端深时地质微生物专业委员会。

深时极端地质环境微生物学还未形成一个独立的学科体系，研究内容分散在古生物学、生物地质学和地球化学等学科领域。在资助力量方面，我国还没有专门针对深时极端地质环境微生物的单独学科体系或资助计划，都是依托其他学科进行的，如国家自然科学基金委员会地球科学部的古生物学、生物地质学和生物地球化学学科；科技部也没有单独的专项计划或其他大型项目来资助这方面的研究；中国科学院也没有这方面的专项计划。

第六节　深空环境的生命

一、国内发展现状及同国际发展的比较

截至2020年9月，通过检索国内外数据库（中国知网、维普、万方和Web of Science），2000～2019年天体生物学方向发表的文章共3000余篇，其中英文文献3069篇、中文文献626篇。随着时间的推移，天体生物学研究成果呈稳定的线性增长趋势。检索结果显示，“生命起源”“极端环境的宜居性”

是最主要的研究内容，其次依次是“冰卫星”“地球类比”“生物标志物”“系外行星”的研究，最后是研究成果相对较少的“生命的定义”课题。对地球极端环境下生命痕迹的研究是天体生物学的主要贡献力量。基于外太空的主要研究对象为冰卫星、火星、系外行星和陨石，少数来自对彗星和星际物质的研究。从数量上看，国际研究是国内研究的4.75倍。从质量上看，国际研究多发表在SCI期刊上；国内发表的文章除少数综述外，原创型的学术论文较少，科普、新闻稿件较多，发表的期刊也更多样化。国际上对天体生物学各方面的研究发展较为均衡、稳定，而国内的天体生物学研究还处在初级水平，较多的研究关注生命起源，诸如行星宜居环境的形成和演化、地外生命的识别和探测等课题研究较少，甚至是处于空白状态。

造成国内外差异的原因可能包括以下两点：①研究基础不足，空间探测稍滞后于欧美国家；②重视程度不够，缺乏针对性的立项研究。

天体生物学是21世纪最前沿的学科之一，2018～2020年在国际上达到发展高峰。学科发展在一定程度上依赖空间探测的结果，随着对太阳系认识的加深和对地球研究的深入，以地球极端环境为研究对象的天体生物学研究是寻找外太空生命的重要方法。我国正在进行火星探测，使得天体生物学成为迫切需要开展的前沿学科，亟须利用现有的科研环境和自然资源开展详细而深入的研究。

二、国内发展态势

（一）总体起步晚，研究较弱

天体生物学在国内的认知程度总体偏低，开展的相关研究还比较少。其中一个重要原因是，基于传统地球生物学的研究方法和手段，天体生物学似乎无处下手，因为迄今还没有在地球之外的天体上发现任何生物。这一观念上的问题，导致从事相关研究的人员少，研究平台层次低，标志性成果少。目前，我国的学科设置中还没有天体生物学这个学科，也没有完善的人才培养体系。

（二）开始了地球极端环境相似性研究、空间生命科学研究、生命的空间传输研究

近年来，我国学者在生命起源与有机磷化学、类火星地貌与环境、极端环境生命、空间生命科学、地磁场变化对地球生物圈和宜居环境的影响、火

星有机质与水和系外行星宜居环境等研究领域取得了一系列新进展。

我国深海探测技术的迅速发展为生命起源的研究提供了基础。潜水器、无人智能装备和原位探测技术的发展和运用推动了天体生物学的发展。例如，我国科学家利用在“发现号”深海 ROV 上搭载的我国自主研发的深海激光拉曼光谱原位探测系统（RiP）在深海热液区（1400 m）首次原位观察到超临界二氧化碳及其中含有的大量氮气和未知有机组分，后者很可能与氨基酸合成相关，这一发现为生命起源以及初始有机质的形成提供了新的启示。

我国科学家在利用柴达木盆地作为火星环境类比研究方面开展了卓有成效的工作。包括中国地质大学（武汉）、香港大学、自然资源部地质研究所在内的多个研究团队开展了柴达木盆地类火星地貌的宏观对比研究，也对该地区几类极端环境的物质组成、形成环境、地表和地下高盐环境的微生物及生物标志物开展了系统性的工作，对与之类似的火星古湖泊盆地的形成历史和环境宜居性开展了重点研究，发表了一些论文和专著。该地区已经成为在国际上认可度很高的火星环境类别研究基地，同时也形成了多支研究队伍，搭建了科研平台，建立了教学科研基地。

在空间生命科学领域，我国对载人航天飞船舱内环境的微生物进行了大量研究，包括样品采集、监测、菌种鉴定、生长生理特性分析、菌种致病性和材料腐蚀性检测等。其中，北京航空航天大学长期从事基于生物再生生命保障系统的微生物研究工作，利用所研制的“月宫一号”实验系统开展了密闭舱室微生物检测与控制研究（图 4-5），阐明了密闭舱室内环境微生物的多样性和发展演替规律，证实了蔬菜引入可减少密闭舱室内空气和表面微生物的量，建立了密闭舱室内微生物监测和控制体系，开展了航天器舱内材料表面微生物消杀的设备研制和方法研究。中国虽然截至 2021 年尚未开展空间环境舱外暴露实验，但是正在建造的近地轨道空间站的科学实验中已经规划了舱外暴露实验。同时，航天微生物数据库也正在构建，该数据库囊括了我国空间站、火星探测器等航天器在建造过程中的环境微生物菌株目录。

在临近空间研究方面，我国从 20 世纪 80 年代就开始利用高空气球搭载多种农作物种子和食用菌菌种开展生物诱变的相关研究。近年来，随着浮空器技术和生物学研究方法的进步，我国学者也开始研究临近空间极端环境中的微生物生命。在中国科学院“临近空间科学实验系统”A 类战略性先导科技专项的支持下，我国学者开展了临近空间的天体生物学相关研究，研制临近空间生物实验载荷并建立了临近空间生物学研究方法，分别开展了临近

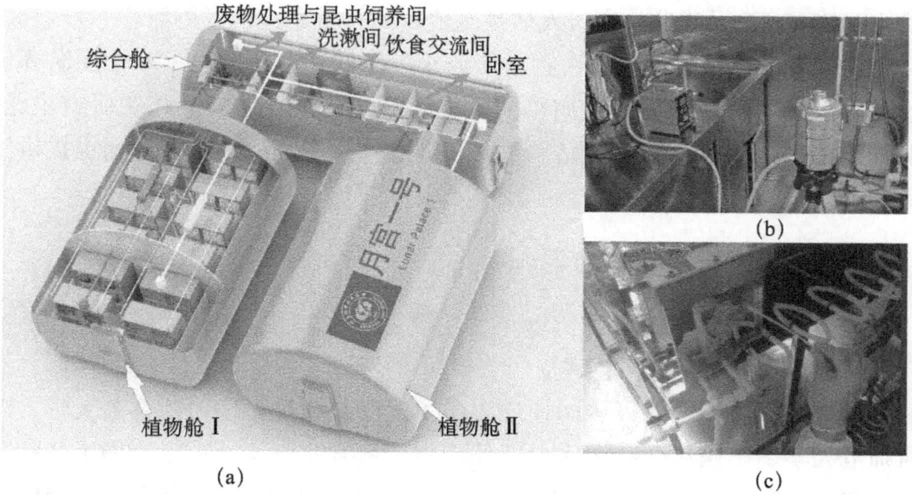

图 4-5　北航“月宫一号”密闭舱室内环境微生物演替发展规律研究

注：(a)“月宫一号”模型图，(b)空气中的微生物监测，(c)关键设备表面的微生物监测。

空间生物暴露和临近空间生物原位收集的研究，以期在临近空间生物的多样性、生命适应临近空间极端环境的机制等方面取得新认识。目前已成功研制出具有自主知识产权的临近空间生物暴露载荷和临近空间生物收集载荷，并成功开展了两次超过 30 km 高度的飞行实验。这些载荷和飞行实验有助于我们更好地认识临近空间极端环境中的生命以及这些生命的生存和适应机制。

（三）有望与深空计划更加紧密地结合，新的机遇已经出现

近年来，我国在行星科学和天体生物学的学科建设和人才培养方面进行了积极尝试。2019 年初，中国科学院大学地球与行星科学学院及中国科学院地质与地球物理研究所联合在中国科学院大学增设“行星科学”一级学科，将“天体生物学”列为“行星地质学”二级学科下的一个重要研究方向（李雄耀等，2019）。2019 年 7 月，中国高校行星科学联盟正式成立，首批 27 所高校中有许多都设有天体生物学研究方向。这些工作有助于促进我国行星科学和天体生物学的高层次交叉人才培养，提升我国天体生物学研究的成果产出和国际影响。同时，许多高校和科研机构成立了天体生物学相关的研究团队，如中国地质大学（武汉）、北京航空航天大学、清华大学、中国科学技术大学、香港大学、澳门科技大学、中国科学院地质与地球物理研究所、中国科学院水生生物研究所、中国科学院地球化学研究所、中国科学院深海科学与工程研究所。我国一些大科学装置的建设，如 500 m 口径球面射电望远

镜（FAST），为寻找地外宜居天体甚至地外高等生命提供了可能。

总体来看，我国的天体生物学研究正处于快速起步阶段，多个团队开展的地球极端环境生物学开始向空间延伸，基金项目开始涉及相关研究主题，学科交叉氛围浓厚，新的研究平台正在搭建，火星等深空探测计划也提供了新的机遇。

（四）国内研究队伍和平台

国内的相关研究团队和实验平台主要分布在高校和中国科学院系统中，已经形成了几支研究队伍，建立了多个研究平台。

行星类比研究团队以中国地质大学（武汉）为主，还包括香港大学、陕西师范大学、中国科学院寒区旱区环境与工程研究所等，重点开展了柴达木盆地与火星地貌和环境的类比研究，类火星极端环境的微生物、生物标志物的种类和保存条件研究等，发表了一系列成果（Xiao et al.，2017；Xiao，2020），已经把柴达木盆地打造成为国际上公认的类火星研究基地。此外，山东大学（威海）建立了模拟火星环境试验舱，可在模拟火星大气和辐射等条件下开展相关科学试验。

在空间生命科学方向，北京航空航天大学建立了空间生物再生生命保障系统地基大型科学实验装置“月宫一号”，并系统地开展了空间舱内微生物污染规律和防控技术研究。

在生命与磷化学方面，厦门大学赵玉芬院士团队及其创建的清华大学生命有机磷化学实验室从20世纪80年代就开始了磷的生物化学研究，探讨磷酰化在蛋白质、核酸、糖的制备中的作用与地位，提出“磷是生命化学过程的调控中心”假说，认为磷酰化氨基酸是地球生命起源的种子，它可以定向组装成多肽、多核苷酸，形成生物膜，从而构成一个生命进化的最小系统。

在地外生命探测方面，运行中的FAST有可能为系外生命探测提供新的平台。正在实施的火星“天问一号”探测任务，可以为火星环境的宜居性评价提供第一手素材。

有建制的相关研究平台主要包括以下几个。

（1）中国地质大学（武汉）行星科学研究所成立于2011年，是国内高校中最早成立的行星科学研究机构，依托生物地质与环境地质国家重点实验室及地质过程与矿产资源国家重点实验室，主要从事行星地质学、比较行星学、天体生物学和天体化学等方向的研究。2009年，该研究所开始设立行星地质与比较行星学博士点，出版了第一部《行星地质学》教材，在利用“嫦娥工程”探月数据及国际月球探测数据研究月球地质方面，取得了显著成就。在火星地质学和天体生物学研究方面也处于国内领先地位，是柴达木盆地类火星研究基地建设的主要贡献者。该高校已经设立了行星地质学本科生专业方向，将是我国主要的行星地质学人才培养基地。

（2）2018 年成立的中国科学院比较行星学卓越创新中心建设有包括行星化学、类地行星探测资料应用在内的 7 个研究单元，构建了天体化学实验平台，利用高精度现代化微量元素和同位素分析仪器，对小行星、月球和未来的火星样品进行了系统化分析研究，揭示了太阳系起源和演化历史。

（3）中国科学院地质与地球物理研究所地球与行星物理院重点实验室研究地幔、地核及衍生的电离层、磁层，并比较研究其他行星的内部与空间环境，以深入理解行星和地球的形成和演化，并满足国家在资源环境、防灾减灾、国防安全、深空探测等方面的战略需求。深空探测、天体生物学、比较行星学等是该实验室的重要研究方向，在行星水的演化、行星电离层粒子逃逸、行星磁场与生命演化、临近空间等方面取得了系列成果。万卫星院士曾担任中国首次火星探测首席科学家，实验室研制了其中的关键载荷——磁强计，用于火星表面的磁场测量。该实验室主持创刊出版了《地球与行星物理》（*Earth and Planetary Physics*）科技期刊，新成立了地球物理学会行星科学专业委员会，并积极推进“行星科学”一级学科的建设论证，吸引和凝聚了一批国内外相关学科的研究力量，解决了地球科学和行星科学面临的重大科学问题。

（4）中国科学院地球化学研究所月球与行星科学研究中心在行星地质方面有较好的研究基础和特色。

（5）中国科学院水生生物研究所刘永定研究员领导的科研团队开发了利用荒漠藻结皮技术治理荒漠化土壤的技术，并在我国北方推广，使 24.5 万亩[①]流沙土地变成了草地。该团队据此提出利用荒漠藻类结皮改造火星地表以促进其地球化的科学设想。该团队还利用临近空间模拟火星的飞行实验，对荒漠藻适应火星环境进行了研究，结果发现荒漠藻适应能力突出，未来可作为火星基地物质循环和环境改造方面的先锋物种。

（6）澳门科技大学月球与行星科学国家重点实验室也在发展行星地质和天体生物学研究方向。

---

① 1 亩≈ 666.67 m^2。

第七节 高压环境微生物

与国际相比，国内在高压环境微生物方面的研究十分薄弱。但一些实验室已经建立了高压取样和培养的实验平台，有望推动我国在这方面的研究，并在国际上产生重要影响。

一、技术平台

截至 2020 年 8 月，在 Web of Science 核心合集数据库中检索到 2000～2020 年深海耐压 / 嗜压微生物（piezotolerant/piezophilic/barophilic/barotolerant）文献 314 篇。其中，2015 年发表的文献数量是 26 篇，2019 年发表的文献数量则达到了 30 篇，是这 20 年来的峰值（图 4-6）。进一步对这些文献的研究方向进行分析发现，超过 70% 的研究领域集中在生物科学（包括微生物学、生物化学与分子生物学、遗传学、生物技术与应用微生物学、环境科学与生态学等），超过 30% 的研究领域集中在地球科学（地质学、地球化学）以及水科学等，其中综述性文章有 41 篇。从世界范围来看，日本、美国参与的研究数量较多，此外参与相关研究较多的国家还有法国、中国、德国等。

国际上 JAMSTEC、CNRS、加州大学系统产出的文章最多。中国科研学者在相关研究领域也做出了卓越的贡献，涉及的研究机构有上海交通大学、中国科学院深海科学与工程研究所、上海海洋大学等。其中，上海交通大学是国内主要的贡献机构。

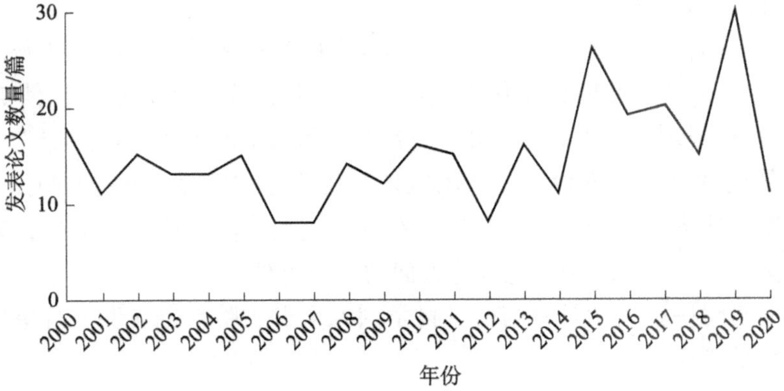

图 4-6 2000～2020 年深海耐压 / 嗜压微生物研究方面发表的论文数量

二、研究进展

高压对生命起源具有独特的意义，但国内开展得极少。国内目前有很好的深海采样设备，如载人深潜器和无人遥控潜水器，以及完善的深海环境模拟培养平台和各种压力规格的高压培养装置，能够通过实验室的模拟培养得到深海微生物。此外，测序水平的不断提高和成本的下降，也有利于进行后续大规模的多组学研究分析。

国内对生命边界的进一步拓展研究较少。目前，国内的研究集中于极端环境中生物资源的挖掘，从而被动地发现生命可能存在的边界，而非主动性地尝试突破现有的生命边界。2009 年发现 *Pyrococcus yayanosii* CH1 是国内外机构合作完成的，国内参与机构包括自然资源部第三海洋研究所和厦门大学生命学院。

对于深海及热液等极端环境，国内有不少机构和平台都开展了积极的探索，但是囿于采样过程耗资高、培养设备设计和维护较难等多种因素，对于极端环境下生命边界的探索较少，且未能取得突破国外纪录的成果，对于生命边界条件下各种生命活动的研究也较少，这直接限制了人们对生命边界温度 / 压力等的进一步突破和认识。真正尝试将模拟培养、代谢组学与微生物功能群紧密结合的研究还处于筹备和起步阶段。但是随着深海采样技术、培养技术和模拟设备的进一步发展，模拟深海甚至海平面以下 10 km 深度，可以匹敌马里亚纳海沟的超高压力条件已经成为可能，这为后续我国科学家对深海高压环境微生物的研究奠定了坚实的基础。

第五章 发展思路与发展方向

第一节 关键科学问题

对于各种不同的极端地质环境，有关微生物的科学问题有很多，也有很大的差异，但总结起来，主要涉及前沿科学问题以及服务社会的重大难题这两大类问题。这些科学问题和重大难题也正是我国在极端地质环境微生物方面的薄弱研究环节，需要在未来5～10年进行突破。

一、学科前沿的基础科学问题

学科前沿主要涉及生物学问题和地质学问题。生物学问题需要回答极端地质环境微生物在生物界总构架中的地位，地质学问题则需要回答极端地质环境微生物在创建宜居环境中所起的作用。

（一）极端地质环境微生物在生物界的总体地位

可以说，极端地质环境微生物在生物界占有举足轻重的地位，具体表现在以下几个方面。

首先，生命的边界都处在极端地质环境，因而极端地质环境微生物决定了地球上生命的边界。目前已经能承受122 ℃最高温度的微生物就是从深海热液口分离出的。在11 000 m深的马里亚纳海沟乃至太平洋海底沉积层的500 m深处都发现了活细菌。在深地3 km坚硬的岩石圈当中，还能发现活跃的微生物。

其次，极端地质环境微生物构建了“生命之树”的关键节点。“生命之树”根部的大多数微生物都是极端环境的微生物，因而现代极端环境的微生物是揭示地球上生命起源及其早期演化的关键载体。“生命之树”的许多重要分支节点往往出现在重大地质环境突变期，体现了极端地质环境与微生物功能群的起源和演化息息相关。

最后，极端地质环境的基因库是人类解码生命起源与演化的密钥，也是人类赖以生存的重要的战略基因资源。有研究认为，位于深海沉积物顶部的10 cm空间，据估算约含有4.5亿tDNA，是地球上最大的基因存库。深海、深地、地表极端环境微生物都有许多特殊的基因和酶，极端地质环境微生物的基因多样性特别值得关注。人类的基因组包含超过32亿碱基对，但其中只有约1.5%编码蛋白质。这说明人类很少有特殊的基因，相当小的基因变化就能导致动物出现高级的组织模式。提供能量、允许蛋白质合成和完成基本代谢机制指令，都是以数十亿年前进化的微生物创建的基因平台为模板的。

当然，由于研究的局限与微生物时空分布的巨大差异，人们对极端地质环境微生物的丰度、种类、多样性等争议较大，这些问题还有待深入研究。回答这些问题更有助于人们进一步认识极端地质环境微生物在生物界举足轻重的地位。

1. 微生物的空间分布、丰度、多样性

微生物的空间分布、丰度和多样性是所有环境都要涉及的有关微生物的最基本问题。如前所述，对极端地质环境微生物的丰度和多样性存在很大的争议，主要是由巨大的空间差异以及研究程度低造成的。对于海洋极端地质环境，需要加强海底沉积物不同深度的微生物研究。对于深地环境，尤其需要加强超深钻微生物的原位监测与研究。对于深时极端地质环境，微生物的空间分布、丰度和多样性研究难度特别大，还需要进行长期的积累。

例如，咸盐环境（咸盐湖泊、咸化土壤）在地表广泛分布，咸盐环境微生物物种、功能多样性直接影响微生物参与的地球元素循环。借助新兴的组学和同位素技术，围绕这一问题开展研究具有重要的生态学意义。又如，以AMD环境为代表的极端酸性环境中微生物多样性和微生物功能群结构具有统一性和特殊性双重特征，不同类型的金属硫化物矿山和不同演化阶段的AMD中微生物群落的生物多样性、细胞丰度、代谢速率等具有不可忽视的差异特征，系统描述这些差异特征并揭示导致这些差异的主要环境因素（如

pH、Eh、温度、硫化物类型、营养状况等），有助于提高对极端酸性环境微生物多样性特征和演化趋势的理解和预测能力。

2. 对单一环境因子的适应及其机制

这方面取得的较为系统性的认识主要来自高压环境，但温度、盐度、酸碱性等对微生物研究影响的规律如何，人们了解得还很少。这也正是极端地质环境微生物需要大力解决的重要科学问题，也是与正常环境微生物研究的一大不同之处。

目前，有关高压对微生物的影响有了一个大致的框架性认识。例如，压力 2～30 MPa 导致微生物运动性降低，30～60 MPa 则影响 DNA 的复制，60～100 MPa 影响蛋白质的翻译过程，100～300 MPa 可以杀死一些微生物，300～700 MPa 则挑战生命抗压力边界。因此，高压是深部生命共同应对的核心环境因子，是生命长期演化的驱动力。以高压环境为例，下面三个问题是比较关键的难题。

1）是否存在特异性的压力适应途径及相应的代谢产物？

选择一个生物标志物，其基础是有针对某一环境压力的特殊代谢途径及相应的代谢产物，而该代谢产物能够在地质历史中得以长期保存或转化为可辨识的次一级产物，所以第一个需要解决的问题是微生物是否存在特异性的压力适应途径及相应的代谢产物与其他环境因子加以区分。基于物理化学参数，压力会影响某些代谢过程，使生化反应朝体积减小的方向进行。即使低至 10 MPa 静水压力，也会极大地影响微生物的代谢方式。10 MPa 的压力足以影响降解烃类的微生物，如食烷菌属和海杆菌属的石油降解和细胞复制；嗜压硫酸盐还原菌和产甲烷菌的分离证明深部环境中高压诱导的硫酸盐还原和产甲烷过程主要是由嗜压微生物引起的；但压力敏感菌株奥奈达湖希瓦氏菌（*Shewanella oneidensis*）MR-1 介导的 Fe^{3+} 还原速率在 40 MPa 的压力下增强，说明压力敏感菌株对深部铁元素循环有更重要的贡献。目前，压力对微生物代谢的影响主要依赖菌株进行分析。通过对嗜压菌 *Photobacterium profundum* SS9 研究发现，在最适压力下氨基酸发酵产能反应——斯提柯兰氏（Stickland）反应和氧化三甲胺还原相关代谢途径基因表现为上调，且负责有机物如几丁质、支链淀粉和纤维素的降解代谢途径基因被激活。对嗜压菌热液口脱硫弧菌（*Desulfovibrio hydrothermalis*）AM13 在其最适压力的分析中发现，最适压力条件上调的基因主要涉及谷氨酸和脂肪族氨基酸的合成，从而使得谷氨酸在高压下积累，且高压下细胞内 ATP 再生水平增高，通过硫酸盐还原进行的呼吸过程导致氧化磷酸化水平增高。通过对嗜压菌嗜压脱硫弧菌（*Desulfovibrio piezophilus*）C1TLV30^{T} 分析发现，高压条件下乳酸作为碳源会有部分流向乙酰-CoA 氧化途径，并通过产生 H_2 和甲酸进而产生 ATP。此外，高压对海底深部生物圈和热液口环境的微生物参与的某些厌氧过程有促进作用，包括硫酸盐还原耦合厌氧甲烷或有机质氧化、有机碳呼吸、铁锰还原和硝酸盐还原等过程。因此，后期对嗜压微生物群体微生物代谢途径的研究分析可进一步揭示是否存在特异的压力适应途径。但是该研究的基础是获得嗜压的微生物群落，原位环境所采取的样品并不能确定极嗜压的微生物群落。因此只有通过高压富集培养的方式才能获得嗜压微生物群落，从而对其进行宏基因组、转录组甚至蛋白质及代谢组学的研究，以分析其特异的高压适应途径。

2）是否在不同微生物功能群间存在对压力的共适应机制？

通常情况下，并不是所有活跃在深海及深部高压环境下的微生物都拥有适应高压环境的机制，部分非耐压 / 嗜压微生物也与耐压 / 嗜压微生物一起出现于深海环境。那么，那些压力敏感菌群是否得益于已经适应了高压环境微生物菌群呢？最有可能的两条途径适应这一目标：①共享相容性物质；②通过病毒等导致的基因横向转移获得压力适应能力。但这些假设仍然需要验证。

当细胞应对冷、热、pH 和高压时，为了维持细胞的膨压并恢复细胞体积，生物体会积累小分子量的渗透压调节剂，主要是有机溶质，亦即相容性溶质。共适应机制的提出是基于微生物在面对不同的压力刺激条件时，可能采取相同的抵抗策略。基于前期的研究，生物在面对不同压力应激时，胞内会积累相同的相容性溶质以保持细胞的完整状态，如海藻糖可作为渗透胁迫积累的相容性溶质，还可保护细胞免受热、冷冻、干燥和其他环境压力。海藻糖被认为是酵母细胞面临各种压力应激发挥作用的相容性溶质。因此，进一步思考，是否在不同群体间存在利用相同相容性溶质来抵抗环境胁迫的现象？例如，面对高盐胁迫时，超盐蒸发池中分层不同的微生物群体均通过在胞内积累甘氨酸甜菜碱来应对渗透胁迫；面对高压胁迫时，深海中的不同动物通过胞内积累氧化三甲胺来应对高压胁迫。目前，对嗜压菌 *Photobacterium profundum* SS9 来说，细胞会在其最适压力下积累 β-羟基丁酸作为相容性溶质；对 *Desulfovibrio hydrothermalis* AM13 和 *Desulfovibrio piezophilus* C1TLV30^{T} 来说，谷氨酸可作为其最适压力下的相容性溶质。由此，对微生物来说，在高压环境下，不同微生物功能群间可能会共享相容性溶质。

3）压力适应（抗氧化）如何影响生态系统的演化？

压力与抗氧化的关系已经有越来越多的证据，包括微生物与宏生物，但至少还有下述几个关键问题亟待解决：①伴随着抗氧化（抗逆）进程，到底发生了哪些代谢组分的显著改变？识别出显著变化的代谢产物是揭示其环境影响的前提和基础。②这些显著变化的代谢产物在元素循环（包括同位素分馏）中到底发挥多大功能？甄别变化的代谢产物与元素循环相关性是解析逆境胁迫对生物地球化学循环的关键。③逆境胁迫下生物学功能的显著改变与重大地质历史事件的关联是什么？

无论是厌氧微生物还是好氧微生物，都面临氧压力，包括外源性氧压力和内源性氧压力。呼吸链是少数容易受到 O_2 和 H_2O_2 伤害的潜在位点，因为黄素、醌和金属酶促反应中心有足够低的电位与氧发生反应。应对氧胁迫的厌氧生物和好氧生物均演化出了不同的机制。典型的氧胁迫耐受酶如 SOD 和过氧化氢酶在好氧生物中比较常见，在厌氧生物中只有少量发现。绝大部分厌氧微生物拥有非血红素类、含铁的超氧物还原酶等。这些酶可清除超氧化物产生的氧毒害而不产生额外的氧气，同时用过氧化物酶替代过氧化氢酶可清除 H_2O_2。此外，新近发现绝大部分厌氧菌含有烃基过氧化氢物还原酶，还原过氧化氢和烃基过氧化物为水或相应的醇，通常情况下抗氧化可以导致 ATP 产量的下降。

可以借鉴现代合成生物学的方法研究在早期地球历史环境中微生物功能群的生存边界与潜在的功能大小。在生物演化的过程中，基本的代谢途径大约起源于 30 亿年前，在极端环境条件下，氧化还原平衡的破坏是对微生物体最主要的生理伤害之一。极端的温度、压力、盐度、pH、氧胁迫、辐射、干旱、寡营养、重金属和其他有毒化合物都会影响生物的代谢过程。代谢网络分析模拟与微生物功能群代谢重构是明确微生物功能群在地质历史时期关键作用假设可行性的核心内容，也是早期地球环境演替重建的分子基础。

3. 对多种环境胁迫因子的生命适应和机制

在许多极端地质环境中，单一的环境胁迫还是比较少的，更多的情况是存在多种环境胁迫因子，导致生物对这些环境因子都能产生适应，如干旱盐碱土壤环境中的微生物物种、功能多样性及环境适应机制。相比其他高盐极端环境（如深海、盐湖等），干旱 / 半干旱地区盐碱土壤不仅高盐，缺水、高 pH 同样给微生物的生存带来了严峻的考验。在多种生存胁迫条件下，干旱 / 半干旱地区盐碱土壤环境中的微生物物种、功能多样性同其他高盐环境有何区别？其嗜盐/耐盐机制与其他高盐环境有何区别？

深地微生物大多要面对深地环境多种极端残酷的生存条件，无论深地微生物是土著的还是外界带来的（地表水的灌入、裂隙水带入、油藏的注水开发等），进入深地环境，都必须经过严格的多种环境条件筛选，最终适应这些极端条件，从而生存下来。但目前对有关微生物对多种极端环境的适应机制与适应速率还知之甚少，这一问题的解决将有助于加深人们对生命极限和起源的认识。

以 AMD 为代表的极端酸性环境往往还是重金属异常富集的流体环境，微生物的长期演化形成了独特的代谢途径和环境适应机制，在形成独特的胞内和胞外电子传递路径的同时，EPS 和周质空间还发展形成拮抗重金属离子侵害和维持高 pH 梯度的生物化学机制，但是强酸性和重金属胁迫条件下环境微生物的代谢机制和环境适应机制还有待在分子层次或代谢组层次系统阐明。开展这方面的研究还有助于筛选并获得独特的微生物资源。

微生物对极端环境的适应机制的研究，已经取得了很大的进展，但需要总结一些规律性认识，特别是对多种极端环境因子的适应问题。例如，基于微生物组学、富集分离培养技术和同位素示踪技术，在群落和纯培养菌株层面，探究高温地质环境微生物具有的特殊能量代谢途径，以及驱动元素循环的分子机制。通过生理学和分子生物学技术手段，探究高温地质环境微生物适应高温环境的适应机制。通过原位及室内模拟实验，探究微生物互作模式对高温微生物环境适应性的影响。通过整合基因组学、有机地球化学、脂类组学、同位素组学、合成生物学等手段，揭示膜脂化合物对高温微生物适应环境的作用。但是，如何在以上研究基础上，进一步深入研究多种环境因子（如高温、高压、缺氧）的共适应问题还亟待突破。微生物首先通过核苷类分子、ROS 等信使分子感应外界环境条件，通过合成酶或分解酶调控，改变一些代谢反应速率。

（二）微生物在创建宜居地球环境中的作用

自生命起源以来，微生物见证了生命和地球的演化历史，为创建适合人类居住的宜居星球发挥了至关重要的作用。在地球早期，能量缺乏的极端环境触发了光合自养微生物的起源，进而向地球生态系统提供了更加充分的能量。同时，微生物的代谢作用使得地球维持在气候宜居的状态。例如，产甲烷古菌的代谢向环境中释放了大量的甲烷，其温室效应缓解了早期太阳亮度不足（faint young sun）的问题，使地球处在一个相对温和的环境下。此时地球表层处于无氧状态，是厌氧微生物的宜居地球，但并不适合好氧生物的生存。随着光合微生物的发展，地球表层环境所能提供的电子进一步降低，进而触发了光合微生物的变革，开始利用地表无处不在的水中的电子，演化出产氧光合微生物。产氧光合微生物的发展彻底改变了地表的环境，使得早期的无氧环境转变为有氧环境，逐渐形成了好氧生物的宜居地球。氧气的出现，极大地提高了生物获取能量的效率和地表物质的循环，为真核生物的出现铺平了道路。真核生物的大量繁盛，加强了有机质的沉降和埋藏，进一步促进了大气和海洋的氧化，逐步形成了适宜后生动物生存的环境。

极端地质环境微生物对宜居地球形成的作用还表现在显生宙生物大灭绝期。化石记录表明，动物自起源以来发生过五次重大的灭绝事件，宏体生物多样性和分异度显著降低，并伴随着地表环境的剧变，如出现高温、海洋缺氧硫化等。许多研究表明，后生动物的大灭绝期同时伴随着微生物的繁盛，如约 2.5 亿年前的二叠纪—三叠纪之交显生宙最大的一次生物灭绝事件。这些微生物的繁盛，调节并改善了地表环境，包括固氮微生物的繁盛向生态系统中输入生物可利用氮、硫酸盐还原微生物的繁盛促进了有机质的矿化和硫化物的埋藏、光合硫氧化细菌减缓了水体的硫化等。这些极端地质环境微生物的作用使得地球重新回到适宜后生动物生存的环境。

因此，微生物在其演化进程中使得地球适合它们自己生存，最终也适合人类居住。极端地质环境微生物存在很多工农业应用价值的种质资源，不仅可以改善人居环境，还可以提升工农业生产活动的价值。可以说，人类离不开微生物，但微生物可以离开人类，功能互补的深地微生物可以形成共生互营养的自维持生态系统。如果人类灭绝了，微生物世界仍将继续进入新的稳定状态，并帮助维持一个宜居的星球，以便新的更高级生物生存和发展。微生物的代谢活动是其作用于环境的途径，查明地质微生物功能群的组成和丰度是深入理解深时极端地质环境微生物作用于环境的必经之路。

1. 地质微生物功能群

极端地质环境微生物的一个关键科学问题是微生物的生态学和生物地球化学功能及其对极端地质环境的改造作用。生物与环境之间是存在相互作用的，一方面，环境影响了生物，生物发展出许多适应机制以应对这些恶劣的环境条件；另一方面，微生物不是被动地适应，而是具有主动改变环境的功能，这主要与它的生态学与生物地球化学功能有关。地下油气藏环境和地表的酸性矿坑水都是微生物作用于极端地质环境的充分体现。

微生物的功能主要与碳、氮、硫、铁等元素循环有关，涉及比较多、进展比较大的物质循环是甲烷循环，但对有关氮循环还知之甚少，对硫和铁循环的了解更少。在深海、深地和地表极端环境下，主要通过研究微生物的功能基因实现。但在深时极端地质环境，必须依托脂类分子和同位素才能有所突破，一些重要的矿物也是一个关键窗口。

陆地深部生物圈的微生物功能群与海洋深地环境相比可能更加多样和复杂。地下水中的微生物可能面临各种污染物的胁迫以及与不同含水层岩石之间的水岩相互作用；洞穴中的微生物面对的则是较大空间、高 CO_2 浓度、低 CH_4，与 CO_2 和 CH_4 这些温室气体循环相关的微生物功能群更加引人注目；石油和煤炭中的微生物功能群更多的则是面对难利用的高分子有机化合物；岩石裂隙中的微生物则必须利用化学反应中产生的各种产物进行化能自养。因此，查明陆地深地生物圈中微生物功能群对深刻理解微生物驱动的元素循环及生态环境作用具有重要意义。

1）微生物功能群的识别和定量化

现代极端地质环境微生物的研究可以通过不同的手段，包括纯培养、基因组学、蛋白质组学和代谢组学等，开展不同层面的研究，如微生物的属种特征、微生物的代谢功能特征等。受技术方法的限制，这些手段目前根本无法应用于绝大部分深时极端地质环境微生物的分析。如何识别深时极端地质环境下的微生物功能群类型及其含量是深时极端地质环境微生物研究的重大关键科学问题。

目前，深时极端地质环境微生物研究群体已建立了多种深时极端地质环境微生物功能群的研究手段，如脂类生物标志物的组成，C、N、S 稳定同位素地球化学，与微生物作用密切相关的矿物学和地球化学等。但目前这些研究还远远不足。一方面，这些记录缺乏针对性，往往具有多解性，如 C、N、S 同位素的组成受多方面因素的影响，不同的生物过程和地质过程可能会形成相似的同位素记录；另一方面，目前的这些指标一般都只停留在定性的层面上，很难开展定量化的研究，进而限制了对微生物作用于环境的认识。下一步亟须加强深时极端地质环境微生物功能群代用指标的构建以及微生物功能群的定量化研究。

2）微生物功能群的能量来源与物质循环

能量与营养是限制深地微生物生长的最基本因子。在明确了微生物功能以后，紧接着就需要进一步解决两大问题。一是不同极端地质环境、不同类型的微生物的能量来源问题。例如，在深地环境，微生物可以从氢气和硫代谢方面获得能量。如此，不同极端地质环境微生物获得能量的过程到底有什么规律？二是这些微生物的地质作用在全球尺度上到底有多大？哪些地质环境就像酸性矿坑水那样明确是由微生物作用造成的？哪些地质突变期的哪些环境主要是由微生物造成的？一些极端环境包含有大量的未知微生物，一些甚至是全新的门类，这些新的微生物类群是否参与了深部重要的地球化学元素循环？贡献大小如何？是否具有新的物质能量代谢机制？

例如，在深地环境，由于地表水与地下水的交互作用，地下水、洞穴、地层破碎带都会间歇性地接收地表有机质的输入，而在石油开采过程中，注入水以及表面活性剂和激活剂的添加也会将部分营养物质带入地下。因此，查明外源有机质及营养物质的输入，以及深地热液或水岩反应中产生的还原性物质，对阐明深地生态系统的物质循环、能量流动以及维持生态系统的功能等至关重要。H_2 可以驱动地下深部的多种生物地球化学循环过程。

2. 微生物与环境相互作用的生态效应与机制及其定量化

深地环境中尽管存在着不可忽视的微生物群体，并且该微生物群体具备从这些极端寡营养的环境中汲取营养、介导生源要素循环的能力，但它们如何与地质过程相互作用，特别是对地质过程的作用，目前知之甚少。譬如，矿物岩石的风化、矿床的形成、元素循环、气候变化等地质过程到底有多少来自微生物的贡献，微生物调控的地质过程速率和通量，微生物调控与非生物过程的区分，微生物活动在地质记录中的保存等问题，将是未来关注的热点科学问题。在不同极端地质环境条件下，微生物与微生物之间、微生物与地质环境之间的物质、能量如何传递与交换？极端环境微生物功能群之间的共生现象可能非常普遍，但是具体调控机制是什么？有哪些生态效应？在全球变化背景下，极端地质环境微生物如何响应气候环境变化？如何反馈作用于气候环境变化？这些问题都有待深入研究。深地等极端环境微生物共生或互营养是一种常用的代谢方式。一种微生物擅长从环境中获取能量，将产生的次级代谢产物传输给另外一种微生物；另一种微生物擅长从环境中摄入营养，经过加工以后，将营养提供给第一种菌，因而形成互惠共生关系（图 5-1）。

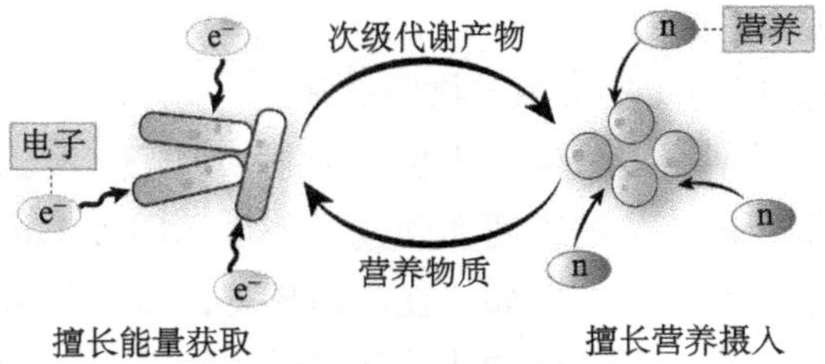

图 5-1　深地等极端环境微生物共生或互营养示意图

尽管人们已经对一些极端地质环境中微生物功能群的结构有了一些了解，初步揭示了多种微生物功能群之间的协同互馈关系，但是，在生物多样性丰富、生物量低的极端地质环境中，微生物种间的互作关系直接影响了Fe、S、As等多种元素的地球化学循环，微生物是如何通过这些生物地球化学过程影响极端地质环境的？理解这些地质过程对一些地质环境的修复和治理具有重要意义。例如，长久以来，人们认为一些干旱区（如沙漠）没有生命存在。但近年来的研究指出，干旱区微生物资源非常丰富，并且对干旱生态系统起着重要作用。探究微生物如何影响干旱区的生态系统十分必要，可以为防治沙漠化进程和土壤恢复提供依据。

微生物与环境的相互作用体现在两个层面上：一是环境变化对微生物组成和代谢的影响，二是微生物组成和代谢对环境和其他生物的作用。这是深时极端地质环境微生物学研究的核心，也是薄弱环节。目前的工作主要停留在定性研究层面上，亟须向定量 / 半定量研究拓展。以C、N、S等元素为主的生物地球化学循环为切入点，将各类环境因子、微生物功能群及其与特定环境条件下的作用机理和代谢融合到生物地球化学循环模型中，开展单个生态、区域范围及全球范围等不同层面的微生物与环境相互作用的定量化研究。

另外，环境参数的定量化重建也是深时极端地质环境微生物研究中最基础的问题，因为环境条件影响了微生物的组成及其作用的机理和过程。不同于绝大多数现代极端环境，几乎所有的环境参数都可以非常精确地测量，深时极端地质环境参数的定量化研究是非常困难的，如地质历史时期温度的变化。虽然一些磷酸盐质壳体化石所记录的氧同位素信息可以定量重建地质时期的温度，但会受很多其他因素的影响，如后期成岩作用对原始信息的改造、同期海水的氧同位素组成不得而知、生物分布的稀少性等。这方面的研究需要完善实验方法体系，并构建更多反映古环境变化的定量参数指标。

除定量化值之外，环境参数还有一个很重要的方面就是变化速率。最显著的一个例子就是现代全球变化领域讨论最多的当今大气中的二氧化碳浓度和温度变化的问题，如果仅就变化幅度来说，当今的变化幅度是微乎其微的，如 pCO_2 从工业革命前到现在增加了约 140 ppm，温度升高约 1.5 ℃。但考虑变化速率的话，当前这两者的变化速率可能远远高于任何一次重大地质转折期。环境参数的变化速率也是影响微生物组成和代谢的重要因素。因此，深时极端地质环境参数的定量化重建还需要高精度年代学的辅助。沉积地层的高精度年代学一直是一个重大的科学难题，需要同位素年代学和旋回地层学研究的共同推进。

3. 微生物与环境的协同演化

总体上来说，目前人们对微生物的起源和地质演化是不清楚的，这导致人们对各类不同微生物的起源和演化的环境背景很不清楚。首先，许多微生物的起源时间十分不明确。尽管“生命之树”根部的微生物大多起源于极端环境，但具体的起源时间很不清楚。哪些微生物是在前寒武纪的极端环境下起源的，哪些是在显生宙的极端环境下起源的，目前还没有框架性认识。其次，虽然大致了解到部分微生物功能群的起源时间，但是在起源以后，同一微生物功能群的不同属种又是如何演化的，这个关键问题很不清楚，甚至处于研究空白，这直接影响了某类微生物功能群演化树的构建。例如，根据硫同位素，人们知道硫酸盐还原菌可能在34.7亿年前就起源了，但在这之后，各种不同的硫酸盐还原菌是如何演化的？如何构建硫酸盐还原菌的演化树？可以说，目前还没有构建出任何一类微生物功能群可靠的演化树。最后，由于一些微生物的起源时间不明确，起源之后的演化关系不明晰，对应地，这些微生物起源和演化的环境背景就更不清楚。这需要地球科学与生命科学的联合才能突破。

随着组学技术的发展和人类对微生物新种研究的深入，微生物代谢的多样性被越来越广泛地认知。在代谢网络相对保守的组成形式下，越来越多的具有物种特异性（organism-specific metabolic differences）的代谢途径被公之于众。不同的微生物在应对不同的环境因子时展现出不同的代谢需求，尤以极端环境微生物为甚，常见环境因子包括营养源、能量源、温度、pH、盐度、压力、氧化还原度等。在微生物与环境因子的协同演化进程中，代谢多样性和代谢网络的演化痕迹被保存在微生物基因组中。环境因子如何影响代谢网络的演化是一个非常重要的问题，而目前关注代谢网络演化的研究相比物种演化的研究少之又少。

二、与人类经济社会发展和国家需求相关的重大科学难题

极端地质环境微生物对人类有利有弊，人与微生物需要和谐相处，犹如人与自然需要和谐相处一样。

（一）微生物对资源能源的影响及其开发利用

1. 微生物的种质资源

极端地质环境的微生物有许多非常重要的特殊基因和独特的生物酶，可以很好地应用到各类工业生产中。低温微生物能在低温条件下生长、繁殖，并产

生低温酶，能使必须在低温条件下进行的众多催化反应得以更高效地进行，在污水处理、工业生产和医药等方面有着广泛的用途。嗜热微生物具有超强的热稳定性以及蛋白质变性的耐受性，可以在高温条件下快速降解淀粉、纤维素、几丁质等物质，具有非常广阔的生物工程应用前景。嗜碱微生物在洗涤剂、食品和医药等工业生产中具有广泛的应用价值。盐湖微生物资源在食品、医药与化妆品、生物电子、生物材料、生物能源以及污水处理等行业具有巨大的应用价值和开发前景。与嗜压微生物和耐压微生物相关的生物技术被广泛应用于食品加工和食材灭菌中，用以改善食品的风味和颜色。耐辐射微生物在食品工业具有广泛的应用前景，并在放射性废物的富集与处理技术方面具有重大的科学意义。油气藏微生物，包括功能基因和酶一类宝贵的生物资源，在化石能源生物开采、生物修复、挥发性石油烃的吸收和降解、原油及其炼制过程中的脱硫和脱氮、微生物腐蚀与生物控制等方面具有潜在的巨大工业应用价值。

然而，各种极端环境微生物的许多特殊基因和独特的生物酶还有待研究和开发，包括模式生物、各种基因和酶的类型、物理化学特性与功能、提取方法、工农业生产应用中的驯化等科学难题。例如，干旱区盐碱土壤的治理和利用一直是研究的热点。不断发掘耐盐/嗜盐微生物资源，并将其合理运用到农业生产、盐碱土壤治理等方面的意义不容小觑。但截至目前，有关干旱区微生物应用的研究还相对较少。已有研究表明，干旱区微生物在药物开发、土壤修复、植物促生等方面具有重要的潜在应用价值，但仍有许多未知领域亟待人们去探索开发。同时，随着宏基因组学和培养组学的快速发展，人们对干旱区“暗物质”资源的认识揭开了新的篇章。微生物“暗物质”是指那些目前可以被识别但尚未得到培养的微生物。

2. 微生物与能源

极端地质环境微生物对能源的形成、勘探和开发利用等均具有重大影响。页岩中的纳米孔隙对页岩气的储存至关重要，而这些纳米孔隙的产生，可能部分地与微生物及衍生有机质有关。在深部缺氧条件下，异化铁还原菌、硫酸盐还原菌和产甲烷古菌等厌氧微生物都具备还原黏土矿物结构铁的能力，从而促进黏土矿物的脱水和溶蚀，并形成纳米和微米级的溶蚀孔。同时，微生物与含铁黏土矿物相互作用可以形成诸如石英的前驱相——无定形二氧化硅，以及长石、滑石等矿物。石英、长石等矿物的存在增加了页岩的孔隙。更重要的是，这些矿物可以增强页岩气的压裂效果，从而对非常规油气资源的开采具有重要意义。水驱采油是油气开采过程中的常规手段之一，但注入的含水流体可能会造成地下储层中的蒙皂石发生水化膨胀从而导致储层渗透率下降，甚至堵塞油气输送通道，给油气勘探开采带来很大的难题，如出现水敏性问题。通过激发油气藏土著厌氧微生物的活性（譬如注入有效碳源），促进微生物介导的蒙皂石伊利石化反应，理论上有助于解决油气勘探中黏土矿物的水敏性问题。微生物不仅可以提供天然气水合物的气体来源，还可以形成一些重要矿物，如自生碳酸盐矿物、黄铁矿等，为示踪深部天然气水合物矿藏提供服务。油气藏微生物可以用来进行生物采油，主要包括微生物驱油、枯竭油藏残余油生物气化开采、稠油等复杂油藏生物改造，以及油藏环境 CO_2 生物转化与利用等。

极端地质环境微生物在地下油气藏、天然气水合物、页岩气等能源的形成、勘探开发过程中的作用还有待深入研究，包括这些微生物的作用机制、驯化条件、作用效率等问题。

3. 微生物与矿产资源

已经证实，在铁、锰、铜、铅、锌、金等矿床的形成和勘探，以及相应矿石的选冶过程中，微生物及其产物都能发挥出色的作用。许多金属矿床与油气矿床、盐类矿床、生物礁、叠层石、热水沉积岩等地质体有重要的空间和成因联系，它们之间的桥梁或纽带就是微生物及其有机质，是生物成矿系统的体现。微生物找矿技术早已为人们所熟悉，其基础是微生物的抗金属毒性能力，即耐金属性。这方面比较成功的例子是利用土壤蜡样芽孢杆菌指示隐伏金矿床，在中国、美国、比利时、加拿大、墨西哥等国家都开展过这方面的研究，并取得了比较好的效果。随着全球矿产资源日益向“贫、细、杂”的趋势发展，以及微生物技术所具有的高效和低污染优势，微生物选矿技术的研究越来越受到重视。微生物选矿技术实际就是微生物浸矿技术，也叫微生物湿法冶金。目前用于浸矿的微生物主要是嗜酸的自养型硫杆菌属、耐热硫杆菌、硫化裂片菌等极端环境微生物。利用微生物技术浸出的矿石种类已由铜、金、铀等扩大到镍、钴、锌、锰、银等十多种金属元素的矿石，由研究微生物对矿石的直接浸出到微生物在矿物表面的吸附、对矿物表面的改性（可浮性）、絮凝以及利用微生物处理浮选药剂等。目前认为，细菌浸矿的机理主要有直接作用、间接作用和复合作用等多种方式。

然而，虽然人们早已认识到在矿产资源各方面应用的微生物主要是适应极端地质环境的微生物，但一些关键的基础科学问题还亟待深入研究。例如，在微生物浸矿方面，基础科学问题包括研究浸矿微生物的生理生化特性、环境条件、生长繁殖规律等，还包括研究这类微生物的来源、培养及驯化的理论和方法，以使浸矿微生物能够适应复杂的浸矿环境，提高浸出速度和浸出率。

（二）深地、深海和深空等重大工程计划中的科学问题

1.深地计划的微生物难题

向地球深部进军是人类必须解决的战略科技问题，除了地下深部资源能源的勘探与开发，还涉及地下空间的利用等问题，特别是地表以下到500 m的这段深度，将是未来地下工程的关键所在。地下空间利用首先遇到的是地下环境的改善问题。地质微生物及其生物地球化学过程是改变环境的一大关键驱动力。地下深部（深地）环境的开发，就像打开一个“潘多拉魔盒”，需要加强深地开发过程中地质微生物的研究与调查。在未被人类利用之前，地下深部大多处于缺氧环境条件，微生物地球化学循环以厌氧过程为主要特征。在缺氧条件下，一些深部生物地球化学循环使得一些元素能够保存在地下，减弱了对地球表层系统的影响，深部地下因此成了许多元素的汇。除了这些生源要素的地球化学循环外，在深部地下还原条件下，一些岩石和矿石中的有毒金属元素也难以释放出来，永久地保存在地下环境。微生物作用具有两面性。如何充分利用微生物好的一方面，规避其坏的一方面，是在利用地下空间时需要考虑的现实问题。需要研究古菌、细菌和病毒等生命系统在地下环境是如何相互关联的，又是如何改善这些恶劣环境条件的，这些都是深地生物圈亟须解决的重大科学前沿问题。在未来深地开发中，需要探索这些环境条件下微生物的适应能力、改造能力以及对人类社会的影响，从而为改善人类的宜居环境服务。

在当前人类活动向大气释放大量CO_2等温室气体的背景下，固定和封存大气CO_2是维持人类社会可持续发展的紧迫任务。其中一种策略就是将CO_2注入地下岩层，以实现CO_2长期与大气隔绝，这个过程称为地质封存。地质封存成本低，潜在封存地点分布广泛，被认为是最具工业应用潜力的CO_2封存方法。然而，地下深部微生物会对地下封存的CO_2及其周围的环境产生重要影响，需要研究微生物与微生物之间、微生物与地下岩石矿物之间、微生物与地下流体之间的相互作用及其对地下CO_2封存的影响。

核废料是一种难以再循环利用且具有长期危害的物质。高放射性核废料含有较多寿命长的辐射体，对处置方法与处置地点的要求相对较高。地质储存是目前公认的、唯一可行的长期安全处置半衰期长、高放射性核废料的方法。正是这个原因，多国政府在可用于核废料储存的地质构造中建立地下实验室，评估核废料地质储存的可靠性和安全性。这些实验室包括加拿大怀特

塞尔（Whiteshell）地下研究实验室、瑞典 Stripa 实验室和阿斯波（Åspö）地下研究实验室、瑞士泰利山（Mont Terri）地下实验室以及日本幌延町地下实验室等。然而，深地微生物会影响核废料储存库的可靠性和安全性，对核废料的地质储存有正、反两个方面的影响，包括降低储存库气压以及诱导储存库金属结构的锈蚀等。在核废料长期管理的规划设计中，除了物理和化学过程，还需要详细了解可能发生的微生物过程。

2. 深海计划的微生物难题

海洋沉积物中有机地球化学指标的精确构建，可以帮助人们指示和重建地球极端气候历史时期微生物与碳循环、气候变化之间的关系，以及地球历史时期微生物对碳循环的作用和可能的机制，这为预测未来气候环境变化提供了参数条件和模型。查明深海及其沉积物的微生物地球化学过程对全球变化具有重要意义，这是深海计划中微生物与全球变化需要解决的科学难题。

大陆边缘通过开放性断裂面和改变沉积物中气体水合物稳定性而具有调整板块、造成液体流动的功能。大陆边缘构造活动是否形成和调控了深部生命活动的多样性、分布区域？地球活动的长周期循环与深部生命活动的关系是什么？这些过程如何影响海底的气体水合物资源？海底的各类金属矿产资源与微生物的关系如何？这些是深海计划中在处理微生物作用与海底资源能源的关系过程中亟须解决的重大难题。

深海极端环境蕴含着许多重要的微生物资源，它们含有独特的基因和生物酶，是重要的种质资源，虽然已经有了一些研究，但还需要进行全方位的深入探索，以期更好地为工农业生产和生活服务。

3. 深空探测的重大科学难题

1）地球生命的边界

人们对生命的认识均基于地球上可观测的各种生命现象，因此对地外生命的探索也受限于地球生命形式。然而，物理和化学规律的普适性使我们可以将生命的基本要素推至其他天体，即生命的存在需要溶剂（液态水）、能量源和基本构件。其中，液态水的可得性是控制地球生命空间（biospace）的最关键因素。在液态水存在的前提下，环境温度、pH、压力、盐度、辐射梯度等条件相互作用，共同影响营养物质和能量的可获得性，限定了生命可以出现的区域及其边界。在不同环境中，不同参数对微生物的影响程度不同。例如，在地热水中温度是更重要的参数，在土壤环境中 pH 的影响可能占主导，而在干旱环境中水含量起着决定性作用。在不同空间尺度上，主导因素可能也不同。在纳微米尺度上，最重要的控制因素可能是水活度和 pH，而在某些宏观环境中，温度可能更重要。

在地球上，几乎所有存在液态水的地方都有微生物。对极端地质环境（图 5-2）及其微生物的研究，可以确定限制生命条件参数的上下限及其联合效应，从而评价地球生命的边界，为地外生命探索提供依据。当前的研究主要侧重于对温度、pH、压力、盐度、辐射梯度等条件的探讨（表 5-1）。

表 5-1　根据培养极端微生物获得的生命边界值

菌株	域	类型	来源生态系统	温度 /℃	pH	压力 / MPa	盐度 /%	水活度 /a_w
坎德勒甲烷火球菌（*Methanopyrus kandleri*）116	古菌	超嗜热菌	深海热液喷口	90～**122**	6.3～6.6	0.4～40	0.5～4.5	—
大岛嗜酸菌（*Picrophilus oshimae*）KAW2/2	古菌	超嗜酸菌	热泉，喷硫火山口	47～65	**−0.06**～1.8	—	0～20	—
嗜压热球菌（*Thermococcus piezophilus*）CDGS	古菌	嗜压嗜热菌	深海热液喷口	60～95	5.5～9	0.1～**125**	2～6	—
西尔弗曼嗜盐砷酸盐利用杆菌（*Halarsenatibacter silvermanii*）SLAS-1	细菌	嗜盐嗜碱菌	碱湖	28～55	8.7～9.8	—	20～**35**	—
嗜盐游球菌（*Planococcus halocryophilus*）Or1	细菌	嗜盐嗜冷菌	海冰芯	**−15**～37	6～11	—	0～19	—
蛇纹石化单胞菌（*Serpentinomonas* sp.）B1	细菌	嗜碱菌	蛇纹石化体系	18～37	9～**12.5**	—	0～0.5	—
帚状曲霉（*Aspergilluspenicillioides*）	真菌	嗜旱菌	古董木制工艺品	—	—	—	—	**0.585**

注：加粗数据表示极限值。

pH 对微生物群落从纳米到宏观尺度均有重要影响。所有微生物都必须将其细胞质 pH 维持于近中性，才可能具备生存和代谢必备的细胞功能（Krulwich et al.，2011）。地球环境中观察到的 pH 范围为 −3.6［美国加州沙斯塔（Shasta）的铁山（Iron Mountain）］～13.3［波兰赫扎诺夫（Chrzanow）地区的高尔卡（Gorka）湖］，二者都为受矿山废物污染的环境。目前已知微生物可生存的极限 pH 为 −0.06～12.5，前者为分离自热泉的超嗜酸古菌

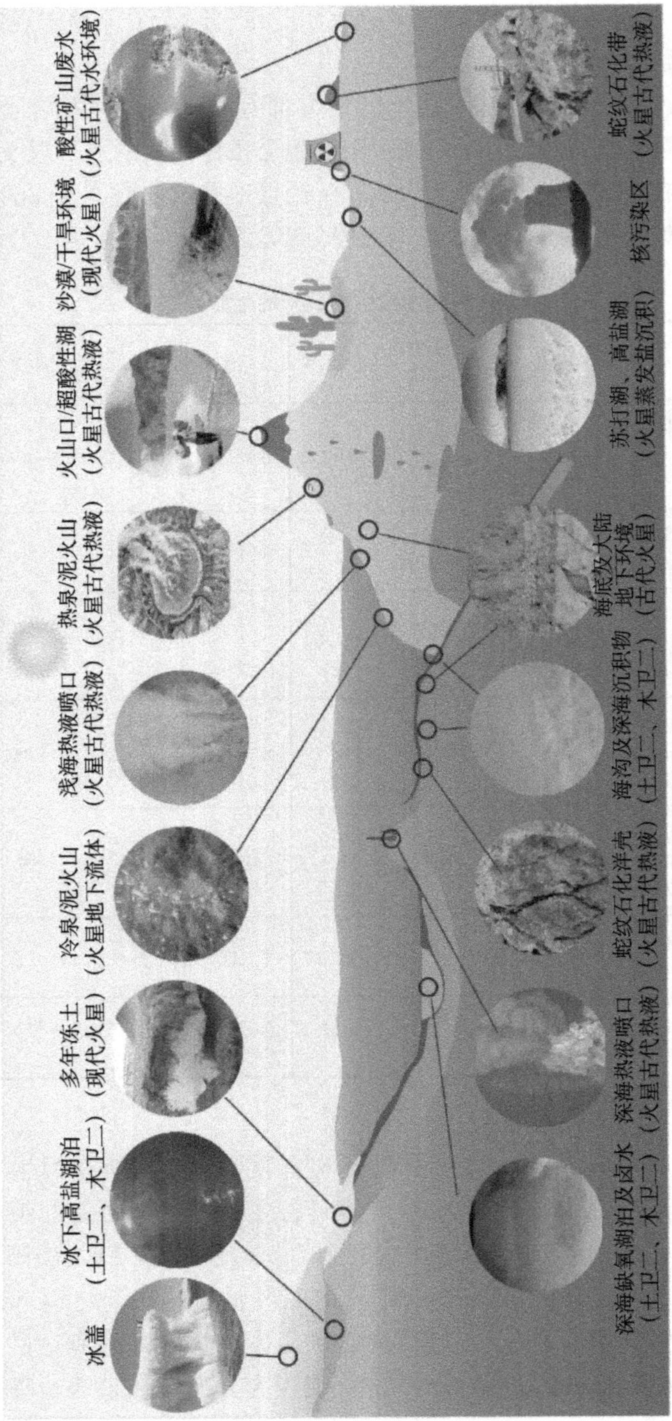

图 5-2 地球代表性极端环境（文后附彩图）

Picrophilus oshimae 和干热土嗜酸菌（*P. torridus*），后者为分离自蛇纹石化系统中的嗜碱细菌 *Serpentinomonas* sp. B1。

地球环境的盐度变化范围很广，海洋平均盐度通常为 3%～4%，热泉中可达 10.5%，而盐碱湖盐度高达 37.1%，在有些盐包裹体中甚至达到 49.7%。可分离培养的微生物可适应的盐度范围为 0～35%。由于盐度及水体离子成分显著影响水活度，因此很多嗜盐生物必须适应低水活度环境。嗜盐菌所需的最低水活度为 $0.611a_w$，但近年来在水活度低至 $0.4a_w$ 的深海超盐度缺氧盆地中发现了微生物群落。另外，盐度的增高可以显著降低水的冰点，从而拓展了寒冷环境下的生命边界条件。

近半个世纪以来，生命的温度边界条件多次被拓展，目前已知生命可以生存的环境温度极限为 −25～130 ℃，而在 −20～122 ℃条件下，微生物可以保持新陈代谢活性。理论研究表明，生命的温度上限可能接近 150 ℃，而下限约为 −40 ℃。随着温度的升高，微生物群落复杂性趋于降低，如在腾冲地热区，随着温度升高（50～90 ℃），土壤微生物逐渐以古菌为主导（Li et al.，2015）。

压力可以影响微生物生长，尤其是在极端温度条件下其影响更加显著。目前已在深海中发现了一些嗜压、耐压微生物，其中嗜热古菌 *Thermococcus piezophilus* 可生存在高达 125 MPa 的压力条件下。某些耐压微生物甚至可短时间内承受 2000 MPa 的极端高压条件。为了适应高压环境，嗜压微生物的细胞膜中含有大量不饱和脂肪酸，从而可增加细胞膜的流动性。海洋中很可能存在大量未发现的嗜压微生物，亟待进一步研究。低压环境主要出现在高山（如珠穆朗玛峰顶大气压为 0.0033 MPa）、太空或近地轨道（10^{-13} ～ 10^{-10} MPa），一些原核生物、真菌和苔藓可以通过孢子形态或形成微生物膜在这些低压环境中存活数月至数年时间。

辐射来源包括紫外线、X 射线、γ 射线和宇宙射线，它们可直接或间接地（如产生活性氧）作用于微生物细胞，导致其 DNA、蛋白质、脂类和 RNA 受到损伤。在地球环境中，极端强辐射环境主要来自受人造核反应堆污染的地区，而地下深部放射性元素的衰变也可产生辐射环境。研究表明，一些微生物可在太空辐射环境中生存数百天（De Vera et al.，2012）。在大氧化事件（约 24 亿年前）之前，地球臭氧层尚未形成，紫外辐射强度可能比现今环境高出几个数量级。早期地球微生物可能通过基因备份冗余、浓缩类细胞核、改进 DNA 修复功能、产生色素等方式来适应极端辐射环境（Merino et al.，2019）。尽管类似的辐射适应现象在古菌、细菌和真核生物中都有发现，但其适应机理的起源和演化还有待确定。

综上，自 1969 年首次发现极端环境微生物以来，地球生命边界就一直在被拓展。目前已经在温度、pH、压力、盐度、辐射梯度等方面对地球生命的边界有了一些认识（图 5-3）。例如，已知生命的温度边界为 −25～130 ℃，而理论预测的温度边界为 −40（代谢速率降至极低水平）～150 ℃（细胞组件变性失稳）（Schulze-Makuch et al.，2017）。地球自然环境的 pH 范围为 0.01～12.5，但被人类活动污染的环境 pH 范围可达 −3.6～13.3，并且在其中观察到了微生物群落。类似地，微生物可存在的压力范围（0.1～125 MPa）也超出了地表生态系统的范围（0.1～112 MPa）。生命的边界是否可以被进一步拓展？答案似乎是肯定的。遗憾的是，尽管目前已经成功培养了一批多重极端微生物，但对它们承受多重环境压力能力的研究还极其缺乏。例如，多重极端条件参数之间如何相互影响、它们会产生什么样的联合效应等科学问题都有待解决。此外，目前还未被培养的微生物在属一级的数量上估计为 7.3×10^{29} 种，它们广泛分布于地下、超咸水环境、海洋沉积物、热泉、热液喷口等地区，极有可能包含更多类型的极端微生物。未来对这些极端微生物的研究将进一步拓展我们对生命边界的认识（Merino et al.，2019）。

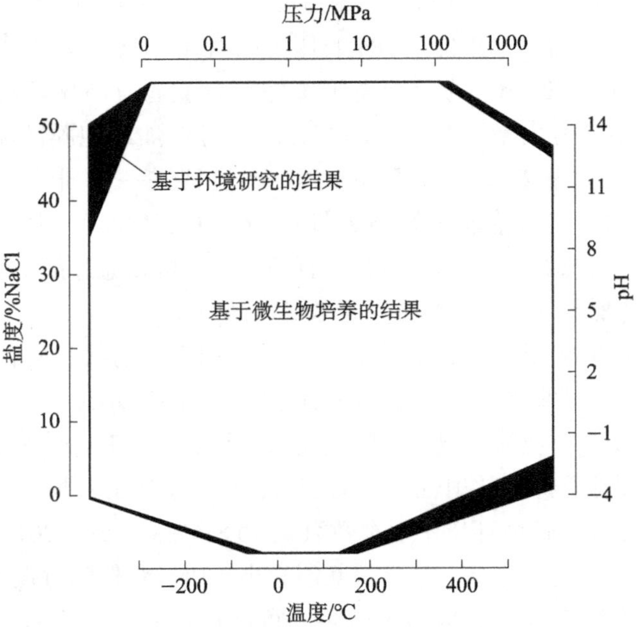

图 5-3 生命的边界条件

资料来源：修改自 Merino et al.（2019）。

2）生命是否可在其他天体上起源、演化和生存及是否存在其他生命形式？

基于对地球生命边界的研究，人们可以对宇宙其他天体的宜居性和可能存在的生命形式进行判别和预测。通过与地球环境类比发现，很多天体具有支持已灭绝或现存生命的潜力。例如，土卫二和木卫二可能具有含盐的地下海洋，火星含有富氯的地下水，嗜盐嗜冷微生物（halo-psychrophile）有能力在这些环境中生存。有些天体如谷神星、木卫二和火星的表面受到强辐射的影响且极端寒冷，很可能不适于生命生存，但在其地下环境中辐射量会大大减少；热液系统通过水的辐射分解或蛇纹石化产生的 H_2 和 CH_4 有可能支持化能自养微生物生存；土卫六地下海洋中的静水压力可能高达 140～800 MPa，远远超出了地球嗜压微生物的边界，然而，已有研究证明，地球微生物能在高达 2000 MPa 压力的环境下短暂存活。生命是否可以在其他天体上起源、演化和生存？哪些天体适合生命的存在？宇宙中的生命有没有可能突破地球生命的边界？宇宙中是否存在碳基生命之外的其他生命？这些都是深空探测面临的重大科学难题。

由于硅和碳属于元素周期表内的同一族，公众对硅基地外生命也很感兴趣。单从硅和碳在类地行星上的物质总量来讲，硅基生命可以像碳基生命一样普遍。但元素硅与氧结合的键能比硅与氢、氮、碳和硅本身形成单键时的键能要高很多，因此硅在有氧存在的室温环境中容易形成稳定分子二氧化硅。与此对比，碳与氢、氮、碳和氧元素在室温环境形成单键和双键的键能彼此相差不大，易于生命所依赖的化学活动的进行。此外碳══碳双键键能是其单键键能的 2 倍，三键键能是单键键能的 3 倍，有利于形成更加复杂的碳基分子；与此对比，硅══硅双键和三键在室温环境都非常不稳定，不利于形成复杂的硅基分子。从这个角度看，在类似地球的系外行星上自然出现硅基生命的可能性不大。

随着计算机技术的发展，人工智能将会越来越先进，可以想象有一天人类将可能创造出智能生命。这种生命形式是否在系外行星上存在，如何观测？这类问题目前还研究得较少。

3）太空环境如何影响地球生命？

太空环境非常极端，具有微重力（$<10^{-6}\ g$）、高真空（$<10^{-14}$ Pa）、高辐射（太阳辐射和银河宇宙辐射）、剧烈的温度变化、原子氧侵蚀等复杂环境条件。由于生命的起源过程不可能在实验室完整复现，因此研究微生物在太空环境下的存活、耐受和适应机制是开展天体生物学研究的主要方法。目前关于太空环境如何影响生命的实验研究主要集中在近地轨道空间（200～

1000 km 高度），对深空环境如何影响生命的研究还处在初级阶段。

太空的独特环境对微生物的影响涉及基因、生物组件、细胞功能、遗传多样性和代谢产物等多个层面，表现为生长速度加快、迟缓期缩短、毒力增加、抗菌药抗性增强、更易形成生物膜等行为。目前，有待解决的科学问题主要包括：太空环境下微生物生长状况的长期变化规律如何？太空环境下微生物群落的物种多样性、适应机理、动力学特征以及维持机制如何？微生物响应太空环境的核心调控因子与生物学特征变化的相互关联及其相互作用机制如何？太空环境下有益的植物-微生物如何相互作用，如何建立抑制植物致病菌的生物方法？太空环境下微生物降解复杂有机物的特征及其机制如何？

我国已于 2020 年 7 月 23 日发射了“天问一号”火星探测器，2022 年前后将完成中国载人空间站建造。在这些深空探测任务中，太空环境如何影响生命是一个重要的科学问题，具有重要的科学意义和应用价值。在科学研究方面，生命能否在太空中生存可以帮助回答地球生命起源和生命星际传播等科学问题。此外，太空环境对生命有何种影响的研究还有助于更好地评估宇宙生命的可能存在形式和分布情况等。在应用研究方面，太空环境对生命特别是高等生物的影响可以用来指导载人空间站、未来载人深空探测以及移民其他星球中与宇航员健康相关的研究和工作，还将促进空间育种和空间合成生物学的发展与进步，将为载人深空探测、地外基地建设乃至未来的行星开发等提供可能的技术支撑。

（三）生物安全及其对策

1. 极端地质环境病毒对人类的潜在危险

病毒广泛分布于深海及深部生物圈环境中，对生态系统具有非常重要的功能。虽然病毒的活动完全依赖于寄生生活，但是病毒与其微生物宿主之间存在着多种复杂的相互作用关系。已有研究表明，除了作为微生物裂解的主要因素，病毒与其微生物宿主之间还存在多种互利互惠的关系，如携带辅助代谢基因、调节宿主的生理代谢等。根据已有的研究结果及深部环境的特点，我们提出“病毒—诱导抗氧化—压力适应”的科学假说，即在深部环境中，病毒的侵染会诱导宿主的抗氧化反应，从而提高宿主对包括压力在内的多种极端环境适应的能力，最终协助宿主更好地适应深部极端环境并在其中生存。考虑到病毒在深部环境中的普遍性，我们认为这种策略可能是深部微生物适应的普遍机制。当然，这一科学假说还需要更多实验的验证。

病毒对国家生物安全存在巨大的潜在的挑战，可能会导致重大疫情的暴发、生化战争，甚至能在短时间内造成全球的流行病疫情等。例如，深地环境的开发利用是否像打开“潘多拉魔盒”一样，使得极端地质环境的病毒等微生物得到大范围散播和繁殖？然而，人们对极端地质环境病毒的空间分布、丰度及其适应机制很不清楚，几乎是空白，需要加强这方面的突破，包括它的宿主、在各类极端环境的分布规律及其受控因素。开展极端地质环境病毒的研究尤其重要，因为它可能对人类构成潜在的巨大威胁。

2. 极端地质环境微生物的有利生物安全因素

极端地质环境的病毒在可能对人类构成潜在威胁的同时，一些微生物也为生物安全提供了极其难得的机遇，可以充分利用这些极端地质环境微生物的种质资源，开展医药、各种生物工程研究。人类开发利用矿井、洞穴等地下空间从事医疗活动的历史久远，利用溶洞 / 地下环境治疗疾病和进行医学研究的历史可追溯到 100 多年前，相关的学科被称为洞穴医疗或气溶胶盐疗法。洞穴医疗中起治疗作用的主要是矿井或洞穴环境中的空气、射线、矿物、微生物等。但是这些空间多处于地下浅表环境，岩层覆盖较浅。除了岩性化学、相对封闭的空间等因素，其余因素因人为干预已经与地面环境差别不大。随着人类向地下空间的进发，掘进的深度会越来越深，与人们习以为常的地面环境和浅层地下环境相比，深地空间是明显不同的。在深地环境下有增重效应，同时具有“三无”（无宇宙射线、无太阳光、无氧气）与“三高”（高温、高压、高湿）等环境特征。这些因素对地下生命体繁衍进化及地表生物的生理、病理和心理的影响程度及其规律，目前相关研究尚接近空白，更谈不上形成相关学科。如何规避不利因素，利用有利因素，创造良好的深地宜居空间环境，是目前亟待解决的世界性公共卫生健康战略科学问题。

3. 极端地质环境生物安全的总体对策

从上面分析可以看出，极端地质环境微生物有利有弊，到底是打开还是关闭地下空间开发的“潘多拉魔盒”，是一个值得深入研究的问题。但是，人类与微生物的关系和人类与自然的关系一样，需要和谐共处。因为人类有人类的空间范围，微生物有微生物的领地。疫病的暴发是因为人类有意（如生化武器）或无意（如猎食欲望）地侵入微生物（包括病毒）的生态圈，打通了它们的传播通道。人类若不克制自己，随意打破人与自然的平衡，瘟疫将永无休止。

因此，人类明智的做法是切断疫源微生物通向人类社会的传播途径，让它们在自己的自然领地中安静地繁衍。人走人的“阳关道”，它们走它们的“独木桥”。不是谁消灭谁，而是和谐共处，只有这样，方可长治久安。

第二节　学科发展总体思路与发展目标

一、总体思路

针对我国的实际情况以及目前的优势和薄弱环节，按照“有限目标、突出重点”的总体要求，未来5～10年需要通过加强以下三方面的工作来开展我国极端地质环境微生物的研究，包括与重大工程计划结合突破关键技术、加强不同学科之间的交叉融合、不同极端地质环境的对比和联合研究等。

（一）与深地、深海和深空等重大科学工程计划相结合，突破“卡脖子”技术

我国正处于科技大发展的黄金机遇期，一个关键标志是围绕一些大科学难题实施了一些大的工程计划。我国的深海计划、深地计划和深空计划正在大力推进，不仅发布了许多国家重点专项计划（如国家重点研发计划），同时实施了许多工程计划（如海底观测计划等），这为开展极端地质环境微生物的研究创造了极其难得的机遇，需要及时抓住这些机会。因此，需要我国的科学工作者积极参与国家大科学工程的前期论证和设计以及后续的研究。我国的地质微生物学家已经参与了正在酝酿的深空计划的科学论证和设计工作，也积极参与了深地计划的论证，深海观测计划已经布设了一些海底微生物观测站，特别是我国地质学家正积极推进利用一系列的超深钻开展野外原位观测与模拟研究。一个科学前沿研究和国家重大工程计划相衔接的时代正向我们走来，将带来有关极端地质环境微生物研究前所未有的重大突破。

极端地质环境微生物的突破需要技术上的革新。如前所述，在钻探和保真取样技术、野外原位观测和模拟技术以及实验室分析技术方面，我国的“卡脖子”技术是在前两项，因此在利用重大工程计划的同时，我国需要在微生物的钻探和保真取样技术、野外原位观测和模拟技术方面开展大量的前

期设计和研究工作，以形成具有自主知识产权的技术体系。这一点对我国来说尤其重要。

1. 海洋极端环境涉及的深海技术

1）大洋钻探技术

大洋钻探是获得深海极端环境微生物样品的一项关键技术。大洋钻探需要钻探船，如美国的“决心号”（JOIDES）和日本的“地球号”。“决心号”是一艘装备动态定位系统的钻探船，可以在75～6000 m水深范围内在海底钻深2000余米。“地球号”是科学界第一艘装备立管的科学钻探船，能够在2500 m水深的海底钻深达7000余米。欧洲的海洋科考船仅具备海底钻探数十米的功能。这些科学钻探船具备在深海和大洋进行多学科综合研究探测（包括深海钻探、海底井下观测等）与样品采集能力，并兼具科学研究和工业应用的能力。

我国发展深海科学，必须在现有国力的基础上建造具有中国特色的第三代大洋钻探船。这种船既要避免像日本船那样规模过大，又要吸收美国、日本、欧盟国家的先进技术而成为高效率、高水平的新设施，成为IODP新阶段与美国、日本、欧盟并驱的第四国，这必将带动我国相关科学、技术和产业的发展，实现国家目标。我国的第三代大洋钻探船能够在不同海区和地质环境下实施钻探，大小约2.2万吨级，介于美国、日本的之间，钻探设备应具有国际最新水平，实现非立管泥浆返回钻探系统、海底井下观测建设等多种功能。

2）保真取样技术

除了钻探，还可以在一些深海极端环境中直接获取样品，特别是热液系统的样品。深海热液保真采样系统适用于科考船或水下运载器，可实现对深海热液的采样，研究热液的物理、化学及微生物梯度变化特性。同时，可实现对热液的分时采样，研究热液活动的变化规律，揭示热液形成机理，探索深海生态环境变化规律。深海保真采样系统的成功研制促进了我国深海科学考察研究的发展，为深海资源勘探和科学研究提供了先进的技术装备支撑。该装备研制至今已有多种系列的深海保压采样器，均获得多次成功的应用。现主要使用单管气密采样器（CGT）和多管序列气密采样器（外国专家称之为Six Shooter）两种型号。目前，深海水体保真采样器的最大作业水深一般为4000～7000 m，最高采样温度达400 ℃，单管气密采样器采样容积为200 mL以内。

3）原位观测和模拟培养技术

人们还开发了一些用于热液中微生物采集和原位培养采样的装置，比如生物采样器（bio-sampler）、闭合系统（seamless system）和新型深海远程实验装置（LAREDO）等热液采样及原位微生物培养装置。浙江大学所研制的深海水体采样器多次参与了国内外深海热液采样工作。

许多海洋和地球科学问题需要通过长期原位实验、取样和观测而得到答案，这就促进了海底观测站的诞生。CORK 海底钻孔井控与观测装置，是海底观测站的一个典型例子与重要组成部分。最初 CORK 的结构和功能均与红酒瓶的软木塞有某种程度的相似性。CO 是封闭钻孔，意即对钻孔进行控制，阻止洋壳地层与海水间发生流体交换；RK 是（重新）进入钻孔，通过再入圆锥筒均可再入钻孔并安装仪器设备进行科学实验和观测。CORK 原位观测系统已经在海底深部生物圈探测中发挥了重要的作用，取得了重要发现。未来的发展是建立电缆连接式 CORK 观测系统网络（Cabled CORKs，CCOs）。这种观测系统网络通过光电连接器和电缆两种控制方式实现 CORK 与 CORK 以及 CORK 与地面站之间的实时联系与数据传输，从而帮助人们获得连续不间断的深地时间序列数据，相当于固定在深地的实验室。它不但连续产出实时数据，还可以进行交互式原位实验。深地 CORK 观测系统网络的建立，为科学家对深地生物圈进行数秒到数十年时间尺度的地质学、地热学、水文学、化学、微生物学与生物地球化学过程基础研究提供了手段。

2. 深地微生物所涉及的深地技术

1）深地钻探与掘进技术

深地生物圈研究需要获取深部地下的地质样品，而钻井取样是其中的关键。钻头在钻探取样过程中与岩石特别是密度较高的大理岩等接触时会产生大量的热量，当这些热量传递至样品后会对其中可能存在的微生物造成致命损伤，因此，可以考虑采用低速钻探工艺，配合高效热交换系统，将产生的热量快速传递至采样设备的其他部位。另一种思路是采用大口径钻头，较大的采样筒口径可以保证热量不会在短时间内传递至岩心中心部位而破坏微生物；较大直径的岩心样品还可以很好地保护位于中心部位的微生物，待岩心样品取出后再在无菌环境中进行样品表面灭菌等后续处理，可以有效避免外源污染的问题。相比钻井技术，掘进技术的最大优势是可以提供深部直接取样的机会。目前世界上的采矿深度已达 4350 m，借助采矿工程的深井，生物学家可以直达深部地下空间，在深达 4000 m 以下的地层中直接观察、取样，对深部生物圈研究具有重要意义。

2）保真取样技术

深地岩层极端环境微生物的生存环境与常规环境差异巨大。当极端环境微生物脱离原始生境后，细胞可能会立即发生破裂，无法正常生长繁殖甚至死亡。在采样过程中维持原有的极端和无菌环境的保真取样技术是进行后续微生物生长、代谢过程研究的基础。这要求样品在采样过程中受到全封闭式采样器的保护，原始物理和化学特征尽可能小地受到外界环境干扰。同时，保真采样系统应具有良好的温度控制能力，确保样品中可能含有的微量微生物不被破坏。更重要的是，全封闭式采样器能够确保样品在采集过程中不受外界微生物污染。

3）原位观测和模拟培养技术

在原位观测方面，目前 CORK 在陆地上基本上还没有使用，但是与海洋一样，应该可以监测与微生物代谢活动相关的指标（温度、Eh、pH、溶解氧、Fe^{2+}、NH_4^+、SO_4^{2-}、NO_3^-等），来推测深地微生物的活动。

在深地环境原位观测微生物有直接和间接的方法。间接的方法主要利用光纤传感器等设备检测由微生物活动造成的流体物理化学性质的变化。近年来开发的一种深紫外（<250 nm）—激光诱发的荧光光谱则可以实现地下原位环境微生物的无损直接观测，可以在从微米到几十厘米的岩石表面拍到单个细胞的图像，这一方法近来被用于深海玄武岩井孔微生物的原位成像研究。

可以把循环式渗透压取样器改装成循环式渗透压培养系统（flow-through osmo colonization system，FLOCS），安装在放有 CORK 装置的钻孔中，用钻孔所处地层位置流体内的微生物作为培养系统中的菌种进行实验。另外，深地流体因含有丰富的营养成分而成为微生物活动的“热点区”，可以在地下流体富集的地区开展地下原位的微生物实验。瑞典的阿斯波坚硬岩石实验室（Äspö Hard Rock Laboratory）已经有成功的经验，基本思路是从已存井孔的井壁打一口斜井（borehole），穿过含有流体的裂隙，斜井与裂隙的交叉部位用橡皮塞将一段含有流体的裂隙隔离，这样用抽水泵可以将地下原位的流体直接抽入培养容器中进行原位培养或进行分子微生物分析，也可以在一口井中加各种营养成分，再让这些营养回流到裂隙，与地下岩石流体反应以后，通过另外一口井抽水造成的压力差，使得这些化学物质迁移到另外一口井中，在这一过程中，观察微生物对这些加入组分的改造。另外，也可以在一口井中加入示踪微生物，在旁边井位抽上来，以此观察微生物的地下迁移。

3. 天体生物学所涉及的深空技术

天体生物学的学科特点是与深空探测任务紧密结合。深空探测在发展前期通常以工程任务为主，但人类并不满足于可以走出地球，而是更想了解地外空间，因此与科学研究的结合甚至以重大科学问题为导向来制定探测任务是深空探测发展的必然趋势。天体生物学家有责任和义务提出重大科学问题，并以此为牵引，规划探测任务并最终共同领导实施深空探测任务。深空探测任务的复杂性和技术的先进性决定了工程师和科学家必须进行深入的交流，合作完成任务。一方面，天体生物学家需要了解航天和深空探测的基本背景知识，这需要在人才培养方面下大力气；另一方面，需要研究人员与工程师进行顺畅的沟通交流，这需要在管理和保障体系方面做足工作。在这方面可以参考欧美的成功模式，成立固定的专家组和工程组，建立例会制度，鼓励科学家深入参与工程任务，通过纪要或其他形式来保护科学家的原创思想。

美国和欧洲天体生物学的成功经验表明，通过顶层设计制定的路线图和战略规划有利于明确科学问题和优先发展方向，推进天体生物学研究和探测任务的顺利实施。例如，NASA 分别于 1998 年（Morrison，2001）、2003 年（Des Marais et al.，2003）和 2008 年（Des Marais et al.，2008）发布了三版天体生物学发展路线图，并于 2015 年（Scharf et al.，2015）和 2019 年（An Astrobiology Strategy for the Search for Life in the Universe，2019）制定了两版天体生物学的战略规划，为美国天体生物学未来 20 年的发展指明了方向。ESA 于 2016 年发布了天体生物学路线图（Horneck et al.，2016），并于 2019 年成立了欧洲天体生物学研究所，总体规划欧洲未来开展的天体生物学研究和主要发展方向。在我国未来的载人空间站建设和“登月探火”等深空探测计划中，天体生物学是一项重要的基础科学研究内容。这需要我国的天体生物学研究者通过深入交流和研讨，尽快制定符合我国国情的天体生物学发展路线图和战略规划，明确优先发展方向，全面提升我国在天体生物学领域的研究能力和国际影响力。涉及的主要技术如下。

1）轨道器探测

轨道器探测是人类探索地外生命最直接有效的方法之一，其传回的图像和其他数据可以帮助人们更加直观地认识地球以外的其他星球。对地外天体的地质环境、大气演化过程以及圈层结构的分析，有助于寻找和判断地外生命产生、维持或灭绝的可能。

例如，NASA 于 1969 年发射的“水手 6 号”第一次成功飞越火星，传回了 75 张火星的照片，让人们第一次看到地球以外的模糊流水作用遗迹。1996 年发射的“火星全球探勘者号”（Mars Global Surveyor）搭载了火星轨道激光测高仪（MOLA）和火星轨道相机（MOC）等仪器，传回的图像上可见火星表面复杂的沟渠和湖网，甚至能追溯其源头和最终流向。1997 年发射的“卡西尼”（Cassini）环绕土星进行空间探测，拍摄了第一幅土卫六的照片，由此分析，土卫六上有疏松的冰、有机物以及冰冻的碳氢化合物，深部存在全球性的地下海洋等。ESA 于 2003 年发射的“火星快车”（Mars Express）轨道器，首次测量并分析了大气中水、臭氧和尘埃的分布，并探测到了痕量可能是微生物来源的甲烷。不同于其他的着陆器，2011 年发射的“好奇号”着陆器携带有收集和处理岩石、土壤样品并进行初步分析的仪器。

2）原位探测技术

由于太空环境极为恶劣、太空旅程耗时太长，天体生物学的原位探测往往依赖于探测器上搭载的相关载荷（与机器人无关）：从 1976 年的“海盗号”、2003 年的“勇气号”“机遇号”、2008 年的“凤凰号”、2011 年的“好奇号”到 2018 年的“火星生命探测任务”（ExoMars）……这些探测器都经历着原地着陆器就位探测——火星车巡视探测的发展阶段。由于火星被认为和地球在早期有着相似的演化阶段，目前行星探测任务又能够企及，所以目前火星是轨道器和原位探测技术的首要应用对象。

例如，“海盗号”重启了寻找火星生命的探测活动，主要目的是搜寻生物标志物和观测火星的气象、地震与地磁，利用生物仪器载荷探测火星表面及亚表面数厘米的土壤中是否存在与地球上类似的微生物。“海盗号”着陆器携带了生物实验舱，其中包括 3 组微生物实验仪器：热解释放仪（pyrolytic release，PR）、标记释放仪（labeled release，LR）和气体交换仪（gas exchange，GEX），目的是通过对火星土壤进行类似地球上的生物培养实验，检测火星土壤中是否存在类似生物光合作用的过程，以及是否存在被 ^{14}C 标记的 CO_2/CO 的光合作用或化学固定产物，用以测定是否存在“火星生命”吸收 ^{14}C 标记的营养液和水蒸气，并发生新陈代谢和交换气体过程（Mckay，1997）。当时的实验结果显示没有生命信号，不过后续的多种分析使该实验结果存在争议。

3）地外样品研究技术

地外样品可分为两大类，即自然坠落到地球上的陨石和人类采用行星探测技术带回到地球的样品。前者由于在行星际滞留的时间极长，研究思路往往侧重于检验是否存在遗迹化石；后者则被认为是最可能存在地外生命的，可对其开展现存或曾经存在生命的各种研究。由于存在疑似火星微生物，ALH84001 是天体生物学研究中最著名的火星陨石之一。

4）地外智慧生命探测技术

地外智慧生命搜寻（the Search for Extraterrestrial Intelligence，SETI），是一个致力于搜索宇宙间智慧生命的科学组织，它试图通过分析射电望远镜采集到的地外无线电信号，主动搜寻外星智慧生命存在的证据。其中一些小组致力于搜寻宇宙中的数十亿种电磁波，辨别地外智慧生命发出的无线电信号；其余小组则观测恒星发出的光脉冲以进行搜索。SETI 针对地外智慧生命进行了主动式探索，但由于存在一系列原因（如宇宙空间极其浩瀚，电磁频段非常宽而导致我们选择的困难，地外生命形式可能与地球上的相差甚大，智慧生命科技太不发达而无法与我们通信，或智慧生命科技极其先进却对我们不屑一顾等），该计划自实施至今，没有接收到任何地外生命发出的通信信号，但随着天文望远镜及信号数据分析能力的不断增强，我们也期望 SETI 能带来好消息。

5）地外生命的相似性研究技术

相似性研究是指将地外天体上地质的、环境的和假定生物的条件放到地球上来研究。目前三个关键主题在相似性研究中占有主导地位：①比较行星地质，包括过程研究和相似性物质的确定；②宇宙生物学；③探测科学，包括仪器的测试和开发、宇航员的训练及行星探测相关活动。

由于天体生物学的研究对象尚不可及，目前主要的研究力量聚焦于在地球上类似地外恶劣“生态环境”地区的相似性研究。从智利的阿塔卡马沙漠到西班牙的力拓河，从南极的干谷到北极的斯瓦尔巴特群岛，还有一些科学家将他们的“漫游者”抛到了遍地乱石的沙漠，行星科学家一直在地球上搜索地球上的“地外”环境。

4. 微生物研究的新技术新方法

最近 10 年来，地质微生物新技术、新方法层出不穷，日新月异，分子克隆、高通量测序、微生物基因芯片都可以用来研究微生物在深地环境的种类多样性，而功能基因芯片可以快速检测特定环境条件下功能基因的丰度，全基因组测序、单细胞基因组学、宏基因组等现代生物学方法可以发现新的功能基因与代谢途径，宏转录组学可以检测功能基因与代谢途径的表达水平，蛋白质组学可以检测介导元素循环的各种蛋白质与酶的表达和活性。

同位素技术与分子微生物相结合发展起来的示踪技术，可以在特定环境条件下，检测某一类微生物的活性与代谢途径，譬如DNA、RNA以及脂类的稳定同位素探针技术（SIP），通过标定碳源或氮源可以用来确定在一个复杂的群落结构中，某些特定功能微生物的代谢活性与途径。同位素标定技术与原位成像技术相结合，则可以在原位条件下确定某一类微生物的特定功能及其与周围矿物岩石介质的相互作用，其中最著名的技术是纳米-二次离子质谱仪成像技术（nano-SIMS），如果把这项技术与微生物荧光杂交技术结合起来，那么可以在原位条件下，确定在一个微生物群落里面，哪一类微生物吸收了同位素标定的底物以及吸收的速度有多快，这对评估低生物量、低活性的深地微生物至关重要。

以上描述的这些方法都属于微观尺度，要在野外尺度上分析微生物活动对地质过程的影响，就要借助于大尺度的地球物理方法，生物地球物理就是在这样的背景下诞生的一门新兴学科。生物地球物理是地质微生物学与地球物理的交叉学科，是基于传统地球物理的延伸，除了检测地质载体的物理化学性质外，主要探测地下微生物的生长、微生物与矿物岩石的相互作用，因为这些过程会在地球物理信号上记录下来。譬如，地下水污染物的降解、生物产生的气体会影响水溶液的电导率、介电常数，因此可以通过电阻率法、探地雷达检测；生物膜的形成与生物矿化，特别是磁性矿物的沉淀会产生界面极化效应、磁化率的变化，因此可以通过激发极化法（简称激电法）/频谱激电法/介电法检测；生物矿化与生物气的产生会影响弹性波速，因此可以通过地震波检测。将地球物理信号反演从而推测微生物过程需要强有力的实验室验证，只有这样，反演得到的信号才会有实际意义。

（二）加强不同学科之间的大交叉和大联合，打通学科壁垒，打破零散的研究局面

在技术突破的前提下，需要加强不同学科之间的交叉联合。我国在极端地质环境微生物方面的一个现状是研究力量分散，学术共同体没有很好地形成，这需要通过不同学科之间的交叉才能实现。实际上，极端地质环境微生物的研究涉及地质学、大气科学、水文地质学、微生物学、生物化学、生态学、地球化学等学科，开展多学科的联合攻关是深入了解极端地质环境微生物的必经之路。地质条件和水文地球化学特征是微生物赖以生存的基础，是无机条件，而生物之间互为环境条件，制约生物之间的相互作用，维持生态系统的平衡，这是生物赖以生存的有机环境。因此，极端地质环境微生物的

研究需要多学科人员的共同协作。

一方面，需要加强科学研究与工程技术的交叉联合，特别是加强与那些“卡脖子”技术的研究人员的联合，包括保真取样技术和野外现场观测与模拟技术。当前，我国投入巨大经费引进了许多国际一流的实验设备，但更缺乏的是自主研制的设备。极端地质环境微生物的保真取样技术和现场观测技术都需要自主研发才能实现，很少有商业化的成套设备可以不加改进就能利用的，这既是挑战，也是我们发展极端地质环境微生物研究面临的重大机遇。我国需要加大这方面的经费投入，促进科学家与技术专家进行不自觉的联合，从而带动我国自主研发的创新和突破，形成原始创新的良好氛围。特别是，深部生物圈的原位监测和保真采样技术至关重要，需要科学家与有关工程技术人员紧密合作，突破采样技术瓶颈的制约，保障深地样品的保真采样，防止污染，获取高质量的深地样品。

另一方面，需要加强地球科学与生命科学乃至与化学等自然学科之间的交叉。这三大自然学科在研究方法和理论体系上都存在极大的差异，地球科学比较注重野外现场和宏观尺度的研究，生命科学和化学多注重实验室和微观尺度的工作。但正是这些自然学科之间的差异，才形成了它们之间的互补性，为交叉联合奠定了基础。实际上，一些技术方法体系可以相互衔接。例如，分子生物学的基因工作可以与分子地球生物学的脂类工作结合，探索脂类分子的合成途径，一些化学领域广泛应用的分析技术可以引入生物地球化学领域，开展地质体中各类有机分子和同位素的研究。又如，开展极端地质环境病毒的研究，不仅仅是生命科学家的工作，更是地球科学家的责任。

生命科学的大量工作都是在实验室完成的，而地球科学特别重视野外现场工作，要实现地球科学与生命科学的交叉融合，最关键的一点是模拟实验研究，具体包括如下工作。

1. 极端环境微生物地质过程的模拟研究

加强极端环境微生物学的模拟研究是深入理解微生物与环境相互作用的机理和过程的重要手段。微生物代谢过程是微生物与环境相互作用的基本途径。因此，对微生物地质作用过程和机制的认识，首先需要从分子层面和细胞层面针对典型微生物代谢功能开展深入研究。通过实验室微生物纯培养及自然环境微生物的富集培养，构建相关的微生物种质资源库和环境参数及微生物代谢参数的大数据库，发现和揭示微生物代谢的主要途径及关键基因，建立微生物特征基因的遗传操作系统，解析具有显著生物地质作用酶的结构和分子作用机理，以及影响其基因表达和酶活性的主要环境因子等。在实验室搭建能够模拟不同古海洋水化学环境模拟平台，查明不同环境条件下典型微生物代谢过程的具体参数，为定量研究深时微生物地质作用提供所需的参数和边界条件。

2. 微宇宙实验研究及极端环境的现场观测与研究

由于微生物通常以大量多样性丰富且功能各异的类群所构成的复杂微生物群落进行代谢活动，并能够形成相互作用紧密的代谢网络，故需要深入研究在生态层面上微生物如何影响地球化学元素循环的过程，从群体的角度揭示生物地质过程，以及如何作用于地球表层环境。因此，一方面，需要加大对自然环境中微生物群落及相应理论的探索研究；另一方面，可以通过人工模拟自然环境，利用微宇宙技术（microcosmic technique）对不同条件下微生物群落的动态过程及相应的环境变化进行监测。两者相互结合，深入揭示微生物之间的相互作用关系及环境效应，建立各种典型环境下的微生物生态学模型。

3. 成岩模拟研究，构建地质微生物地质 / 地化标识

受微生物自身的局限性（个体微小、难以形成化石及鉴定困难），如何识别深时地质微生物是深时地质微生物地质过程研究中的一个瓶颈问题，但未来取得突破的机会和提升空间很大。第一，基于现代地质微生物过程的研究，构建微生物功能群不同层面的地质 / 地化标识，如微生物代谢过程的同位素分馏特征、微生物代谢过程所形成的特殊产物，包括生物成因的矿物；第二，从微生物生化组成角度入手，建立典型微生物的脂类生物标志物及碳同位素组成特征，结合埋藏学实验，构建构型稳定的生物标志物。

除了以上的模拟研究，深时极端地质环境微生物学的发展还需要加强与现代分析技术方法的结合，从空间和时间两个层面提高分辨率。例如，与二次离子质谱相关技术的结合，可以实现形貌和化学成分的结合，以及原位同位素和脂类生物标志物组成分析的结合；与旋回地层和高精度年代学的结合，可以探讨环境因子和生物因子的变化速率，进而为定量化的模型研究提供更多参数。

（三）实现不同极端地质环境的对比和联合研究，以形成带动整体突破的一体化工作

在我国，针对各类极端地质环境都开展了点上的微生物工作，但没有进行很好的对比研究，没有形成一体化的系统性研究，导致各类环境的研究难以衔接，没有达到相互借鉴、形成合力的目的。深海极端环境与深地环境的结合可以同时了解微生物对温度、压力等极端环境的适应能力及其差异，从而形成规律性认识。陆地地表环境和深海海底的两套热泉系统的联合研究，可以阐述在同样的温度条件下微生物的适应策略，了解两套系统在适应机制上的异同点。

其中，深时极端地质环境微生物的研究是关键，它既是连接现代极端环境微生物的桥梁，也是连接现代生物学与天体生物学的桥梁。要进一步加强深时极端地质环境微生物学与现代极端环境微生物学、天体生物学的交叉融合，深入利用现代极端环境微生物学的技术方法手段，并吸收其研究成果，深入揭示深时极端地质环境微生物的地质过程和环境效应，进而服务于地球系统科学和天体生物学。

1. 从对多种环境因子的共适应机制入手，开展微生物的分离培养

微生物具有普遍适应性策略来响应多种不同的环境因子胁迫。这里以许多极端地质环境存在的高压微生物的分离培养为例，通过协同控制多重环境因子，提高环境中的能量和降低氧化还原度，能够帮助微生物应对高压，进而高效富集、分离多样的嗜压微生物。这一培养策略要求建立稳定的高压流动体系，同时独立控制多重环境因子，以维持稳定的能量输入和稳定的氧化还原度。上海交通大学深部生命国际研究中心肖湘教授团队攻克了诸多工程难题，例如，通过低流量高压恒流泵精确控制物质输入（对应底物种类和浓度），通过机械先导式背压阀精确维持系统压力（对应水深），通过增加高压配料单元制备高压下的气体饱和溶液（对应甲烷、氢气等在深海的喷发），通过双阀取样口的设计实现不泄压取样检测，通过电势能和化学势能的同时调节步进式控制环境中的氧化还原梯度等，实现了在实验室模拟开放的高压深海环境，成功研制的0.1～200 MPa、0～200 ℃的静置压力培养装置和0.1～40 MPa、0～150 ℃气液混合高压连续流动培养系统，使我国在压力培养装置上居于国际领先地位。目前从事嗜压微生物分离工作的研究人员仍非常缺乏，深海高压环境还有大量的嗜压微生物资源有待我们发掘和研究。

除取样、模拟等工程技术难度外，制约对高压下微生物功能群理解的关键还是可培养嗜压菌资源的匮乏。据不完全统计，全球曾经获得培养的嗜压细菌 / 古菌有 400～500 株，但得以保存下来并正式发表的不到 100 株。总结现有的研究发现，我们推论高压对微生物群落的塑造可以是同一营养类别微生物的亚类水平，并决定其亚类分布和发挥功能的水深范围，该推论为认识深水区至深渊区碳氮储库的形成与转化提供了理论支持。

嗜压菌的分离由于依赖深海样品和特殊的压力培养装备，工作难度较大，研究进展较慢。目前世界上已分离的嗜压菌共有 78 株，其中有 10 株嗜压菌是由我国科学家分离的，共发表 8 篇 SCI 论文。其中嗜冷希瓦氏菌（*Shewanella psychrophila*）WP2 和耐压希瓦氏菌（*Shewanella piezotolerans*）WP3 于 2007 年分离自西太平洋，标志着国内嗜压菌分离培养工作的开端，并且成功在这两株嗜压菌中构建了遗传操作系统，后续开展了一系列与压力适应相关的工作。*Pyrococcus yayanosii* CH1 于 2009 年分离自大西洋中脊的热液口，是世界上第一株获得纯培养的严格嗜压古菌，该菌株已经成功获得常压突变株，并构建了成熟的遗传操作系统。*Thermococcus eurythermalis* A501 于 2015 年分离自瓜伊马斯盆地的热液口，是一株条件嗜压菌，嗜热祝融菌（*Zhurongicum thermophilus*）3DAC 于 2018 年分离自东太平洋中脊的热液口，是目前已知嗜压细菌中生长温度最高的菌株。上海海洋大学深渊科学技术研究中心的方家松教授团队分离了 3 株嗜压菌：沉积物盐厂盐单胞菌（*Salinimonas sediminis*）N102 在 2018 年分离自新不列颠海沟的沉积物，生长压力范围是 0.1～70 MPa，最适压力为 10 MPa；枯草芽孢杆菌（*Bacillus subtilis*）29R7-12 和泛酸枝芽孢杆菌（*Virgibacillus pantothenticus*）19R1-5 于 2017 年分离自西北太平洋的沉积物。自然资源部第三海洋研究所海洋生物遗传资源重点实验室的邵宗泽研究员团队分离了两株嗜压菌：发酵厌氧杆菌（*Anoxybacter fermentans*）DY22613 于 2015 年分离自东太平洋中脊的热液口，生长压力范围是 0.1～55 MPa，最适压力为 20 MPa；太平洋古老球菌（*Palaeococcus pacifcus*）DY20341 于 2013 年分离自东太平洋中脊的热液口，生长压力范围是 0.1～80 MPa，最适压力为 30 MPa。近年由于我国对深海研究的重视及对深海装备研究的投入，我国科学家成功研制了 7000 m 载人潜水器“蛟龙号”、4500 m 载人潜水器“深海勇士号”，同时 11 000 m 级载人潜水器也在研发中，一系列深海装备的进步保障了我国在获取深海样品上处于国际前沿地位。

除了适应能力的对比，还可以进行功能群、地质演化方面的对比。例如，病毒在不同极端地质环境中是否存在差异，受控因素是否不同？各类极端地质环境的微生物是否都有代表最早起源的一些共同微生物类群，有哪些共同的特征？现代极端环境的分子钟分析得出的微生物起源的地质背景如何？这需要把现代与深时的研究进行结合。

2. 从已知的地质环境条件入手模拟微生物作用

海洋环境的变化直接关系到全球气候的波动，气候的剧烈变化在地球历史上会引发诸如“雪球地球”、强温室效应、海洋硫化缺氧等重大环境的变迁，对地球生命产生极大影响。这些影响不仅与地球表面有关，而且与海洋深部过程相关联。例如推测距今 2.5 亿年二叠纪—三叠纪之交的环境重大变化和生物大量灭绝就与古海洋水化学条件（如缺氧、硫化等）的变化有关。因此，对重大地质历史时期海洋环境演变规律的清晰认识，能够让我们合理预测和应对当前气候与环境的变化。微生物是驱动海洋环境中地球化学元素循环的核心，从地质历史来看，原核生物占据了超过 35 亿年之久的时间，因此在调节和改造全球气候的过程中具有重要的作用。然而，微生物（古菌、细菌及单细胞真核生物）不仅个体的代谢过程非常复杂，它们还通常以大量多样性丰富且功能各异的类群构成复杂微生物群落的方式进行代谢活动，并形成相互作用紧密的代谢网络。要研究这些微生物及其形成的群落在不同地质历史时期海洋环境中的演替特征，以及它们是如何影响海洋中地球化学元素循环的过程，进而影响全球气候变化的，就需要人工模拟环境，对不同条件下微生物群落的动态过程及相应的环境变化进行监测。因此，只有在实验室搭建能够模拟高压情况下的古海洋环境模拟平台，从而模拟地球早期、重大地质历史时期及极端条件下海洋环境甚至其他类地行星潜在的海洋环境，才能通过实验来验证地球不同时期及不同条件海洋环境或类似环境中的特征微生物在群落、细胞及分子层面上相互作用的机制与协同演化的过程。同时，根据分子钟进化理论，估算某特定地质历史时期出现微生物重大演化过程及形成全新代谢通路的时间，进而推断发生此类地质历史事件的可能性。最终，基于大量实验和模拟数据，提出可靠的科学假说，建立特征微生物与海洋环境相互作用的数学模型，并对远古时期地球环境进行反演，以及对未来环境进行预测。以上所提出的思路与目标，希望能够在地球生物学学科开辟新的研究领域，进一步扩大国际学术影响力及引领性。

3. 多种环境条件的室内模拟和野外现场原位模拟相结合，建立全面的监测网络和大数据平台

自然生态系统十分复杂，影响因素多样，很难评估各种环境因子对生态过程的控制和贡献。室内模拟实验可以通过单一变量的控制以及同位素示踪技术等，对某一环境变量对生态过程的影响以及某类微生物功能群对某一过程的贡献等进行示踪和定量，再结合动力学模型建立深地微生物的生态模型，定量评估生物量、生长速率、参与元素循环的速率以及煤炭和石油的转化效率。

地面实验室很难模拟深地原位环境的天然条件，这一问题一直制约深地科学的发展，包括深地微生物的发展。利用现有矿井建立深地实验室，或者在地面建立深地模拟舱，有助于更深刻地认识深地微生物在深地原位环境的行为。深海、深空在这方面已经积累了不少的经验，值得深地生物学家参考。

野外监测是了解深地环境时空动态变化的基础，能够揭示生态系统与环境条件相互作用的机制，也是从长时间尺度考察深地生物圈演替及生态功能演化不可或缺的途径。监测网络的建立需要大量的资金和人员的投入，需要在科学问题驱动的前提下，周密选择监测的地点以及监测的环境参数和微生物参数，以便后期更有效地利用监测数据，并产生能服务于公众的产品。目前深海的监测网建设比较超前，值得陆地深地科学家借鉴。另外，规范数据采集和记录格式，构建大数据平台，建立共享机制。呼吁有关部门建立合理有效的资助机制，鼓励更多不同领域的科学家广泛参与，使得监测网络和监测数据发挥最大的效应。

二、发展目标

针对国家科技发展和人才队伍建设需求，本学科领域的发展目标既要在极端地质环境微生物领域产生重要的国际影响，又要在技术、人才等方面为我国的重大战略目标提供全方位的支撑。

（一）研究成果和技术

1. 技术方法层面

在技术方法层面，力争突破极端地质环境微生物的保真取样技术、野外现场观测和模拟技术等“卡脖子”技术，努力提升我国深海、深地和深空等国家重大工程计划的科学研究效应。努力提高实验室模拟深时极端地质环境

的技术，推动单化合物碳-氢同位素的分析技术以及原位同位素和元素的分析技术的发展。

利用各种极端地质环境，分离具有广泛应用前景的微生物种质资源并进行储备，为实现我国的一些重大战略目标提供支撑。借助最新的培养技术，继续加强有价值的陆表极端环境微生物资源发掘，建立极端环境微生物资源库，并结合微生物生态学成果在工农业生产、生物工程应用、生态恢复、污染治理等方面开展应用研究。

深时极端地质环境微生物的研究首先需要获取各方面定量化的参数，这需要技术方法的突破。需要综合利用岩石地层学、生物地层学、年代地层学、沉积学和同位素地球化学等各种技术方法，深入揭示重大地质转折期的环境变化的类型、空间分布特征、环境参数变化的幅度以及持续时间。同时，需要构建刻画深时极端环境相伴随的 C、N、S 等生物地球化学循环的技术方法体系。

2. 研究成果层面

在研究成果上，既要在海洋极端环境、深地环境、地表极端环境、深时极端地质环境和深空环境等各领域形成微生物研究的一些高峰成果，又要在极端地质环境微生物方面形成一体化成果，只有这样才能产出真正国际领先的系统成果。

基于微生物组学、富集分离培养技术和同位素示踪技术，在群落和纯培养菌株层面，通过生理学和分子生物学技术手段，探究地表极端地质环境微生物特殊的能量代谢途径与驱动元素循环的分子机制及其嗜极机理。精细研究地表极端环境的微生物群落结构及其演替规律，利用生物信息学技术揭示控制微生物群落多样性的基本特征和主控因素，进而扩展和丰富对极端微生物驱动地球化学循环、环境质量演化的生态学机制。通过大尺度高通量测序技术手段，分析不同地表极端地质环境下微生物物种组成特征和潜在功能特性，揭示地表极端环境微生物物种和功能多样性对环境因子的响应规律、微生物驱动元素物质地球化学循环与微生物代谢活动对环境的反馈及其对全球气候环境变化的影响。

阐明深地环境生态系统的能量来源、转化途径和生态系统维持的机制；构建微生物之间、微生物与其他生物之间以及微生物与环境之间的相互作用网络模型，揭示不同微生物类群之间的相互作用关系，定量评估深地生物圈的地质作用和资源环境效应。

充分利用我国在深时极端地质环境记录及地质学研究方面的优势，加强薄弱环节，以形成系统性的国际影响，在显生宙重大生物灭绝期极端地质环境微生物与古气候环境、新元古代“雪球地球”冰期及冰后期的极端地质环境微生物与古气候环境、极热时期极端地质环境微生物与古气候环境方面取得全面发展。重建典型微生物功能群的起源时间和演化过程，以及这些生物事件的环境背景；构建一些典型深时极端地质环境微生物和环境两方面的边界条件和变化幅度；揭示微生物与环境之间的相互作用，定量 / 半定量地分析地质微生物的生态环境效应；重建深时极端地质环境微生物与环境的协同演化历史，为全球变化领域方面的研究服务。

结合极端地质环境地球化学特征并塑造地球的重大地质事件的环境条件，开展类比研究，阐明特殊地球地质历史时期（如地球早期、生命演化过程）的微生物种群构成与功能多样性维持机制及其生物地球化学过程对生态系统演化、环境变化与矿产形成的影响。结合极端地质环境地球化学特征和地外星系存在的环境条件，开展类比研究，阐明外星生命存在形式与潜在生理生态功能及其适应机制。

（二）学科体系和人才队伍

1. 地球生物学

地质微生物是地球生物学的重要研究对象。在极端地质环境微生物各分支领域，培养一批具有国际领导能力的杰出科学家，特别是在极端地质环境微生物领域的战略科学家，促进科学共同体的形成，促进地球科学家、生命科学家与化学家之间不自觉地联合，形成具有自主创新的研究氛围，由此推动地球生物学的发展。

地球科学与生命科学的交叉联合具有广阔的发展前景，可以形成许多新的学科生长点。相对于现代生命科学来说，地球科学研究的古生物仅是地球生命的“冰山一角”，地质微生物的研究有待突破。由于地球生物学本身还是一个新兴学科，学科体系还有待进一步完善。应尽快加强深时极端地质环境微生物学的研究，促进学科体系的发展，进而促进地球生物学学科体系的完善。

当前，地球生物学的研究队伍还是非常弱小的，亟须加强人才队伍建设，培养一批具有国际视野的从事深时极端地质环境微生物研究的青年科学家，尤其是极端地质环境微生物领域的战略科学家，加强地质学和现代极端地质环境微生物学的交叉融合。

2. 天体生物学

天体生物学是一门多学科交叉的新兴学科，该学科的建设和发展需要地球科学、生命科学、天文学、行星科学、空间科学、物理学、化学等学科作为支撑。欧美等国家和地区自 20 世纪 60 年代起就开始发展天体生物学，目前已经建立了完备的学科和人才培养体系，培养了一大批活跃在深空探测领域的天体生物学家。我国的天体生物学起步晚，虽然初具规模，但相较于欧美等国家和地区，还缺乏专门的学科和人才培养体系，这极大地限制了我国在天体生物学方面的研究和发展。随着我国载人航天工程和深空探测任务的逐步实施，以及深空探测中科学与工程并举的发展趋势，当前在我国建设和发展天体生物学势在必行。在我国“行星科学”一级学科的建设中，天体生物学是作为“行星地质学”二级学科下的一个研究方向，该方向的主要研究内容包括宜居环境的形成和演化、生命的起源与演化、极端环境生命、地外宜居环境和生命的探测、生命星际传输、行星开发和保护等。积极推进我国的天体生物学学科建设，在若干天体生物学基础研究较好的高校和科研院所开设天体生物学相关课程，并逐步授予天体生物学的学士、硕士和博士学位，将有助于我国自主培养天体生物学高层次人才，服务我国的深空探测国家战略。

近期目标是力争建立和完善天体生物学的学科方向和培养体系，进一步完善已有课程体系，并制定相关教材；引进国内外优秀人才，壮大科研和师资力量；依托我国深空探测战略规划，提出天体生物学相关科学问题，并积极参与探测任务，服务于我国深空探测的战略需求。建成成熟的教学体系、人才培养模式和科学研究平台；明确重大科学问题，并据此规划新的深空探测任务，力争达到国际先进水平。

远期目标是培养天体生物学国际领军人才，实施若干次以天体生物学科学问题为主导的深空探测任务，发展天体生物学新理论，引领国际天体生物学发展。

第三节　重要研究方向

根据目前我国的发展状况以及未来发展的需要，结合国际发展趋势，未来 5～10 年应着重开展六大方向的研究工作。前三个方向分别与深海计划、深地计划和深空计划相关，服务国家重大工程计划，第四个方向则服务于地球系统科学的发展，第五个和第六个方向分别与微生物种质资源和生物安全有关，服务国家的重大需求。这六大方向既符合国家自然科学基金委员会“三深一系统”的长远规划（第一、第二、第三、第四个方向），又体现了基础研究与国家重大需求结合的国家发展战略（第五、第六个方向）。

一、深海计划的微生物及其资源环境效应

在这个方向，需要继续开展海底沉积物、热泉和冷泉、深渊、俯冲带和大洋中脊、缺氧带等核心地区的微生物研究。针对海底沉积物、热泉和冷泉等已经开展了较多的研究，需要继续加大力度。近几年，海洋缺氧带因全球变暖而出现扩张的趋势，该问题受到全球科学家和政府部门的重视。深渊是连接海洋表层和深部的关键纽带，俯冲带和大洋中脊是连接地球表层与地球深部的重要抓手，这三个地区的研究对地球系统科学尤其重要，但目前有关地质微生物的研究基本处于空白，需要加大支持力度。具体内容如下。

（一）海洋极端环境微生物的培养分离技术和极端环境勘探技术的开发

目前在海洋极端环境里，人们所观测到的大量微生物，特别是古菌类还未得到有效的分离和纯培养，微生物代谢各方面的解析主要依靠对其基因组的测序解读。如果能够有侧重点地加强对技术手段的开发，提高这些古菌的分离效率，将极大地增加我们对沉积物中微生物的了解和认识。

同时，随着国家对海洋的大力投入，我们每年都有不少航次进行海洋极端环境的研究工作。勘探工作是我们一切研究工作的起始，为我们的科学实验提供最初的研究样本，其重要性不言而喻。对勘探采集技术的精细化发展，并进行有目的、有规划的进一步整合与细化，会大大地提高我们开展科研工作的效率。

当前受海洋脱氧影响的海域面积与欧盟面积大小相当，而且预测到 2100 年缺氧水体面积还要增加约 7%，因此，缺氧研究具有现实紧迫感，它主要面临三个方面的挑战：①目前海洋的氧观测数据仅有 60 多年历史，缺乏更长时间尺度海洋氧库变动、趋势及驱动因子的信息；②从河口到大陆架区再到大洋，低氧如何影响微生物群落呼吸及生物地球化学循环；③评估及减缓脱氧对海洋资源、生态服务及生物多样性的影响及威胁。氮、磷、铁等生物地

球化学循环对低氧的响应机制已经比较清晰，但缺氧下的碳循环研究则显得明显滞后。因此，待攻关的关键科学问题是气候变化和人类活动双重压力下缺氧对海洋碳循环的影响机制和环境效应。

（二）微生物对极端环境的适应及其种质资源

海底沉积物的极端环境使微生物进化出各种生存策略以适应不同的环境。其在极端条件下的生态功能、生理特征以及分子遗传适应机理、物质循环和信息间的流通等，都有助于我们更加深入地了解这个独特的生物圈，进一步帮助我们开发微生物资源，也是我们了解生命起源进化的一个天然的实验平台。

（三）微生物对全球气候、海洋环境和矿产资源的影响

海洋中储存着大量的甲烷，我国在南海也陆续发现了多个天然水合物与冷泉区，对其机理的解读一直是国际上相关领域的热点问题之一。针对其中甲烷产生、消耗、循环的机制，对相关微生物代谢特点的认识将有助于我们对全球元素循环与气候变迁有更多的理解。

海底沉积物埋藏着大量的矿产资源，一方面，在海底寡营养区域，微生物的生存有时要依赖这些矿物质提供物质能量；另一方面，微生物也通过与矿物质的相互作用促进矿物质的形成、转化。集中解决微生物与矿化相互关系的问题有助于我们提高对海洋矿物质的认识和开发应用。

生态系统是否发展为缺氧状态取决于氧气消耗与物理输送。物理作用对脱氧贡献较为明确，而微生物对脱氧的贡献较难评估，这是因为呼吸作用的量化存在技术性难题。测定微生物呼吸的方法众多，急需标准化；通过颗粒有机碳的矿化测定，呼吸易受到压力和取样的影响；已有的生物地球化学Argo浮标和水下探测设备缺乏原位氧传感器。

可通过加强国际合作（如SCOR工作组）参与微生物呼吸测定方法学标准制定和比对工作；利用已有的保压采样设备等测定颗粒有机碳矿化；对于我国近海已布局的浮标和探测设备，通过添加氧传感器、整合航次及历史数据等，形成氧的监测网络及数据库。

有机物驱动的微生物呼吸产生内源CO_2，会促进海水酸化，加强脱氧和酸化的联系；低氧下微生物有机质分解的速率和途径会发生变化，通常使有机物埋藏得到加强。酸化和缺氧都威胁近海生态健康和渔业养殖等，对其产

生的影响研究比较多，但缺乏对微生物参与的脱氧过程和机制的研究，而且微型生物碳泵在缺氧环境中如何形成碳汇也不清晰。

可通过加强对我国脱氧程度不同的海域开展微生物参与的脱氧过程与机制和碳汇效应研究，加强缺氧环境碳循环与氮、硫循环的耦合研究，提出缓解近海脱氧又能增加碳汇的理论和方法。

（四）典型极端海洋环境的微生物

1. 冷泉区微生物特殊生态功能与代谢特征

一方面，通过分析不同沉积物及水样中的微生物种类和群落结构特点，尤其关注可能参与甲烷代谢的微生物的多样性和分布差异，比较同一冷泉生态系统不同深度及不同冷泉间微生物群落结构的异同点，认识形成不同冷泉微生物种群结构的环境因子。

另一方面，利用宏基因组、转录组和代谢组分析不同冷泉区微生物群体代谢差异，重点分析碳、氮、硫循环。评估此类微生物在有机质代谢成甲烷途径中的作用以及可能对甲烷代谢的贡献。添加甲烷和不同电子受体（硫酸盐，硝酸盐，铁、锰氧化物等）富集甲烷氧化微生物群体，并结合特异的甲烷代谢基因的荧光探针，通过单细胞显微手段结合不同荧光强度建模，定量分析参与甲烷代谢微生物的活性。解析冷泉区微生物代谢途径和能力，探明冷泉区碳元素循环与其他元素循环偶联的机制，以及甲烷缺氧氧化与硫酸盐还原偶联外其他可能的代谢途径。

2. 洋壳岩石圈微生物

在过去近 30 年的时间中，虽然深部生物圈的研究取得了巨大的进步，但针对深部生物圈的研究仍然十分有限，特别是对栖息于海底洋壳岩石中的微生物研究就更少了。受制于取样技术，海底深部生物圈研究还主要依赖于 ODP 和 IODP 的钻探取样，相关研究大多数集中在海底沉积物，为数不多的洋壳岩石生物圈的研究则仅局限于胡安・德富卡、亚特兰蒂斯地块（Atlantis Massif）、亚特兰蒂斯浅滩（Atlantis Bank）、北大西洋中央海岭（North Pond）和路易斯维尔海山（Louisville Seamounts）等区域，占地球面积 70% 的海洋底部洋壳岩石圈的绝大部分地区还处于未探索状态。目前，对洋壳岩石圈中栖息的微生物的规模还没有取得一致的认识。虽然最新的估算表明，洋壳深部生物圈中微生物的细胞总量为 4×10^{29} 个，然而由于洋壳岩石圈中微生物的分布具有很强的异质性，微生物细胞的丰度变化很大，差别可以超过 8 个数量级，因而给洋壳中栖息微生物数量的评估造成极大的不确定性。洋壳岩石深部生物圈的相关研究更多地集中在上洋壳的玄武岩中（Orcutt et al., 2013），对于更深部的辉长岩质的下洋壳研究几乎处于空白状态。迄今，下洋壳生物群系的研究仅有两例，显示其生物量和微生物多样性、生存策略等与上洋壳截然不同（Mason et al., 2010）。从这个角度，对于平均厚度为 6 km 的洋壳岩石圈来说，我们对洋壳在垂向剖面上的微生物分布知之甚少，对洋壳微生物群系所能够触及的深度空间极限也不清楚。再者，虽然研究已经表明在洋壳岩石圈中栖息着生物量巨大的微生物，然而，这些微生物类群的生理生化特性和潜在代谢策略却是未知的，它们对洋壳元素地球化学循环过程的潜在影响几乎没有被研究过（Orcutt et al., 2013），微生物栖息于这些极端环境时的生存策略、适应机理以及与洋壳极端环境条件的互相制约关系等目前也不清楚（Jungbluth et al., 2016）。所有这些问题都需要我们综合多学科的研究手段对洋壳生物圈进行全面的评估和研究。

3. 深海与深渊环境微生物

揭示深渊海沟沉积物和水体中微生物（尤其是专性微生物）群落的明显差异和调控因子，研究微生物（细菌、古菌、病毒）在介导深海 / 深渊生物地球化学循环中的互作关系和演替，研究活性微生物和总体微生物在调控深海 / 深渊生物地球化学循环中的作用，研究细菌芽孢的丰度和在深海 / 深渊生物地球化学循环中的作用，研究地质作用在深渊沉积有机质的分布和微生物群落结构和活性之间的关系，研究深渊生态结构和食物网控制深渊沉积物中各种污染物（包括持久性有机污染物、汞等）聚集的关系。

二、深地计划的微生物及其资源环境效应

陆地深部地下生物圈除了黑暗，还具有特殊的环境特征，如地下水的流动性以及与地表水的相互作用、污染物对地下水微生物群落的影响等。洞穴有较大的地下空间，通过一个或多个洞口与外界连通，有的还具有地下河或季节性溪流，可以持续或间歇性地向洞穴输送营养物质；或者接受地下热液带来的还原性物质，孕育着相对丰富且高度特异化的生物（盲鱼、钩虾、桡足类、鞭毛虫类等），因此该生态系统的营养级与其他以微生物为主的营养级显著不同。对石油微生物和煤炭微生物来讲，面对的又是极难降解的液态和固态的有机质，油藏是一个独特的生态系统，在时间尺度上，跨越从地质

时期石油生成、油藏形成、成熟油藏到现代开发与利用；在空间尺度上，以深度为纵向参照，油藏纵深从距地表几米（露头油藏）至4000 m以下，存在着温度和压力等环境因子的梯度变化；在横向上，随油藏与圈闭环境及其与地下水赋存关系、地层流体分布及化学组成变化等，形成了相应的溶解氧和营养等的梯度分布。特别是经历开发以后，油藏与外部环境之间同时存在能量与物质交换，因而成为开放系统。正因为如此，深地微生物是解决诸多重大科学难题的关键，需要开展典型油气藏、地下煤层、洞穴、深湖底部及其沉积物、泥炭沼泽深部、地下水、超深钻、深大断裂带、地下CO_2储层、地下核废料储层等关键地带的微生物研究。

（一）深地微生物功能群对地质环境的改造作用及生态安全

微生物与环境之间是相互作用的，过去人们关注更多的是环境对微生物的影响和选择作用，但微生物对环境改造作用的研究相对薄弱。微生物对岩石或者有害元素的改造作用会影响元素的毒性，从而影响生态安全。譬如地下水是我国居民饮用水的重要水源，但天然劣质地下水的形成、微生物对地下水有害物质的迁移转化以及在治理中的应用将对供水安全造成重要影响。将来的发展方向应该侧重在深地功能微生物对地质过程的微观机制与宏观效应层面。地下CO_2储层和地下核废料储层等关键地带往往涉及国家安全，需要开展地质微生物对地下环境的影响能力评估。

（二）深地微生物的资源环境效应

深部煤炭资源开采利用将在未来成为常态，深部煤层气资源将是未来能源开发利用的重点，深部煤层微生物功能群特征和资源效应将是重点发展的方向。同时我国许多油田已进入开采后期，枯竭油藏的资源化利用，如残余油气化开采、复杂低品位油藏微生物改造、枯竭油藏环境CO_2生物转化与利用等均是利用微生物的作用来实现的。最近甚至有人提出，微生物有可能对金属矿床的流态化开采起到关键作用，其基本原理是利用特殊功能的微生物将金属元素从矿石中浸出，从而可以将富含金属的淋滤液直接抽到地表。但目前大多处于概念模型阶段，尚缺少先导性实验的报道，对参与这些过程的功能微生物群的特性、运转条件和效率等还亟待研究。

洞穴、深湖和泥炭沼泽底部微生物往往与全球气候变化密切相关，需要加强微生物地质作用的研究力度。

（三）不同深度微生物的地质作用

值得注意的是，深地计划在不同地下深度有不同的目标，因而就会涉及微生物的不同地质作用。地下 500 m 以上，主要涉及微生物的生物地球化学过程对地下空间利用的影响。地下 1000 m 以上，主要涉及微生物对地下水系统水质的影响。地下 3000 m 以上，主要涉及微生物对矿产资源、能源的影响。实际上，人们对油气藏和地下水环境开展了大量的地质微生物研究，已经分离出了许多微生物功能群，并部分地应用到工业上，加强了微生物对不同极端环境的适应研究，特别是对多种极端环境因子（如缺氧、高温和高压）的适应能力的研究。

三、深空计划的天体生物学

天体生物学是典型的交叉学科，需要多个学科团队、多个实验室的通力合作，使用来自不同团队的方法和技术，以降低研究成本。同时，结合不同方向（地质过程、生物保存、行星演化、气候变迁等）的研究进展，多角度共同探索地外宜居性环境，为地外生命探测提出可靠建议。中国已经获得一些自己的行星探测，更需要结合国外已有数据，吸收国外已有的研究经验，开展国际合作，并建立合作共享数据库，加快研究进程，并运用于本国的空间探测和地外生命探测任务。具体研究内容如下。

（一）地球极端环境（微生物）的天体生物学意义

空间环境是研究天体生物学的重要工具，但对空间的探测是有限且相对困难的。利用地球极端环境同太阳系内其他天体进行类比研究可以加快研究周期，降低成本，而且研究的可复制性更强，样品量更大。发展方向主要包括：①探究寻找极端环境微生物在多极端环境下的适应策略，拓宽对生命边界的定义；②寻找古老岩石中保存的生物标志物。

自极端地质环境微生物被发现以来，持续的研究使我们对生命边界的理解在几乎所有层面都得到了扩展。研究发现，目前已培养的极端微生物绝大多数都是多重极端微生物，但对微生物在多重极端环境下耐受能力的研究极其缺乏，阻碍了我们对生命边界的进一步认识。过去 50 年对极端微生物的研究告诉我们，当面对同时出现的多种极端环境条件时，生命的边界会发生改变。例如，当高压条件存在时，生命的温度边界可越过 100 ℃的限制。因此，将来的研究需要进一步聚焦于多重极端环境下各种参数条件的相互作用研究。当考虑生命的基本需求时，生命的边界实际上是由这些需求在实际环境中的影响因素决定的。例如，目前生命温度、压力和盐度的理论边界都直接与水活度或生物分子在这些条件下的稳定性相联系。因此，在探寻生命的真实边界时，需要特别考虑能在群落和亚细胞尺度直接控制水可得性和大分子稳定性的参数。

自有人类历史以来，人们尽管几乎一直都在对地球开展研究，却总能在预想不到的地方发现生命。因此，我们可以预期生命边界会被进一步拓展。对照地球环境的参数范围和生命存在的环境范围，可以设想生命确实已经适应并占据了地球上所有可能的生态位。总体来看，目前的观察表明，地球生物群的真正形态还有待确定。此外，地外天体具有极大的多样性，很可能会扩展目前已知的环境空间，因此地外生命的形式（如果存在）可能会与地球生命大为不同。在评价生命在其他行星体上起源和存在的可能性时，必须注意地球局部环境条件相对于全球平均环境的变化范围。局部和短时环境条件可能足以支持生命的存在，如火星表面暂时性液态卤水的出现可能给生命创造了宜居环境。因此，仅仅通过短时间低分辨率的观察，很难完全排除地外生命存在的可能性。

生物标志物是一种复杂的分子化石，其来源是生物化学物质，特别是脂类生物。生物标志物的检测在对地球以外是否存在或曾经存在生命的研究中发挥着核心作用。对某一特定行星上的矿物学研究可反映其过去和现在的环境条件，通过与地球上极端条件下的生命进行比较，可以对其宜居性进行评估。

由于地外样品获取困难，因而进一步研究地球上极端环境下生物标志物的保存至关重要。地球生物标志物的保存受到矿物类型的显著影响，包括潜在丰度、类型、有效提取和检测。比如西班牙力拓河，该河流系统极端酸性、富 Fe 的环境由岩石-水-生物相互作用产生。从力拓河阶地古老沉积物（2.1 Ma）中检测到了大量微生物和矿物相互作用的产物。火星的部分地区过去可能是高酸性和高盐的，对火星着陆区的矿物学研究将是寻找过去生命痕迹的重要课题。同时，对地球矿物-生物标志物相互作用更系统的研究，有助于了解生物标志物保存机制，更合理地设计各种生物标志物的高效提取方法并应用到地外生命探测中。对类比环境生物标志物的研究中阴性结果（低于检测限）的解释也至关重要，分析其是源于提取过程中被矿物吸附或破坏而导致的大量丢失还是由于其他因素限制了生命痕迹的保存。因为火星上即使存在生命，也可能维持低丰度状态，需要避免由于设备污染造成的假阳性事件。

（二）地外生命探索

探索地外生命是天体生物学的重要研究方向之一，也是深空探测的重要科学目标。认识生命的本质，即生命的定义，是寻找地外生命的基本前提。目前地球生命是我们唯一已知的生命形式，它们的共性是具有有序结构，可以生殖、生长发育，利用能量，对环境做出反应，保持体内平衡，能演化适应等，但目前尚不清楚如果存在地外生命，这些生命是否与地球生命形式类似，抑或是具有不同的形式。NASA 在开展地外生命探索时，提出了对生命的工作定义，即生命是能够进行达尔文式演化的一种自我维持的化学系统（self-sustaining chemical system capable of Darwinian evolution）。但该定义需要从演化的角度进行研究，限制了其在实际探测任务中的应用。因此，生命的准确定义以及具有更好操作性的工作定义还有待进一步厘定。此外，包括如何区分生命和非生命信号、选择何种生物标志物或生命信号来指示地外现在存在或过去存在过生命、如何检测和识别这些标志物或信号等，也是未来重要的研究方向。

基于地基和天基望远镜的遥感探测是目前探索地外生命的最主要手段。对于太阳系内的天体，遥感技术可以在一定程度上实现对大气、天体表面等的探测，而对系外行星的探测主要集中在对其行星大气进行遥感分析上。ESA 正在开展的阿里尔（ARIEL）空间望远镜计划以降低光谱分辨率为代价得到上百颗类木和类冰系外行星的大气的普查信息。为研究宜居行星，大气国际科学家提出两个可能的空间望远镜概念：专门的宜居行星观测者（HABEX）和多用途大型空间望远镜（HDST/LOVIOR）。这两个方案的预想实现时间都是在 2035～2040 年。目前正在建设中的国际 30～40 m 级地基望远镜和将在 2021 年以后发射的下一代空间望远镜 JWST 也都有明确的针对系外行星生命信号的观测计划。

随着我国载人空间站的建设以及浮空器技术的发展，未来部署在空间站和临近空间的探测载荷将成为探索地外生命的有力手段。例如，中国科学院“鸿鹄专项”将突破临近空间浮空器平台关键技术，开展球载行星光学遥感研究，这将为包括地外生命探测在内的行星光学遥感研究提供先进新手段（何飞，2020）。降落到地球的地外样品（如陨石等），也为研究地外可能生命提供了直接研究对象。截至 2020 年 5 月 31 日，国际陨石命名委员会数据库已收录了超过 63 000 块陨石，这些样品为认识地外生命的可能传输途径和分布等提供了条件。人类已实现了月球样品的取样返回，中国、美国等国家也已经规划了火星取样返回的深空探测任务，无疑这些返回样品将为认识火星是否存在过生命提供第一手材料。此外，到目标天体近距离或原位进行探测也是寻找地外生命的重要手段。20 世纪 70 年代的“海盗号”火星表面探测任务和 2011 年发射的“好奇号”巡视器均搭载了多个科学载荷，用于研究火星是否可能存在生命以及火星环境的演变。我国 2020 年发射的“天问一号”火星探测器已经成功实现环绕、着陆、巡视 3 种探测方式，这为我国火星原位探测奠定了坚实的基础。

（三）行星环境的实验室模拟

部分天体生物学相关研究需要在行星上进行原位分析和实验，但地球和其他天体之间存在生物和有机物的交叉感染风险，导致地球和其他天体的生态系统或生物圈遭到破坏。同时，在真实行星环境下开展科学实验研究还无法轻易实现且成本极其昂贵。因此，行星环境的实验室模拟是目前开展空间生命科学实验研究的必要手段。地面行星环境模拟装置不仅可以在一定程度上降低污染风险，而且面对太空实验关于空间、检测和分析手段等多重因素的限制，其提供了一个经济、快捷、有效且具有前瞻性的研究方法去支撑天体生物学研究任务所需的大量实验。

行星环境是由空间高能粒子辐射、超重力 / 低重力、高磁场 / 低磁场、高真空、温度骤变等复杂因素组成的独特环境。空间高能粒子辐射（主要来源是银河宇宙辐射和太阳粒子事件）是制约人类开展长期深空探索的关键因素。对微生物而言，空间高能粒子辐射能量高、贯穿力强，可以直接作用于 DNA，DNA 双链断裂是空间高能粒子辐射造成的最严重损伤。由于没有地球大气臭氧层对 290 nm 以下射线的阻挡，在近地轨道存在全光谱紫外辐射，除了长波紫外线（UVA）（320～400 nm）、中波紫外线（UVB）（280～320 nm）的辐射，还包含被大气层屏蔽的短波紫外线（UVC）（200～280 nm）和真空 UV（<200 nm）辐射，其中，UVC 波长短、能量高，对地球生命体有强烈的刺激作用（袁俊霞等，2020）。近地轨道飞行过程中微重力在 10^{-6}～$10^{-3}\,g$，微重力环境中自然对流、沉积等现象的消失，将诱使生命体产生不同于重力环境的生命特征。此外，不同轨道飞行任务中压力和温度变化也不同：在近地轨道任务中真空度为 10^{-7}～10^{-4} Pa，火星上的大气压力约为 700 Pa，在星际空间真空度可达 10^{-14} Pa（袁俊霞等，2020）；空间站外直接暴露在太阳下的温度在 −120～120 ℃波动，火星表面温度变化范围为 −143～35 ℃。

基于上述参数，行星环境的实验室模拟发展方向为：构建模拟空间辐射、磁场、高真空和重力效应的复合空间环境生物实验平台，利用行星环境模拟装置模拟和重现目标行星上的特定环境条件，如辐射、压力、大气、温度、湿度等。为了更好地支撑和服务深空探测工程和科学研究，越来越多不同类型的行星环境模拟装置被研发出来，并成功应用于天体生物学的相关研究。德国放射生物学系的天体生物学小组在德国科隆航天航空中心（DLR）的航空航天医学研究所拥有 7 个模块化的行星和空间模拟设施（PSI），在一定程度上可模拟低地球轨道（LEO）、火星表面、深空等环境（Rabbow et al.，2016）。西班牙天体生物学研究中心主要有 3 个模拟装置，分别是星际天体化学模拟装置（ISAC）、行星大气和表面模拟装置（PASC）、高压行星环境模拟装置（HPPECs）。ISAC 主要研究在裸露的碳质尘埃与冰相互作用的过程及产物。其原理是，含碳的冰被辐照加热后会产生大量与天体生物学相关的分子，如氨基酸、糖、N-杂环、羧酸等。PASC 可模拟大部分行星的大气和表面温度，同时具有多种辐射源和原位分析技术（红外光谱和拉曼光谱）。HPPECs 用于研究深层环境，尤其是隐藏在太阳系卫星冰面之下或之内的含水储层。

为了探究行星环境对生物的整体性影响，往往将各种耐辐射或耐高温高压的物种暴露在行星环境模拟装置后，分析其损伤诱导、修复能力、适应性和适应性策略、生存率、蛋白质、新陈代谢等方面的变化（Martins et al.，2017）。当前，火星上是否存在生命是天体生物学的热点问题，一大部分天体生物学实验都是在火星环境模拟装置中进行的。火星地表环境恶劣，尤其是存在对生物伤害极大的强辐射，如今对火星宜居环境的研究和生命的探索已逐步从火星地表转入地下。人们将 DNA 保护或修复基因有缺陷的两组等基因的枯草芽孢杆菌突变菌株的孢子分别暴露于有和没有紫外线辐照的火星大气模拟环境中 24h。结果显示，没有经过紫外线辐照的孢子的存活率明显高于经过紫外线辐照的孢子。这表明在火星浅表生存需要躲避辐射对生命的伤害，而被灰尘、岩石等遮蔽的环境更有利于生命的存活。

随着执行首次火星探测任务的“天问一号”的发射，我国的深空探测计划正在稳步开展。同时，鉴于深空探测的发展以及技术的成熟，行星环境模拟装置所使用的参数将会越来越完善，考虑到的环境因素会越来越全面，功能也会越来越齐全，这将大大提高实验数据结果的可靠性，为开展天体生物学相关研究提供有力的技术支撑。通过建立和使用行星模拟装置，我们得以在地面就能进行相关天体生物学实验，这不仅推进了我国天体生物学研究的快速进展，也为未来的太空实验项目提供了辅助，为地外生命探索指引了方向。

四、深时极端地质环境微生物与地球系统科学

深时极端地质环境微生物很好地记录了地球系统各圈层之间的相互作用，是发展地球系统科学的关键抓手。需要开展两次大氧化时期、“雪球地球”时期、显生宙生物大灭绝时期、极热时期等关键时期地质微生物及其地质演化的研究。

深时地质微生物研究的难度极大，分子生物学难以发挥作用，只能依托脂类分子和同位素等分子地球生物学的方法。但深时地质微生物是研究地质微生物对地质环境作用以及地质微生物演化的不可替代的记录。同时，还可以开展地质微生物与动植物危机之间的关系研究，从而可以从整个生态系统角度来看地球上的生物危机及其复苏过程。特别重要的是，一些地质突变期往往是多种极端地质环境因子综合作用的结果，这与一些单一环境因子的影响明显不同，可以带动许多理论突破。

针对我国深时极端地质环境微生物发展的实际情况，以及目前的优势领域和薄弱环节，选择若干我国发育良好的不同类型的深时极端地质环境作为切入点，围绕深时极端地质环境微生物研究所面临的关键科学问题，开展深入研究。未来5～10年，拟选择不同尺度重大气候变化事件作为我国深时极端地质环境微生物研究的发展方向，重点集中在气候发生明显变化的显生宙后生生物大灭绝期（晚奥陶世赫南特期和二叠纪—三叠纪之交）、新元古代“雪球地球”冰期及冰后期和显生宙极热事件（早三叠世）等气候事件，对比研究揭示地质微生物与极端气候环境之间的相互作用，形成系统理论。

（一）生物大灭绝期的地质微生物与气候环境关系

显生宙发生过5次重大的后生生物大灭绝事件，分别发生在晚奥陶世赫南特期（约445 Ma）、晚泥盆世弗拉期—法门期之交（约372 Ma）、二叠纪—三叠纪之交（约252 Ma）、三叠纪—侏罗纪之交（约201 Ma）和白垩纪—古近纪之交（约66 Ma）。除白垩纪—古近纪之交的生物大灭绝事件可能由地外因素（大型陨石撞击）诱发外，其他几次重大的生物灭绝事件都与地球内部因素有密切关系，其中气候波动及其环境效应被认为是最可能的因素。例如，晚奥陶世赫南特期的生物大灭绝事件与冰期的发育有很好的一致性，晚泥盆世弗拉期—法门期之交的生物大灭绝事件也与温度的降低在时间

上有很好的对应关系，而二叠纪—三叠纪之交和三叠纪—侏罗纪之交的生物大灭绝与高温事件之间有密切的关系（Joachimski et al.，2012）。这些表明，生物大灭绝事件与气候环境之间的内在联系非常复杂，需要开展更进一步的工作。

需要注意的是，很多研究都表明，在这些后生生物大灭绝事件中，微生物可能有完全不一样的发展趋势。以二叠纪—三叠纪之交为例，显生宙这次最大的宏体生物灭绝事件伴随着微生物的大繁盛，出现了巨厚生物岩沉积。需要回答的科学问题包括：哪些微生物功能群出现了繁盛？是什么环境因子导致了微生物的繁盛？微生物的繁盛对气候变化有什么影响？微生物的繁盛对后生生物的灭绝和复苏有什么影响？

前期的初步工作表明，在生物大灭绝期，微生物功能群的变化可能对气候有重要影响。例如，二叠纪—三叠纪之交，伴随着海洋缺氧程度的加剧，氮循环会出现显著异常，反硝化微生物功能群丰度升高，进而提高温室气体 N_2O 产生的通量。特别是，当海水中的 Cu^{2+} 因缺氧加剧而显著降低时，N_2O 的通量会更加显著上升。此外，动物的大灭绝和缺氧程度的加剧，可能会导致海水硫酸盐浓度显著下降，进而影响 C、S 循环，提高温室气体 CH_4 的形成通量（Luo et al.，2010）。这些地质微生物过程都对气候有着重大影响，但这还需要更深入的研究。一方面，目前这些研究都还只是停留在定性分析上，需上升到定量 / 半定量程度，来探讨地质微生物过程对气候的影响；另一方面，目前对其他几次后生生物大灭绝的地质微生物过程还不了解，也亟须开展相关的对比工作。

我国拥有非常好的地层古生物资源，并在后生生物大灭绝领域的研究中处于国际领导地位，其中以奥陶纪赫南特期生物大灭绝和二叠纪—三叠纪之交生物大灭绝研究最为突出。一方面，我国的地质学工作者构建了非常高分辨率的时间地层格架，赫南特阶底界和三叠系底界的全球界线层型剖面和点位全部位于我国（Zhang et al.，2006）；另一方面，我国古生物学家经过持续多年的艰苦奋斗，已经深入揭示了这两个重大转折期后生生物的危机过程。下一步，将以此为基础，重点开展这两次后生生物灭绝期的微生物地质过程及其对比研究，探讨不同气候背景下地质微生物过程对气候环境的作用。

（二）“雪球地球”冰期与其他冰期微生物与气候环境关系

新元古代成冰纪（约 720 Ma～635 Ma）发生了两次全球性的冰期事件，被称为“雪球地球”事件（Hoffman et al.，1998）。虽然有关“雪球地球”的成因目前还没有达成共识，但一般认为与硅酸盐风化加强导致的 pCO_2 的显著下降有关。此外，在每一次冰后期的地层中都发现了碳酸盐岩碳同位素组成（$\delta^{13}C_{carb}$）的强烈负偏，如在第一次“雪球地球”的间冰期地层中发现的特雷佐纳漂移（Trezona excursion，负偏幅度达约 18‰）和在第二次“雪球地球”的间冰期地层中发现的舒拉姆漂移（Shuram Excursions，负偏幅度达 15‰）（Fike et al.，2006）。学者们推测这些 $\delta^{13}C_{carb}$ 的强烈负偏可能与深部海洋存在的一个超大型 DOC 库的氧化密切相关，这一推测得到了有机质碳同位素组成（$\delta^{13}C_{org}$）记录的支持（Rothman et al.，2003）。

由于晚新元古代生物泵的重要执行者——后生动物尚未在生态系统中占据主导地位，以我国学者为首的研究团队提出这一超大型 DOC 库的发育可能与“微型生物碳泵”的强烈活动有密切关系（焦念志等，2013）。然而，目前还有很多问题需要解答。例如，是什么微生物地质过程形成了这一超大型 DOC 库？超大型 DOC 库的形成与“雪球地球”冰期之间有什么关系？超大型 DOC 库的形成对微生物的地质过程又有何影响？是什么微生物地质过程导致了超大型 DOC 库的消失？对气候又有什么影响？显生宙其他次一级的冰期会不会伴随相似的微生物地质过程？

无论是国际还是国内都还没有对上述这些问题开展实质性的工作，这些问题是未来国内外科研力量争夺的重要研究制高点。我国拥有非常好的地质资源，发育了一系列从浅水到深水的沉积地层，并且已经建立了非常好的地层格架，在后生生物演化和古海洋环境背景方面做出了非常出色的工作（Li et al.，2010），比如中国地质大学（武汉）李超教授团队已在此方向上开展了相关的研究工作。这些为深入探讨冰期及冰后期微生物地质过程及其与气候之间的关系打下了很好的基础。

（三）极热时期的地质微生物与气候环境关系

很多学者认为地球是一个恒温器。然而，地质记录表明，地球曾经经历了一些极端的气候事件。除上述讨论的“雪球地球”冰期事件之外，地球还经历过一些极热气候事件。目前讨论最多的地质历史时期的极热事件主要集中在新生代早期，如古新世—始新世之交（PETM）的极热事件以及始新世初期的极热事件。这些极热事件持续的时间相对较短，一般只有几十万年，与恒温地球的假说是相符的。

相对来说，早三叠世的极热事件是一次非常特殊的事件。牙形石磷灰石的氧同位素记录表明，这个时期表层海水温度一般在 35 ℃以上，并且持续了500 万年的时间。对于地球如何维持如此长时间尺度的高温气候，目前的认识还是非常有限的。查明这一特殊极热事件的地质微生物过程，对理解上述问题有重要价值。

我国华南地区有发育非常好的地层记录，且具有非常好的研究基础，如建立了早三叠世的生物地层和年代地层格架，揭示了与早三叠世极热事件相伴的碳同位素组成和海洋氧化还原环境的频繁波动等现象。这为极热时期地质微生物与气候环境的研究打下了很好的基础。

五、极端地质环境微生物与种质资源

陆地热泉、雪冰和盐碱湖的微生物主要与生物的种质资源及其工业利用有关，可以重点聚焦在关键功能基因的研究上。酸性矿坑水、重金属污染区、农药及抗菌药污染区等可以聚焦在微生物对各类污染物的降解或分解作用方面，以便为地质环境的保护和治理服务。开展沙漠地区的微生物对比研究，则可以服务于深空环境微生物的工作。具体内容如下。

（一）极端酸碱环境微生物

全面开展咸盐湖泊与盐碱土壤极端环境微生物多样性（包括物种及功能多样性）研究和耐盐 / 嗜盐机制研究、微生物驱动元素循环的机制与环境效应研究，丰富现有认知，开发研究手段；利用新兴技术不断加深对咸盐湖泊与盐碱土壤极端微生物生态功能、生态价值的研究；不断发掘咸盐湖泊与盐碱土壤微生物资源，并将其合理运用到环境治理、生态修复工程当中；拓展咸盐湖泊与盐碱土壤耐盐 / 嗜盐微生物研究与其他新兴科研领域的交叉和合作。

发展并综合利用宏基因组学技术和纯培养技术，揭示极端酸性环境微生物的多样性、群落结构、功能群结构演化特征和主控因素，以及极端酸性环境微生物的代谢机制和环境适应机制。极端酸性环境具有强酸性、寡营养、富重金属等特征，揭示微生物代谢的分子机制和电子传递路径是阐明环境适应机制的基础。从细胞、菌株和群落层次揭示微生物对 AMD、尾矿和其他酸性生境元素地球化学循环、环境属性演变的控制作用，针对地球演化历史上重大环境转折机制，开展极端酸性微生物生态学类比研究，以及极端酸性环境微生物资源的发掘和开发。瞄准生物冶金和重金属污染修复等重大需求，在多种极端酸性环境中筛选分离出具有重要应用价值的微生物菌株，形成种质资源库。

（二）热泉、雪冰等极端温度环境微生物

需要开展大尺度范围内高温地质环境微生物物种和功能多样性调查：含高温环境微生物膜脂化合物合成基因研究、原位活性检测、未培养微生物的可培养化、高温微生物菌种资源库的构建。

随着宏基因组学及高通量测序的快速发展，这些技术也被用到低温微生物的研究中，从而为嗜冷微生物冷适应机制的研究提供了技术支撑。研究嗜冷微生物的冷适应机制可以为它们的分离培养提供更适宜的条件，并且可以进一步探究嗜冷产甲烷菌对低温的响应特点。嗜冷微生物对低温的适应涉及细胞全局性反应，认识微生物嗜冷的物质基础及其机理，将为其开发应用奠定坚实的基础。另外，微生物的分布呈现一定的生物地理模式，而低温环境为研究微生物的生物地理提供了理想的生态系统。宏进化、微进化等综合研究，有助于理解和认识低温微生物的分布规律，有助于深入了解全球微生物的生物多样性格局及其形成机制。

（三）极端干旱环境微生物

针对极端干旱区微生物的研究，我们需要进一步扩大免培养研究的调查范围，以获取我国不同沙漠土壤生境中微生物的群落结构和多样性，揭示极端干旱区微生物的群落功能轮廓及对各种环境因子的响应规律；建立针对沙漠土壤放线菌的高效分离、传代培养方法；基于多相分类研究，明确放线菌新分类单元的系统发育关系，初步建立新类群的精确分类地位；同时基于活性功能研究，从新培养的菌株中筛选得到具有特定功能（如微生态修复功能）的活性菌株，还可以结合抗菌、抗肿瘤细胞等生物活性，筛选出具有显著活性的新化合物，并解析其生物合成基因簇，为新型微生物药物的研发奠定基础。

六、极端地质环境微生物与生物安全

（一）各种极端地质环境的病毒与生物安全

深地微生物多样性高，可能蕴含许多未知的新物种，而我们对这些新物种的认识极为匮乏，比如对深地生物圈中病毒的认识还十分不足，但病毒致病性及对人类健康的危害有目共睹。因此开展此类环境中病毒的多样性以及

它们在生态系统中的作用研究，进而评估它们潜在的致病性等，将极大地提高人们对开发地下空间的生物安全意识，合理科学地规避和控制风险，为地下空间的开发和生物安全指导提供科学依据。

（二）病毒与气候环境的关系

病毒的暴发与宿主的发展有关，而宿主的生长代谢活动又往往与气候环境变化相关。尽管这方面的研究工作较少，但越来越多的证据表明，病毒暴发与气候环境有密切关系，需要地球科学与生命科学的交叉融合，更需要地球科学家与生命科学家的联合。例如，2002～2004 年在北美地区暴发的西尼罗河病毒，其宿主为蚊子，而蚊子的生长繁殖明显与气候变化密切相关，导致病毒出现季节性和地区性的差异。病毒宿主与气候环境之间存在的密切关系，造成了病毒引发的区域性乃至全球性的疫情大暴发，但这也有助于人们根据气候环境的变化来预测未来疫情的走向。

（三）极端地质环境微生物与地质医学

开发地下空间是一个有步骤、有计划的科学系统工程，与人们习以为常的地面环境相比，不同深度的地下空间环境是明显不同的。地下空间无季节天气变化、自然光线缺乏、色彩单调、视觉受限，容易产生空间定位障碍、幽闭感；对空气流通、气压、湿度、温度、重力围岩性质变化，低本底宇宙辐射，微生物及其他未知地下生态圈环境等尚缺乏科学客观深入的认识。更重要的是，地下空间环境，尤其是深地环境对人类生理、心理以及病理的影响规律及其机制研究尚属空白。深地资源开发利用和地下空间环境如何实现卫生安全保障？对于深地生命体（人）的心理、生理承受能力、极限深度如何？深地环境未知生命与疾病发生发展规律所涉及的医学健康科学问题亟待探索研究。深地微生物已经在深地环境中长期生存繁衍，对这些环境已经有了适应能力，因此深入研究深地微生物的极端环境适应机制、揭示深地微生物特有的生理生态特征，对深地医学有直接的借鉴作用。

第六章 资助机制与政策建议

第一节 学科领域的总体政策建议

如第一章所述，极端地质环境微生物研究具有重要的科学意义和战略价值，亟须教育部、科技部、生态环境部、自然资源部、中国科学院等多部门从战略高度进行前瞻性布局，实施系统工程，在多学科交叉与跨学科人才培养、重大科学研究计划的实施、条件平台建设、科学普及四个关键环节上给予全方位的重视和前瞻性的投入，以提升我国在这一关键领域的国际地位，保障我国的生物安全、生态安全和能源安全，为我国重大战略目标的实现做出不可替代的贡献。

一、不遗余力地推进学科交叉和跨学科人才培养

对于极端地质环境微生物的发展，首先必须建立不同于传统学科的交叉学科体系。生命科学、地球科学、行星科学等单一学科体系都很难在这一相关领域开展深入的、全方位的探索工作，难以取得重大的、具有全局性的科学突破。为此，需要不断推动地球科学、生命科学与行星科学的交叉融合，建立地球生物学、天体生物学等新型交叉学科体系。

交叉学科体系的构建是一项长期的系统工程，首先需要从高等院校的课程设置与人才培养入手。地球生物学的课程体系不仅要涉及生命科学领域的主干课程、地球科学领域的核心课程，更重要的是要建立真正体现这两大学

科交叉的课程体系，如涉及从分子生物学到分子地球生物学衔接的课程、示踪从活生物体到沉积埋藏等一系列地质过程变化的同位素技术方法课程等。后者是否成功开设是决定学科交叉成败的关键所在。天体生物学则需要建立涉及地球科学、行星科学和生命科学真正交叉融合的课程体系，而不是这三大学科核心课程的“拼盘”。

涉及学科交叉的课程体系建立以后，需要投入人力和物力进行跨学科人才的培养。目前，真正具有跨学科背景的人才极少，许多涉及跨学科研究的项目主要是通过不同学科领域的专家学者聚集在一起而实现联合研究的，但不是真正意义上的、具有跨学科背景的专家在开展跨学科研究。因此，当前的学科交叉研究还处于第一层次，即不同学科领域的人员进行联合研究，而不是交叉研究。要把这类学科交叉研究尽快提升到第二层次，需要具有跨学科素质的专家加入，这需要国家加大这方面人才培养的力度。因此，跨学科人才队伍的培养是学科交叉课程体系建设的终极目标，亟须加强，而这需要教育部门和各高等院校的重视。

二、加大基础科学研究的资助力度

教育部门和各高等院校在加强交叉学科体系建设和人才培养的同时，科技部等相关政府部门则需要加大学科交叉研究的投入力度。极端地质环境微生物研究能够引发一些重大科学发现，促进地球科学、生命科学与行星科学之间的交叉融合，对推动地球生物学和天体生物学等新学科体系的发展起着至关重要的作用，需要在重大科学研究计划方面加大支持力度，以促进学科交叉研究，推动地球科学的转型发展。

同时，在深空探测计划、深地计划、深海计划、全球变化等一些已经实施的国家重大研究计划中，加大对极端地质环境微生物的研究，以推动相关核心技术，特别是与钻探计划相关的保真取样技术的突破。

三、大力推进条件平台建设

极端地质环境微生物的研究同时需要学科交叉的研究以及核心技术方法的突破，除了推动重大科学计划的实施外，还需要加大一些条件平台的建设。其中，室内实验平台需要建立能开展生命科学与地球科学领域交叉研究的仪器设备，不仅需要分子生物学的常规设备，还需要各类同位素分析仪器、示踪各类代谢过程的模拟实验设备、模拟地质过程的实验设备，以实现分子水平代谢过程与长时间尺度地质过程的全方位衔接。目前，真正能够实现从代谢过程到地质过程衔接的实验平台还极度缺乏。

除了实验平台，还要大力推进野外研究基地的建设。深地计划实施了许多超深钻孔，为原位观测研究深地微生物提供了难得的基地。深空计划开展地外生命探测，需要对地球极端环境进行类比研究。深海计划发现了许多特殊的生态系统，需要开展长期的现场观测研究。地质剖面清晰地记录了地质微生物的起源、演化及其与恶劣环境之间的相互作用，为探索深时极端地质环境微生物提供了相当珍贵的野外研究材料。目前，所有这些研究基地都没有单独纳入对微生物的原位观测系统，需要前瞻布局这些野外研究基地，以实现系统化。

四、加强微生物的科学普及工作

如果说上述的三项政策建议需要依托教育部、科技部等相关政府部门的推动，那么科学普及工作则需要相关企事业单位、各类学会以及科学家群体与个体的联动。2019 年，《环境微生物学》（*Environmental Microbiology*）发表了一篇由全球 32 位微生物学家联合撰写的论文，提出了“提高社会的微生物素养势在必行”的倡议，认为微生物远比互联网对社会的作用大：微生物是人类的“第二皮肤”和“第二分泌系统”，总体上是“人类的朋友而非敌人”，是“生物圈运转的齿轮”，是“生物经济的核心”，是“地球上最大的废物回收者和再生者”，是“这个星球上人类唯一可以依赖的能解决污染的朋友”，是“地球上第一个也是最后一个生物”。正因为如此，从决策者到普通百姓都应该了解微生物学知识，人们需要在社会决策和个人行动中权衡与微生物的利害关系，需要在幼儿园、小学、中学和大学都开展微生物学教育。

特别是，极端地质环境微生物在资源能源、生态环境和生物安全领域做出了一系列突出的贡献，是微生物领域极其关键的一环，因而需要将科学普及工作贯穿到重大科学研究计划实施、野外研究基地设立、学科建设和人才培养等一系列工程建设中。依托这些项目、学科、基地等平台，既要进行极端地质环境微生物的深入研究，服务国家重大需求，又要进行极端地质环境微生物的科普宣传，使其真正成为人类的朋友，实现人类与微生物的和平共处。

第二节　实施地质微生物的重大科学研究计划

一、重大科学研究计划实施的必要性、紧迫性和可行性

（一）必要性

在科学层面，如第一章介绍的，极端地质环境微生物研究引发了许多重

大的科学发现，诸如深部生物圈、黑暗生态系统、生命极限等。同时，它还促进了生命科学、地球科学与行星科学的交叉融合，极大地推动了地球生物学、天体生物学的大发展。极端地质环境微生物的研究因而具有重要的科学意义。此外，极端地质环境及其微生物群落还是解剖宜居地球形成与演化以及地球系统科学的一个关键抓手。例如，海底洋脊的扩张与板块俯冲的拉伸作用等构造活动造成 / 促进了海底洋壳的破裂，破裂洋壳以及其中所含的海洋流体构成地壳海洋（crustal ocean），储存了大约相当于 2% 的全球海洋水量的流体。随着这些流体的循环，地球内部的能量与物质也发生了重要交换，这些过程也支撑并孕育了丰富而独特的海底生态系统，影响并改变着全球的生物地球化学循环。洋壳生态系统的形成与演化过程为解剖地球系统科学提供了关键场景。

更重要的是，极端地质环境微生物还是极其重要的种质资源，拥有独特的功能基因和生物酶，在环保、冶金、能源、电子、材料、医药、食品等各类工业的生产上得到了很好的应用，其巨大潜力还有待进一步挖掘。这些微生物还影响天然气水合物、页岩气等其他资源能源的形成、勘探和开发利用。除了众所周知的地表极端环境，海底洋壳生物圈也蕴藏着丰富的基因资源，这些基因通过遗传表达参与了包括深海成矿在内的众多生物化学过程，提供了巨量的潜在药物、材料资源，可以为国家长期可持续发展提供重要的战略储备，因而对我国的资源能源安全具有举足轻重的意义。

极端地质环境微生物更能对核废料的地下储存、地下 CO_2 封存、地下空间开发利用等产生一系列正面或负面的影响，能够影响干旱等极端环境的生态系统，能够调控和影响气候变化。极端地质环境的病毒还会带来生物安全隐患，造成疫情的暴发，它与气候环境的关系则对未来疫情暴发的预测具有至关重要的现实意义。

正因为极端地质环境微生物具有重要的科学意义以及潜在的资源能源、生态环境和生物安全方面的战略价值，因此实施一些重大科学研究计划以开展极端地质环境微生物的研究极其必要。

（二）紧迫性

极端地质环境微生物的研究具有重要的科学意义和难以估量的战略价值，然而我国极端地质环境微生物研究面临巨大的挑战，亟须尽快突破。

第一大挑战是极端地质环境微生物的研究如何更好地服务于国家安全。能源资源安全、生物安全、生态安全都是影响国家整体安全的关键环节，无论是深地的资源能源还是深时的资源能源，都有待对微生物的研究，因为它关系到诸如天然气水合物、铀矿床、页岩气等一些重要资源能源的形成，还影响这些资源能源的勘探开发以及有效利用等诸多环节。一些极端环境的微生物因具有特殊的功能基因和生物酶，有助于我们治理西部地区不断扩大的干旱生态环境以及东部经济发达地区不断加重的生态环境污染，这些对我国现阶段的生态环境保护来说都尤为紧迫。来自洞穴等极端地质环境的病毒已经给人类带来了诸如 SARS 等区域性乃至全球性的疫情大暴发，需要深入研究这些病毒生存和传播的气候环境条件，以便从源头上阻断自然传播途径（图 6-1）；需要深入研究地下空间开发利用过程中地下微生物的利弊关系，以便了解地下微生物对生态环境的影响。这些深地微生物对我国的生物安全具有极其重要的意义，我国已经将生物安全列入国家安全体系，要系统规划和建设国家生物安全风险防控和治理体系，而极端地质环境微生物是其中的关键一环，对其进行研究显得尤其迫切。

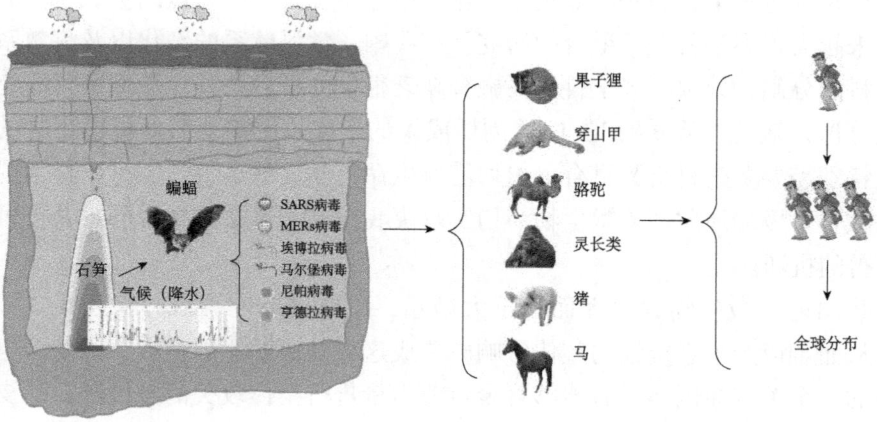

图 6-1　洞穴极端环境蝙蝠携带的各种病毒通过宿主对人类安全构成危险

注：MERs 为中东呼吸综合征。

为防止新型冠状病毒蔓延，人们可以捕杀中间宿主。例如，荷兰政府已捕杀数十万只水貂。然而，蝙蝠种类众多且数目庞大，想要大量捕杀极其困难，难以操作。其次，蝙蝠作为唯一的飞行兽类，多数地区蝙蝠的种类、食谱比较单一，所处的生态位比较独特，没有替代者，所以捕杀蝙蝠会对生态链造成多大的影响实在难以预料，但绝对不乐观。蝙蝠体内发现的一百多种病毒中有 1/3 可以在人类体内生存，比如狂犬病毒。因此，仅靠捕杀动物不是长久之计，关键是要查清气候环境变化是如何影响蝙蝠的活动的，人类活动又是如何影响蝙蝠病毒传播的，如何妥善处理人与微生物之间的关系等。

第二大挑战是重大工程计划的实施急需前瞻性的多学科交叉融合研究，以推动“卡脖子”技术的突破。得益于我国经济的快速发展以及科学研究实力的不断增强，我国在深地、深海和深空领域不断地推出了重大的工程计划，这些工程计划亟须进行一些前瞻性的科学部署，以推动一些“卡脖子”技术的突破。例如，深地和深海工程中的钻探技术、微生物样品的保真取样技术、原位观测和模拟培养技术，深空计划中的轨道器技术、地外生命探测技术等需要极端地质环境微生物科学研究的带动。在极端地质环境微生物的科学研究带动技术创新方面，我国明显落后于美国、德国、日本等一些发达国家，因而具有紧迫性。

当今科学的发展和突破除了需要一系列的技术创新来带动，多学科交叉融合也越来越紧迫，一些诸如全球变化、地球系统科学等领域的重大科学突破越来越需要多学科交叉融合的研究。在我国，学科体系的细化以及科教系统学科的分割，导致了多学科交叉融合缺乏很好的平台条件，从而使得研究力量分散，缺乏“从 0 到 1”的原创性成果的形成。极端地质环境微生物的研究特别需要多学科交叉融合，迫切需要生命科学、地球科学、行星科学的交叉融合。极端地质环境微生物为切实解决我国多学科交叉融合的难题提供了难得的机遇。

我国地质微生物的研究面临重大难题，那就是研究分散、综合集成欠缺，从而难以产生系统性的整体影响。造成这种局面的原因是多方面的，但其中的一个关键原因是没有能够凝聚重要力量进行集体攻关的重大项目的支撑。很显然，到目前为止，在国家重大专项、国家重点研发计划专项、国家重大科学研究计划等方面都没有涉及极端地质环境微生物的研究。尽管国家自然科学基金委员会资助了“水圈微生物”计划，但真正涉及极端地质环境微生物的项目及“卡脖子”技术突破的研究很少。

（三）可行性

实施极端地质环境微生物重大科学研究计划具有鲜明的必要性和紧迫性，在我国也具备可行性。

首先，我国极端地质环境微生物的一些关键领域已经在国际上产生重要影响，有望在一些前沿领域起重要的作用，发出“中国声音”。在深时微生物领域，得益于我国的优势地层资源以及在地球科学领域的重要成就，已经在一些关键地质时期的地质微生物研究方面发挥了重要的作用，无论是显生宙关键时期还是前寒武纪的一些重要时期，有关地质微生物功能群及其生物地球化学循环的研究均具有重要的国际影响，大氧化时期的微生物及其碳、氮循环，生物大灭绝期的微生物功能群及其碳、氮、硫循环，地质微生物与气候环境变化等领域均体现了中国科学家的杰出贡献。

其次，地质微生物领域已经开始在工程、技术与科学方面结合。深地、深海和深空等重大工程计划已经开始实施，极大地促进了我国一些地质微生物研究领域的特有技术的形成和突破，包括高压、热液环境的保真取样技术和实验室培养技术，深地微生物样品的保真取样计划。我国科学家分别在我国西部和东部地区的深钻孔中设计原位观测和模拟培养技术，开始尝试建立深地实验室的研究，分别在诸如洞穴系统、南海深海底建立长期的原位观测系统。所有这些都为重大科学计划的实施创造了极佳的条件，一个地质微生物研究的新局面正在形成，需要我们及时抓住这一难得的机遇，实现一些重大突破。

最后，我国在地球生物学、天体生物学方面的学科体系已经建立，人才队伍在地质微生物的关键领域都有很好的分布，可以凝聚力量来组织一支强有力的队伍进行协同攻关和技术突破。在发展地球生物学这一学科体系中，我国是国际开拓者中的重要力量。我国的行星科学发展极其迅速，在国际上的影响力不断增强。在极端地质环境微生物的一些关键领域，诸如深海微生物、深地微生物和深时微生物均有在国际上有重要影响的科学家以及相应的国家级研究群体，中国地质大学（武汉）研究深时微生物的地球生物学群体和上海交通大学研究深海微生物的群体均已经纳入国家自然科学基金委员会资助的国家创新研究群体。

二、具体研究内容、拟解决的关键科学问题和预期目标

（一）具体研究内容

重点开展以下五方面的科学研究，涉及地球科学领域的“三深一系统”以及极端地质环境微生物的种质资源和生物安全。

1. 深海微生物及其资源环境效应

开展海底沉积物、热泉和冷泉、深渊、俯冲带和大洋中脊、缺氧带等核心地区的微生物生态功能与代谢特征的研究，海洋极端环境微生物的培养分离技术和极端环境勘探技术的开发，微生物对极端环境的适应及其种质资源的研

究，以及微生物对全球气候、海洋环境和矿产资源影响的研究。例如，通过研究深海微生物对高压的适应（图 6-2），人们可以从中获得嗜压微生物的种质资源。

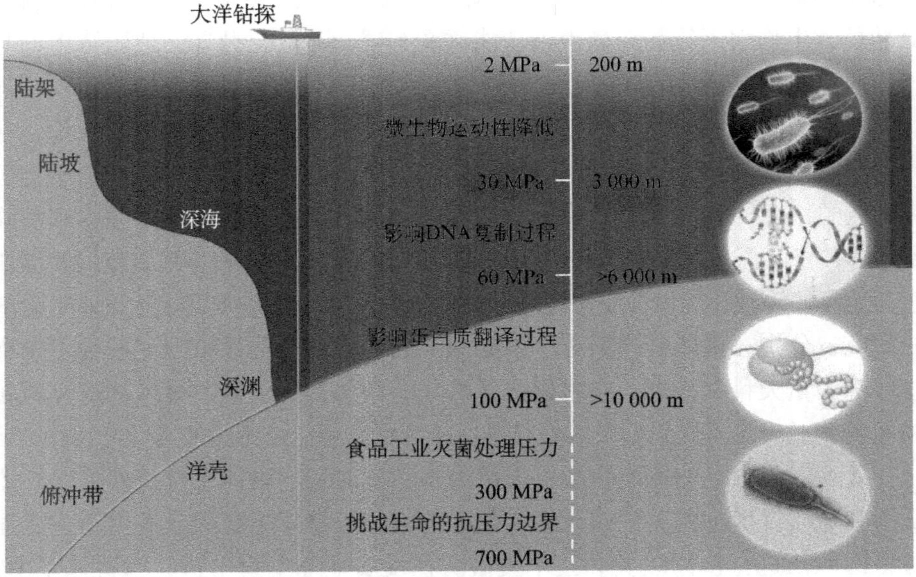

图 6-2　压力条件对深海微生物的影响

2. 深地微生物及其资源环境效应

开展地下水系统、洞穴系统、油气藏系统和地下煤层、深地流体和岩石等深地环境的微生物研究，包括深地微生物功能群对地质环境的改造作用及生态安全、深地微生物的资源环境效应、不同深度微生物的地质作用。例如，深地生物圈可以从氢气或含硫化合物中获得能量，从而对深地环境产生重要影响（图 6-3）。

3. 深空环境的生命与天体生物学

研究微生物在多种极端环境下的适应策略，寻找古老岩石中保存的生物标志物，基于地基和天基望远镜、载人空间站以及浮空器等开展地外及临近空间生命的探索，利用地外样品研究地外可能的生命，通过构建模拟空间辐射、磁场、高真空和重力效应的复合空间环境生物实验平台开展行星环境的实验室模拟。例如，现在地球上的生命主要是碳基生命，除此以外，地外是否存在氨基或硅基生命（图 6-4）？

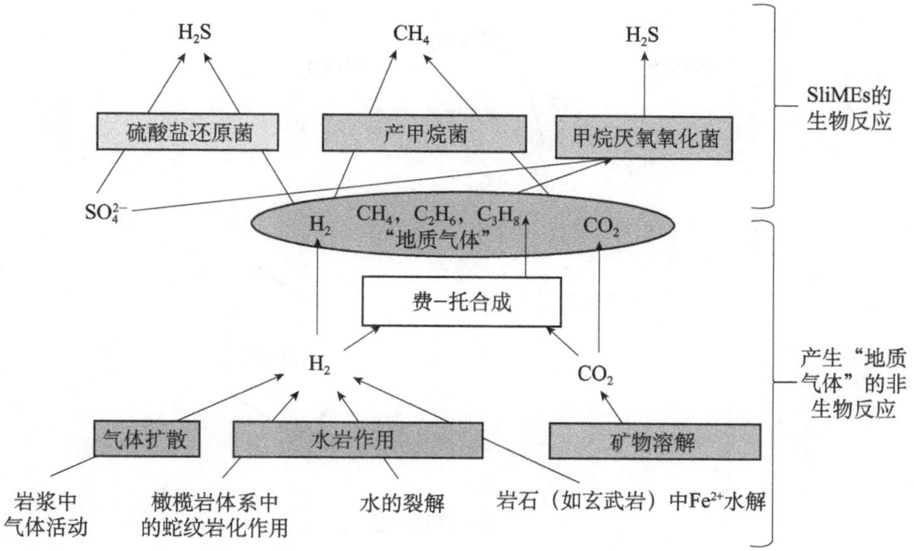

（a）氢驱动

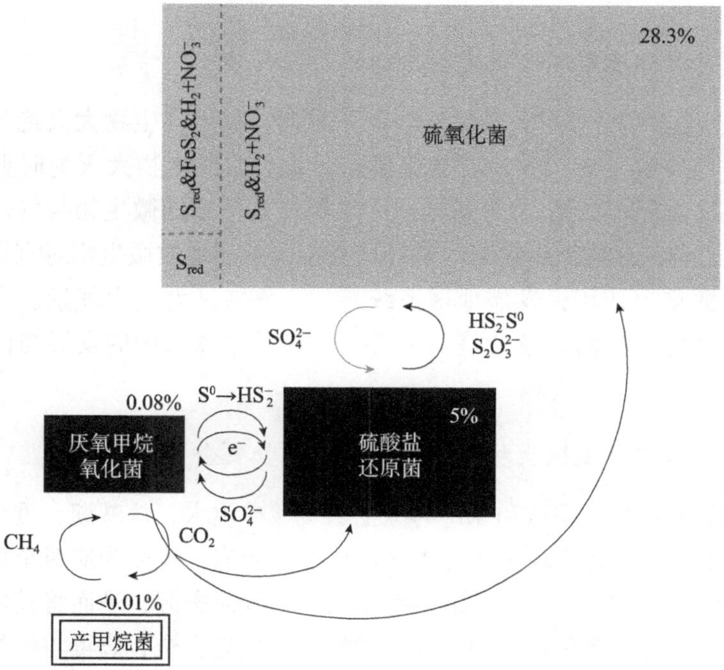

（b）硫驱动

图 6-3　氢驱动和硫驱动的深地生物圈

资料来源：修改自 Kieft（2016）；Lau et al.（2016）。

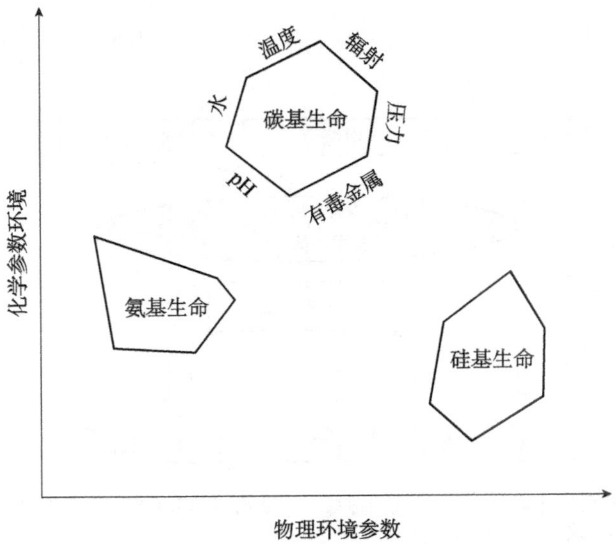

图 6-4　天体生命存在的可能形式及其与环境参数的关系

4. 深时极端地质环境微生物与地球系统科学

开展两次大氧化时期、“雪球地球”时期、显生宙生物大灭绝时期、极热时期等地质微生物及其地质演化研究，重点关注生物大灭绝时期的地质微生物与气候环境关系、“雪球地球”冰期与其他冰期微生物与气候环境关系、极热时期的地质微生物与气候环境关系，阐述地质微生物对宜居地球形成与演化的重要作用，发展地球系统科学。例如，大气含氧量、光合微生物和海洋环境之间的关系就是一个典型的涉及地球系统的发展与演化问题（图 6-5）。

5. 极端地质环境微生物与种质资源、生物安全

开展陆地热泉、雪冰和冰川与冻土、酸性矿坑水、盐碱湖、重金属污染区、农药及抗菌药污染区等重点地区的微生物研究，可以为应对全球变化和生态环境治理提供支撑，并获取微生物的各种种质资源，进而将其应用到工农业生产中。开展各种极端地质环境的病毒与生物安全、病毒与气候环境的关系、极端地质环境微生物与地质医学等一系列研究，保障人类活动的安全以及人与微生物的和谐相处，实现长治久安。

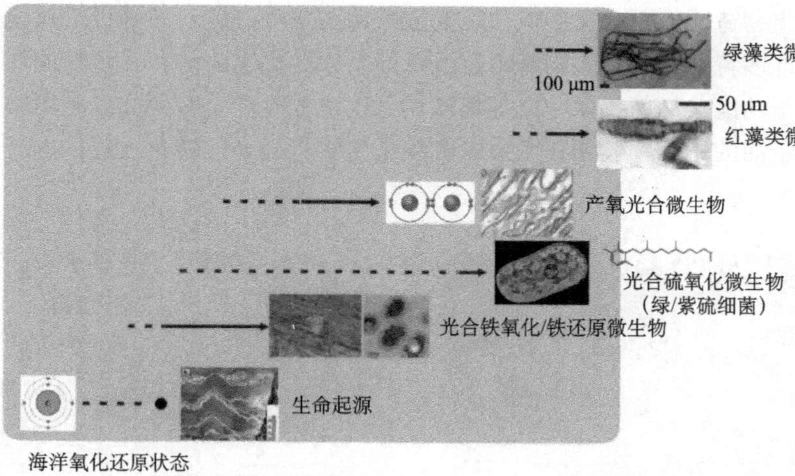

(a) 光合微生物与生命起源

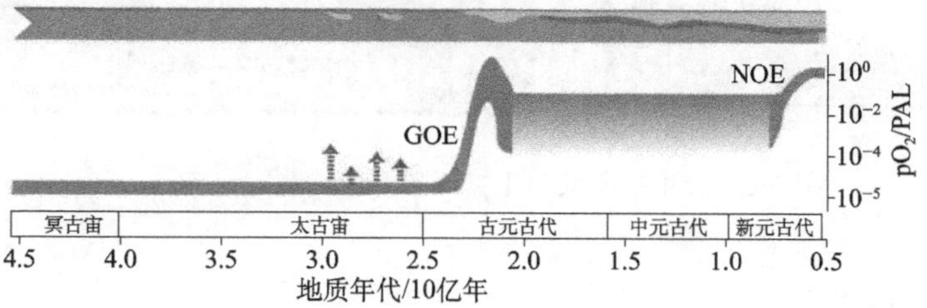

(b) 大气含氧量

图 6-5　地质时期光合微生物、大气含氧量和海洋环境之间的演化关系图（文后附彩图）

资料来源：大气含氧量的图来自 Lyons et al.（2014）。

（二）拟解决的关键科学问题

1. 科学前沿问题

在科学前沿方面，需要回答极端地质环境微生物在生物界总格局中的地位，主要涉及微生物的分布、丰度、多样性、代谢、适用机制等。同时，需要回答极端地质环境微生物在创建宜居环境中到底起多大作用，主要涉及微生物的功能、演化及其与地质环境的相互作用。

2. 国家需求方面的难题

在国家重大需求方面，极端地质环境微生物对人类有利有弊，需要回答人类如何与微生物和谐相处，以及人类的重大工程计划需要如何考虑和权衡地质微生物造成的正、负影响才能实现长治久安等问题。各种极端环境微生物具有许多特殊基因和生物酶，这些特殊基因和生物酶有什么特殊的物理化学特性与功能，如何提取，又如何应用到工农业生产上？例如，利用微生物降解剩余油的微生物驱油技术或者利用微生物将残余油转化成气体进行开采等（图 6-6）。

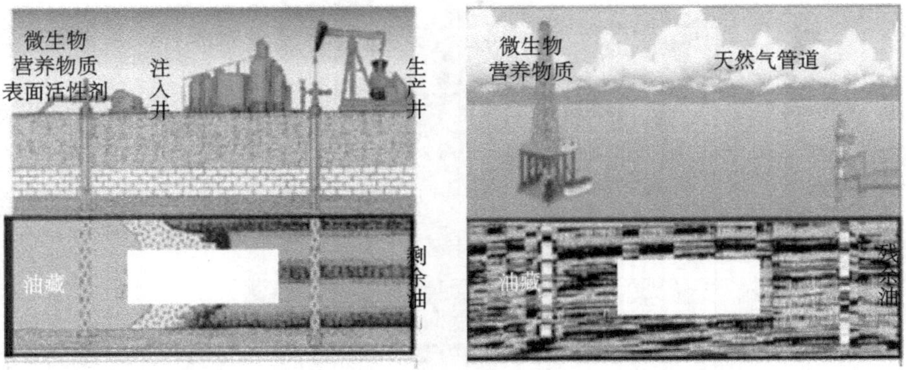

图 6-6　微生物驱油与生物气化开采油藏

（三）预期目标

通过地质微生物重大科学研究计划的实施，预期可以实现如下目标。在科学研究上，阐述地质微生物在宜居地球形成与演化方面的关键贡献，实现我国在地质微生物方面从某些单一领域的影响扩展到整体领域在国际上的重要影响，在国际上的相关学术组织中不断有中国科学家的身影和声音，推动地球生物学和天体生物学的发展。在服务国家重大需求上，获得和积累一些能广泛应用于工农业等诸多领域的关键微生物种质资源，查明深海、深地等一些极端地质环境的病毒分布及其与环境条件的关系，为我国海洋利用和地下空间开发过程中出现的生物安全服务，推动深地、深海和深空重大工程计划的一些重要技术的突破，建立现场原位研究的野外实验室。

三、可能的实现途径

可以通过国家自然科学基金委员会重大科学研究计划、科技部“十四五”重大科技专项、自然资源部重大科技专项三种途径实现极端地质环境微生物的研究。实施极端地质环境微生物重大科学计划，不仅可以形成合力，有机会提升我国在这方面的国际地位，产生许多科学前沿理论新突破，更重要的是可以发现一大批能适应各类极端环境的微生物种质资源，实现从科学前沿到工业应用的全方位的科技成果转化，从而为我国的科研成果转移这一难题提供范例。

首先，无论是拟解决的关键科学问题还是5项具体研究内容，均涉及“三深一系统”，这非常符合2019年国家自然科学基金委员会制定的地球科学领域的总体发展战略和顶层设计的主题内容。国家自然科学基金委员会未来在地球科学领域将重点发展“三深一系统”，即深地、深海、深空与地球系统科学，而地质微生物研究计划的5项研究内容的前4项就是“三深一系统”研究主题在地质微生物领域的最好体现，可以从地质微生物角度入手实现“三深一系统”的顶层设计。特别是，各类极端地质环境是一个相互支撑的有机整体，可以将“三深一系统”通过地质微生物统筹起来。例如，深时极端地质环境微生物学需要来自现代深海、深地极端环境微生物学研究获得的生物与环境相互作用机理和定量化参数。同时，深时极端地质环境微生物学的研究成果能很好地服务于深空的天体生物学、深海极端地质环境微生物学等。这些研究都是一个有机的整体，并不是“1+1=2”的简单组合，而是“1+1>2”的有机融合。

其次，可以从科技部“十四五”重大科技专项中建立地质微生物国家重点研发计划专项。地质微生物研究计划不仅涉及上游的科学前沿研究，还涉及下游的微生物种质资源的开发利用、微生物的生态安全和生物安全，以及野外现场微生物原位研究基地的建设、干旱环境和污染环境的生态治理，从而实现从科学前沿到示范建设的全链条连接，符合国家重点研发计划的专项要求。通过科技部积极促进基础研究-技术研发-工程示范应用的全链条研究。鉴于极端环境微生物有着重要的工程应用价值，并且已经在部分产业中发挥了重要作用，有必要、有条件地创建相关的工程技术示范中心，对接产业，推进极端环境微生物研究的产业化越加必要和迫切。

最后，自然资源部于2018年开始实施以“一核两深三系”为主体的重大科技创新战略，其中的“两深”即深海与深地，因此地质微生物研究计划符合自然资源部的重大科技战略，可以从地质微生物入手来探索深海和深地的各类资源能源，包括微生物的种质资源、天然气水合物资源、页岩气等非常规能源等。

第三节 统筹布局建设国家级地质微生物野外研究基地

一、野外研究基地建设的必要性、紧迫性和可行性

如果说重大科学研究计划的实施凝聚了一批科学共同体围绕共同的科学目标开展长期的科学攻关工作，那么，研究基地的建设则是地质微生物长期发展、持续发力的最根本保障，同时也是进行科普宣传的重要渠道和载体。

（一）必要性与紧迫性

1. 依托野外研究基地通过微生物科普宣传来提高社会的微生物素养

微生物早已和人类及地球水乳交融，是地球上最早也是最后一个“客人”，没有微生物，万物皆不复。然而微生物学作为一门专业课而非通识课程，在大学课堂上才闪亮登场。2019年开始的新型冠状病毒肺炎疫情还在持续，所有人都应了解和学习一定的微生物知识，提高社会的微生物素养已经势在必行。除了各类与微生物相关的学会，建立一些国家级的微生物科学研究基地进行科普宣传也是一种重要途径和载体。

首先，必须树立“微生物是人类的朋友而非敌人”的思想认知。微生物群落在生物圈与环境中所有大生物（动物和植物）的表面形成了“第二皮肤”，形成了动态的生理生态屏障。就细胞数量来说，人类生物群系的50%是微生物。肠道微生物能消化食物并释放各种营养元素，提供人体自身不能生成的微量元素和氨基酸，充当第二分泌系统分泌类激素物质，并在人类健康和疾病中扮演不可或缺的角色。微生物在食品和饮料生产、面包发酵、材料黏合、土壤肥力保持、家庭和工业废弃物处理、洁净饮用水供给等方面做出了巨大贡献。微生物已成为生物经济的核心，第四次工业革命（4IR）悄然而至，微生物过程因无需极端条件、高能耗和有毒物质，非常符合4IR内涵。微生物不仅广泛而深刻地影响着我们个人和集体，还广泛而深刻地影响着整个生物圈。在面临重大挑战和实现可持续发展的漫长道路上，很多措施和计划都将涉及微生物，在制定政策时需要了解相关的微生物知识以及如何利用微生物实现最大利益。

其次，一些全球性安全问题和环境危机的出现是由我们对微生物不了解

而自觉或不自觉地破坏了微生物的环境造成的。微生物是地球和生物进化的核心参与者和利益相关者，缺乏对微生物及其功能的认知，会使一些制定和推行的政策（国际、国家或地区等）充满风险、效果欠佳或无效，甚至适得其反，从而出现全球性的问题。人们的微生物恐惧症（细菌恐惧症）越来越严重，而错误的广告宣传，即“所有微生物都是坏的”，进一步加剧了人们对微生物的错误认识。我们必须采取一定的措施，传授给大家相关的微生物实践知识，在减少病原体与保持正常微生物群健康方面协调取舍。抗菌药耐药问题已成为人们面临的最大挑战之一，导致一些疫情重现。这些都是人们改变了微生物对环境适应性的结果。微生物既可以产生温室气体，也可以消耗温室气体，仅仅知道微生物参与环境过程是不够的，重要的是要知道它们擅长和不擅长什么。尽管微生物能够代谢多种多样的物质，但有些化合物的代谢速率比较慢，造成了污染物在环境和食物网中的积累。目前人们正面临一系列与微生物有关的严重问题，包括耐药病原体的传播、水土流失、海洋塑料污染、微塑料在食物网中的积累等。其实这些问题在很大程度上是可以避免的，这取决于决策者和社会公众对微生物的认识。因此，基于微生物知识的某些决策可防止某些重大甚至全球性灾难的发生，这显示出提高社会的微生物素养势在必行。

2. 依托野外研究基地开展全方位的科学研究和种质资源保护

关键带监测与研究已经引起发达国家的高度重视。特别是，关键带监测已经在全球建立了 60 多个野外观测网络，并开始对社会发挥重要的服务功能。我国许多部门都很重视野外科学观测研究，科技部、中国科学院、教育部、自然资源部、中国气象局等都先后建立了许多野外台站。中国科学院建立了“中国生态系统观测网络”，部分地方和科研院所也建立了以湖泊、湿地、水土、植被等为主题的野外观测站，这些工作极大地推进了我国关键带的研究。

然而，不同于以上这些常规环境的监测与研究，极端地质环境微生物的野外监测与研究具有鲜明的特点。这些极端环境的微生物具有特有的基因和生物酶，是极好的种质资源，而要充分了解和利用这些种质资源，最需要强有力的涉及关键核心技术与方法的平台的支撑，否则不可能实现极端地质环境微生物研究“从 0 到 1”的突破，更不可能通过科学前沿研究带动国家重大需求的发展。

首先，地质微生物研究不仅需要借鉴常规微生物研究的技术方法，更需要特殊微生物的保真取样技术、原位培养技术等一系列技术方法的原创性突破。深海微生物、深地微生物的研究首先要从保真取样、原位培养技术的突破开始，而深时微生物的研究首先需要在识别地质微生物功能群的技术方法上取得突破。建立种质资源库，还需要筛选分离特征微生物菌株的技术方法。所有这些技术方法的创新需要有强有力的平台为依托，进行长期且持续的工程技术人才与科学研究人员的联合攻关。正因为极端地质环境微生物研究最能体现工程技术与科学研究的完美结合，所以它才最需要国家级平台的支撑。从这一点上来说，它与关键带的野外监测有很大的不同，需要把科学与技术很好地进行结合。

其次，极端地质环境微生物的研究除了需要技术的支撑以实现科学与技术的结合，更重要的是需要依托一些重要的野外研究基地，实现科学研究与野外监测的完美结合。特别是，诸如干旱、热泉、洞穴、冰川、盐湖等极端地质环境具有十分珍贵的工农业可以广泛应用的微生物种质资源，为建立长期的野外观测与研究基地创造了极好的条件。同时，由于我国地下深部油气资源和金属矿床资源的勘探开发，我国许多地方都留下了各种深达数千米的深钻孔，这为开展深部地质过程和微生物作用的原位监测、建立深部地下微生物实验室提供了难得的机会。

实际上，许多地质微生物的重要研究成果来源于极端地质环境的野外长期研究基地。例如，在深时极端地质环境微生物方面，有关全球最大生物灭绝期地质微生物的暴发与海洋环境的变化就来源于对二叠纪—三叠纪之交全球界线层型剖面和点的研究（图 6-7）。又如，在现代极端环境方面，基于对酸性泥炭地和洞穴系统的长期监测与研究（图 6-8），发现了微生物与气候、碳循环三者之间的联系。这些地质微生物方面的重要科学发现都与长期野外研究基地的监测密不可分。

最后，极端地质环境微生物的研究既涉及很多现代环境观测和实验室模拟研究，又涉及地质时期的研究工作。随着多学科交叉、新技术方法的应用，未来将产生海量的数据。因此，将知识数据化、数据标准化、数据可视化，建立深时极端地质环境微生物学领域的共享数据平台势在必行。今后，需要加强高精度、全方面极端地质环境微生物学大数据库的建立，实现全球不同学科、不同时间和空间尺度的数据库智能整合分析，让数据共享平台成为科研人员的强大科研助手。借助大数据、机器阅读、人工智能等新技术手段，改变极端地质环境微生物学研究范式，为认识地球系统的演化过程和发展变化规律、预测地球与人类的未来奠定重要基础。要完成这些艰巨任务，需要有一系列的研究中心进行统筹计划。

图 6-7　二叠纪—三叠纪之交生物大灭绝与环境的全球重要研究基地

图 6-8　现代极端环境微生物的重要研究基地——洞穴系统

因此，在全球都在大力建设各类关键带野外观测与研究基地的今天，在全球都在大力开发各类生物资源和挖掘大数据资料的今天，在全球不断出现新型病毒造成的疫情暴发等生物安全问题的今天，在我国建立极端地质环境微生物野外观测基地显得尤为迫切。

（二）可行性

第一，我国拥有可以建立野外观测与研究基地类型十分丰富的极端地质环境，不仅具有各类容易达到的地表极端环境，而且有一些深达数千米的深钻孔。我国还拥有非常好的地质资源，广泛保存有各类深时极端地质环境的地质记录。所有这些极端地质环境覆盖面广阔、类型多样，全球少见，具有很好的地质地理优势，并可以通过野外研究基地的建设转化为生物资源优势、生物安全优势等。

第二，我国已经建设了各种不同类型的野外观测与研究基地，包括生态系统领域的、地质学领域的、地理学领域的，等等。这为建立极端地质环境微生物野外研究基地积累了非常丰富的经验。

二、具体建设内容和预期目标

（一）主要建设内容

需要建设极端地质环境微生物的一系列野外研究基地，具体建设清单如下。

1. 深海微生物野外研究基地

由于在海洋领域已经建立了一些野外研究基地以及海底观测系统，这里重点涉及深渊与海沟、热液区、冷泉区、海洋沉积物、海山与多金属结核区、洋壳岩石圈等深海极端环境，并在这些地区建立研究中心和相应的野外研究基地。需要建设的具体野外研究基地包括舟山缺氧环境监测站、台湾龟山岛浅海热液监测站、台西南冷泉区、神狐水合物区、海马冷泉区。

2. 深地微生物野外研究基地

依托地下水、油气藏、煤层、岩石与流体，以及洞穴等典型环境建立野外研究中心和研究基地。这些生境以黑暗、低氧、寡营养为特征，可以分成两类：一类是现场可以直接到达的，如洞穴系统、地下煤矿区或金属矿区等；另一类是需要依托深钻孔的，包括油气藏、地下岩石、地下水系统等。需要建设的具体野外研究基地包括贵州双河洞洞穴、湖北清江和尚洞洞穴、四川盆地永胜 1 井、黑龙江“松科 2 井”。

3. 深空生命野外研究基地

深空生命野外研究基地主要针对四类基地进行建设，包括干旱与低温等地球类比极端环境、临近空间、月球、火星。其中以地球类比极端环境为主要的地外生命探测研究基地，而临近空间、月球和火星主要依托我国的一些重大深空探测计划构建科学研究中心。需要建设的地球类比极端环境的具体野外研究基地包括柴达木盆地西部干盐滩、腾冲酸性热泉、海南玄武岩、克拉玛依蛇绿岩。

4. 深时微生物野外研究基地

主要依托古元古代/新元古代大氧化时期、新元古代"雪球地球"冰期、早三叠世和古新世—始新世极热事件、显生宙后生动物大灭绝期等这些深时极端地质环境突变期建立科学研究中心和野外研究基地。需要建设的具体野外研究基地包括湖北神农架深时极端低温环境、安徽巢湖深时极热环境、湖北宜昌深时大氧化环境、河北蓟县缺氧硫化环境、浙江长兴最大生物灭绝期。

5. 地表极端环境微生物野外研究基地

主要依托盐碱与干旱环境、极端酸性环境、高温环境、低温环境等地表极端环境建立科学研究中心和相应的野外研究基地。需要建设的具体野外研究基地包括新疆五家渠极端干旱环境、云南腾冲热泉环境等。

（二）预期目标

第一，有步骤、阶段性地推进国家级地质微生物野外研究基地的建设，最后形成地质微生物领域的整体实力和国际影响力。

第二，实现地质微生物的科学研究与野外长期监测的结合，实现微生物种质资源的科学研究与实际利用的结合。

第三，提供来自极端地质环境微生物，特别是病毒的野外第一手数据，构建大数据系统，为实现国家的生物安全提供地质微生物方面的宝贵资料。

三、可能的实现途径

可以通过国家自然科学基金委员会、国家发展和改革委员会、科技部和自然资源部四种途径建立野外研究基地。这些途径各有不同的侧重点，从而可以构建不同的地质微生物研究平台。

第一，地质微生物国家级研究中心可以通过国家自然科学基金委员会的国家基础科学中心来建立。国家自然科学基金委员会从2016年开始建设国家基础科学中心，地球科学部每年最多可设立两个国家基础科学中心。尽管国家自然科学基金委员会以建立基础科学中心为主要目的，但地质微生物研究中心也可以通过这一途径建设，同时可以增加研究中心对野外研究基地建设的要求，从而实现基础科学研究中心与野外研究基地的连接。特别是，极端地质环境微生物涉及地球科学与生命科学两大学科的交叉融合，可以依托国家自然科学基金委员会的地球科学部和生命科学部进行一些跨学部的国家基础科学中心建设，这是一大特色。以上这两点都可以促成国家自然科学基金委员会国家基础科学中心的建设。

第二，地质微生物的野外研究基地可以依托国家发展和改革委员会的重大基础设施建设进行。《国家中长期科学和技术发展规划纲要（2006—2020年）》明确将“全球环境变化的监测与对策”“生态脆弱区域生态系统功能的恢复重建”列为环境领域的优先主题，反映了近地表环境研究与监测的重要地位和作用。如前所述，极端地质环境微生物的研究必须要有一系列技术设施条件的支撑，包括钻探技术、保真取样技术、原位观测和模拟培养技术、微生物研究技术等，这些技术需要纳入重大基础设施来进行建设。特别是，微生物对资源能源、生态环境以及生物安全的重要影响，既能解决重大科学前沿问题，又能极大地服务重大国家需求，很符合国家发展和改革委员会重大基础设施建设的需求。

第三，地质微生物的野外研究基地可以依托科技部的野外研究基地等途径实现，从而与国家发展和改革委员会构建不同的平台模式。科技部已经在全国范围内建立了许多野外研究基地。截至2019年，科技部已经建立了97个国家野外科学观测研究站，绝大部分是生态系统观测站，但没有一个真正涉及微生物的观测站，更没有涉及极端地质环境微生物的观测站。

第四，地质微生物野外研究基地可以依托自然资源部进行建设。我国地质调查工作将从传统的面向矿产资源的基础地质调查向面向生态环境调查转变。在《2015年国土资源公益性行业科研专项项目指南》的“前沿探索与创新”领域，将“生态环境脆弱区地球关键带地质环境研究”列入项目指南。在《中长期科学和技术发展规划纲要（2006—2020年）》中，明确将地质环境与地质灾害领域的“区域（流域）水循环与水土资源合理配置”、基础地质领域的“典型岩溶区地质环境综合整治”等直接涉及地下水和岩溶洞穴系统等极端地质环境的内容列为优先主题。特别是，无论是深地、深海的研究还是深时和深空的研究，不仅需要有大量的技术支持，还需要地质学、物理学、化学、生物学等多个学科同时进行，这些可以通过加强同自然资源部的合作，突破现有的采样和分析技术瓶颈，不断推进极端地质环境微生物研究中心和野外监测基地的建设与发展。2011 年，国土资源部设立了第一批野外科学观测研究基地，共 84 个，2019 年自然资源部公布了 15 个野外科学观测研究站（表 6-1），但这些研究站都没有涉及极端地质环境微生物的主题。

表 6-1　2019 年自然资源部野外科学观测研究站名单

序号	名称	依托单位	运行时间
1	青藏高原盐湖野外科学观测研究站	中国地质科学院矿产资源研究所	1990 年 11 月
2	西南岩溶环境野外科学观测研究站	中国地质科学院岩溶地质研究所	1986 年 3 月
3	大陆科钻深孔地壳活动野外科学观测研究站	中国地质科学院地质研究所	2007 年 2 月
4	京津冀平原地下水与地面沉降野外科学观测研究站	中国地质环境监测院、中国地质科学院水文地质环境地质研究所、首都师范大学、北京市地质环境监测总站、天津市地质环境监测总站、河北省地质环境监测院、河北省地矿局第四水文工程地质大队	2008 年 6 月
5	北京地壳应力应变野外科学观测研究站	中国地质科学院地质力学研究所	2009 年 12 月
6	黄土高原地质灾害野外科学观测研究站	中国地质调查局西安地质调查中心	2008 年 10 月
7	长三角海洋生态环境野外科学观测研究站	自然资源部第二海洋研究所、南通海洋环境监测中心站、宁波海洋环境监测中心站、温州海洋环境监测中心站	2009 年 7 月
8	北部湾滨海湿地生态系统野外科学观测研究站	自然资源部第三海洋研究所、自然资源部第四海洋研究所、广西红树林研究中心	2013 年 9 月
9	北方滨海盐沼湿地生态地质野外科学观测研究站	青岛海洋地质研究所	2006 年 4 月
10	黄河三角洲土地利用安全野外科学观测研究站	山东省无棣土地利用野外科学观测研究院、浙江大学、山东农业大学	2011 年 12 月

续表

序号	名称	依托单位	运行时间
11	渤海海峡生态通道野外科学观测研究站	自然资源部第一海洋研究所	2014 年 8 月
12	海峡西岸海岛海岸带生态系统野外科学观测研究站	自然资源部第三海洋研究所、自然资源部海岛研究中心、厦门海洋环境监测中心站、河海大学	2014 年 1 月
13	南沙珊瑚礁生态系统野外科学观测研究站	自然资源部南海环境监测中心、自然资源部第三海洋研究所	1988 年 8 月
14	杭州全球海洋 Argo 系统野外科学观测研究站	自然资源部第二海洋研究所	2003 年 5 月
15	北极黄河站地球系统科学野外科学观测研究站	中国极地研究中心	2004 年 7 月

第四节　实现地球生物学与天体生物学的全方位连接

地球生物学与天体生物学都是地球科学与生命科学交叉结合的产物，两者有许多共同的研究内容、技术方法和研究目标，不仅可以相互借鉴，还可以相互支撑。没有地球生物学的发展，天体生物学很难深入，天体生物学因而是地球生物学在深空极端环境这一内涵的延伸。同时，两大学科体系均处于初创期，有待共同发展、相互促进。目前，在国际上，从地球生物学到天体生物学的全方位连接还未很好地建立起来，这给我国科学家提供了一个难得的发展窗口期。建议的举措如下。

一、加强天体生物学的地球相似性研究基地建设

深空环境的生命探测，由于研究对象遥不可及，目前主要的研究力量聚焦于地球上类似地外恶劣“生态环境”地区的相似性研究，即将地外天体上地质的、环境的和假定生物的条件放到地球上来研究。认识地球生命的起源、演化及其与环境相互作用是天体生物学的重要研究领域，从这种意义上说，地球生物学是天体生物学研究的基石，因而实现从地球生物学到天体生物学的发展具有极其重要的意义。

在第五章已经述及，地球类比研究涉及三类环境，分别是类似行星环境的早期地球极端环境、类似行星环境的现代地球极端环境、将地球样品置于模拟的太空或行星环境。截至2017年，全球共建立了37个天体生物学类比研究点。在我国，至少可以建立用于类比火星的现代极端环境、类比木卫二和土卫二的环境以及深时极端地质环境这三类类比研究基地。其中，柴达木盆地具有高剂量的紫外辐射和干旱少雨的特点，并发育大型干盐滩和残留部分浓缩型盐湖。这些特征使柴达木盆地被列为火星古湖泊及现代干涸湖泊的类比研究基地。

位于西太平洋的马里亚纳海沟，海沟底部的静态水压是正常大气压力的1071倍，水的密度也增加了4.96%，化能自养型微生物是该区域的主要栖息者，这种生命形式对可能存在于冰卫星深部海洋中的生命迹象是很好的示范。

我国拥有许多地球早期演化产生的地质剖面，以及与地质极端环境相关的显生宙地质剖面。对于这些剖面，已经开展了大量的地球科学领域的诸多前沿研究，包括地质环境条件、海洋环境以及地质微生物及其生物地球化学的研究，所有这些都为建立野外研究基地奠定了基础。

二、依托各类深空计划加强太空环境生物学模拟实验研究

近年来，我国深空探测在工程技术方面发展迅速，已成功实施5次月球探测任务，以火星为目标的“天问一号”也已经发射，空间站正在有序建设中，这标志着我国空间探测技术进入世界前列。与深空探测在工程技术方面取得的进步相比，我国在行星科学和天体生物学基础科学研究方面还有所欠缺。天体生物学是深空探测基础科学的重要研究内容，也是未来深空探测有可能取得重大突破的领域。NASA和ESA都把天体生物学列为其主要发展方向予以优先支持，分别制定了天体生物学的发展路线图和战略规划。建议结合我国深空探测的战略总目标，借鉴美国、欧洲等成功经验，通过顶层设计，总体规划并制定符合我国国情的天体生物学发展路线图和战略规划，凝练重大科学问题，明确优先发展方向，首先实现在若干个研究方向上的突破，再以点带面，全面提升我国在天体生物学领域的研究能力和国际影响力，服务国家深空探测重大战略需求。

在进行顶层设计时，需要考虑依托先进的研究平台和实验载荷的支撑开展天体生物学的研究，了解太空环境对生命系统的影响，这是实现从地球生物学到天体生物学的最关键一环。一方面，依托已有的国家大科学装置和

重大科技基础设施，如FAST、空间环境地面模拟装置、郭守敬天文望远镜（LAMOST）等开展天体生物学研究；另一方面，依托各类探测工程进行生物学实验，包括空间站及深空航天器内太空育种、空间站舱内与舱外微生物暴露实验等，探索微生物在大温差、不同辐射强度和辐射源下的适应及应变机制等。

三、加强地球生物学与天体生物学学科体系和队伍及平台的建设

地球生物学与天体生物学的两大学科体系在我国都处于初创期，但都发展迅猛。国家自然科学基金委员会于2007年将“生物地质学”新增为地球科学领域的二级学科，对应的英文是geobiology（“地球生物学”），重点研究生命系统和地球系统的相互作用，包括其机制、过程和演化历史等。无论是从具体内涵还是英文表述都与现在学界普遍接受的“地球生物学”一致。2000年以来，geobiology在国际上得到了迅猛的发展。美国国家研究理事会于2001年将geobiology列为21世纪地球科学基础研究六大机遇之一，并于2008年再次将其列为固体地球科学十大科学问题之一。与此同时，在英国、德国、法国等国家的21世纪地学研究框架中，也都列入了生物圈与地圈相互作用的主题。同时，geobiology成为国际地质大会、美国地球物理联合会年会、美国地质学年会等国际地学重大会议的主要议题之一。2004年，美国国家科学基金会地球科学部正式将geobiology单独列为一个计划予以稳定的支持。2016年成立了国际地球生物学会，并每两年召开学术会议，戈登地球生物学研究会议（Gordon Research Conference of Geobiology）也每两年召开学术会议。

近十年来，地球生物学在国内迅速发展。2009年，中国科学院地学部将地球生物学列为21世纪基础研究重点之一，2015年出版了《中国学科发展战略·地球生物学》，这是国内第一次将地球生物学作为一个独立学科来编写发展战略报告。从2010年起，我国已经连续举办了四届地球生物学国际会议，国家自然科学基金委员会、科技部和中国科学院等部门也已经多次组织地球生物学前沿科学与技术论坛。2018年成立中国古生物学会地球生物学分会。目前，中国科学院和部分高校已经设置了地球生物学本科或硕/博士招生专业，开设了相关课程，出版了一批关于地球生物学的成果和教材。我国已初步形成了一支具有一定规模的地球生物学研究队伍和国家

级实验平台。地球生物学作为一个学科的研究对象，内涵越来越清晰，有望成长为与地球化学和地球物理学平行而独立的学科体系，分别从生命运动、化学运动和物理运动三个方面研究地球的系统过程，是地球系统科学的重要支柱学科之一。大力加强地球生物学学科体系建设的一些条件已经成熟。

与地球生物学相比，天体生物学在学科体系、人才队伍和学术组织方面还有待尽快提升，天体生物学还没有在传统的基金资助体系中列出专门的资助方向。美国 NASA 天体生物学研究所在建立之初就设立了专门的天体生物学基金，鼓励本国和少数国家的科学家进行申请。目前美国 NASA 天体生物学研究所已经建成了一个成熟的天体生物学资助体系。中国国家自然科学基金委员会近年来在资助交叉学科和行星科学方面已经做出积极尝试，但在目前的资助体系中尚没有天体生物学的方向，建议尽快设立该资助方向或者在生物地质学（即地球生物学）中增加天体生物学的内容。这样一方面可以对天体生物学基础研究形成稳定支持，另一方面可以实现从地球生物学到天体生物学的连接与发展。

我国的深空探测已经取得了举世瞩目的成就，目前我国正处在深空探测由大变强的关键时期，规划中的“登月探火”等探测计划也越来越注重科学和工程技术的并举，行星科学基础科学研究将发挥重要作用。目前中国科学院和部分高校正在积极推进我国的行星科学一级学科建设，规划中的行星科学一级学科包括行星物理学、行星地质学和行星化学 3 个二级学科（吴福元等，2019）。国家自然科学基金委员会也自 2018 年开始新设立了行星地质学方向。国家航天局正在与国家自然科学基金委员会协调，共同规划和支持深空探测及相关科学主题的研究。天体生物学则是行星地质学的主要研究方向之一，应以此为契机，设立硕士和博士培养点，培育本土的天体生物学青年人才。

随着我国载人航天和深空探测的发展，天体生物学研究在我国日益受到重视，一些全国性学会也成立了与天体生物学相关的委员会，如中国空间科学学会的空间生命专业委员会和空间生命起源与进化专业委员会、中国宇航学会的航天医学工程与空间生物学专业委员会、中国矿物岩石地球化学学会的陨石及天体化学专业委员会、中国微生物学会的地质微生物学专业委员会等。然而，目前还没有专门的天体生物学学会或专业委员会，建议在中国科学技术协会地球科学领域的一级学会下成立天体生物学专业委员会。

第五节　在资助策略、人才和平台建设方面的其他建议

一、在深地、深海和深空等重大工程计划和科学计划中加大对地质微生物研究的支持力度

在我国的一些关键领域，加强科学与技术联合的时期已经来临。一方面，极端地质环境微生物研究需要突破保真取样、野外原位观测和模拟等“卡脖子”技术，这需要依托一些工程计划开展科学家与技术专家的合作才能实施；另一方面，我国正在实施面向国家重大需求的一系列重大工程计划，需要提升工程计划的科学前沿研究效应。这两方面的需求正是我国在深海、深地和深空等重大工程计划中加大对地质微生物支持力度的极好机遇期。

在深海，开展冷泉区和沉积物的地质微生物研究，可以查明天然气水合物等能源的成因机制，服务这些非常规能源的勘探。开展热泉区的地质微生物研究，不仅可以提供强大的能够适应高温环境的种质资源，为工业生产服务，还可以为深空微生物的研究、地球早期生命起源的探索提供关键依据。深渊微生物的研究可以发现一些嗜高压的微生物种质资源。

例如，在国家自然科学基金委员会“西太平洋地球系统多圈层相互作用”重大研究计划中，西太平洋作为全球海山系统分布最为集中的海域，还有诸如马里亚纳海沟这一全球最深的海沟海底。这里的地质运动非常活跃，不仅是富钴铁锰结壳等矿产资源的重要分布区，还拥有丰富的生物资源，如许多生长慢、生命周期长的生物；不仅可以开展西太平洋海山的生态环境、海洋地质、水文物理、化学等多学科联合调查研究，还可以推动深海极端环境微生物基因资源调查研究，这些将有助于理解海洋地质、洋流和海洋微生物之间的作用联系，有助于深入海洋生物起源与演化研究，有助于海洋矿产资源和微生物种质资源的可持续性开发利用。

又如，以全球变暖、海洋酸化及缺氧研究为突破口，加强长时间及多点观测数据，脱氧过程与机制研究更需要海洋物理、海洋生物、海洋化学、海洋地质等多学科交叉，可以通过设置专项的形式加强资源的整合。依托我国海底观测网计划，并在出现季节性近岸缺氧地点建立野外观测站等。通过国

家自然科学基金委员会、国家重点研发计划等设置国际合作项目，更有利于我国融入全球变化研究中。

在大陆深部地下环境，油气藏微生物的研究将为生物驱油提供强大的支撑，还可以发现耐高温的缺氧微生物资源。雪冰和洞穴微生物也可以提供许多种质资源，还可以服务全球气候变化。开展地下水微生物的研究，为查明地下水的水质提供科学依据，同时也为污染地区的地下水治理提供微生物学方法。地下深部基岩微生物的研究，为地下空间的安全利用提供科学决策依据，为地下环境 CO_2、核废料等安全储存奠定基础。

在深空，开展外太空环境影响生命系统的研究，为未来人类移居外星球奠定基础，也为太空育种等技术的突破创造条件。

特别重要的是，深海、深地和深空这三大工程计划可以相互结合，都可以开展各种不同地质微生物对极端环境条件的适应研究，开展地质微生物的功能及其地质作用的研究，从而不仅可以提升大科学工程的整体效率，而且可以提升地质微生物研究的整体性、系统性突破，从而形成合力。

综上，尽管极端地质环境微生物具有重要的地质作用，对提升一些重大工程计划具有极其重要的意义，而且我国也已经形成了 150 余支研究团队，包括 5 个国家创新研究群体，然而诸如深地微生物至今还没有被列入国家级深地研究指南。建议我国在实施深海、深地和深空等大科学工程时，加大对地质微生物研究的资助力度，加大对地质微生物专家参与大工程的科学论证、实施的支持力度，加大对地质微生物相关技术研发的资助力度。

二、在平台建设、人才培养和国际合作方面加大对学科交叉联合的倾斜支持

我国科学发展面临的一大难题是多学科的交叉与联合。尽管科学发展到今天，淡化学科界限、面向大学科问题的态势越来越明晰，我国科技工作者和管理部门都已经认识到多学科交叉联合的重要性，但目前还没有形成有效的机制，也没有独立的管理和资助部门。尽快寻找一个合适的抓手是突破这一难题的关键之一。

极端地质环境微生物就是进行多学科交叉联合的关键抓手之一。一方面，它涉及科学与工程技术的联合，只有那些“卡脖子”技术突破以后，科学才能实现质的飞跃。这只有依托我国正在实施的深海、深地和深空等重大工程计划才能更有效地实现。另一方面，它涉及地球科学、生命科学与化学三大自然科学的交叉联合，特别是地球科学与生命科学的交叉融合。这需要

国家自然科学基金委员会和中国科学院给予倾斜支持与资助。

地球科学与生命科学的交叉联合是一项长期且艰巨的任务。一百多年前，国际学者就倡导这两个学科之间的交叉，可以说每次古生物学质的飞跃，都和生命科学的理论方法突破有关。当前，分子生物学的快速发展再次激发了古生物学的活力，一门新的交叉学科——地球生物学应运而生，带动了占“生命之树”绝大部分地质微生物的研究，带动了生物对环境作用这一薄弱环节的研究。地球科学从庞大的时空尺度也为生命科学的发展注入了活力，让人们更加关注生物圈的未来走向，关注当今地球环境的大保护。

有鉴于此，建议国家自然科学基金委员会和中国科学院在一系列平台建设、人才培养和国际合作方面加大对地球科学与生命科学交叉联合的倾斜支持力度。特别是，在优秀青年科学基金、杰出青年科学基金和国家创新研究群体方面加大对地质微生物和地球生物学的支持与资助。在一些重大国际合作项目方面，加大对深海、深地、深空和深时地质微生物研究的资助力度，以尽快提升我国在某些领域的领导地位，并最终实现在极端地质环境微生物研究领域的整体国际地位。在国家自然科学基金委员会的生物地质学、生物地球化学和微生物学等项目申报学科领域，加大对各类极端地质环境微生物研究的资助力度。

任何学科的发展都离不开人才队伍的建设和进步。海洋极端环境的研究需要具有海洋学、微生物学、地质学等多学科的基础，而且要借助生物、化学、物理等相关分析手段。因此，海洋极端环境微生物学的人才需要具有宽广的知识结构，并且要熟悉海上采样过程，了解行业发展的制约因素。建议加大相关学科在目前学生培养过程中的比例，大力支持交叉学科的教学和对学生研究兴趣的引导，并且为学生提供出海实习的机会，让他们更加全面地认识学科特点。另外，建议加大对相关领域青年研究学者的支持力度，促进青年人才的成长，通过重大研究项目等培养未来学科带头人和具有国际影响力的科学家。最后，全球海洋面积巨大，无论是采样还是分析都离不开国际合作，建议我国科学家积极参与国际研究项目（如国际大洋钻探计划等），争取领导一些重大研究计划，并且争取重要国际会议的举办权，不断提升我国在海洋领域的科研实力和国际影响力。

通过国际合作，在建设各类相关的全国研究中心的同时，需要根据条件成熟情况分阶段地不断推进国际研究中心的建设。成立于 2018 年的上海交通大学国际深部生命研究中心，旨在形成汇聚深部生命科学领域全球顶尖科学家的人才教育和科学研究高地，将成为深部生命原创性重大成果的发源地。若以此为基础，建立更全面系统的极端微生物学研究中心，必将会对我国在深部生命和海洋极端环境研究上的国际影响力的提升具有巨大促进作用。

第六节 关于建立地质微生物常设联络中心的建议

在国内，目前有多个相关的学会或机构涉及地质微生物的研究工作。一是中国古生物学会地球生物学分会，它兼顾地质时期和现代过程的地质微生物研究，偏重地球科学领域，理事长是谢树成教授。二是中国微生物学会地质微生物分会，它聚焦现代微生物研究，偏重生命科学领域，主任委员为董海良教授。三是中国海洋湖沼学会微生物海洋学会，它偏重海洋科学，理事长为焦念志院士。

为了凝聚各学会的力量，统筹推进国内地质微生物的发展，特别是本书提出的未来有关地质微生物的重大科学研究计划的申报、野外研究基地的建设以及与一些重大工程计划的衔接等事项，需要建立常设的联络中心来继续推进这些事关地质微生物长远发展的事务性工作。

殷鸿福院士先后负责了地球生物学、深地生物圈、极端地质环境微生物这三大学科发展战略项目的研究，他来自中国地质大学（武汉）生物地质与环境地质国家重点实验室。以上的第一、第二各分会主要负责人和秘书长也来自中国地质大学（武汉）生物地质与环境地质国家重点实验室。本书涉及的五类极端地质环境微生物中，深时极端地质环境微生物（罗根明负责）、地表极端环境微生物（蒋宏忱负责）、深地生物圈（董海良负责）、深空环境生命（肖龙负责）的负责人均来自该实验室。同时，生物地质与环境地质国家重点实验室主要聚焦在地球生物学，涉及方方面面的地质微生物研究，时空跨度大，研究队伍庞大。因此，建议地质微生物联络中心常设在中国地质大学（武汉）生物地质与环境地质国家重点实验室。

谢树成教授主要从事地质微生物的研究，是生物地质与环境地质国家重点实验室副主任。他也是地球生物学、极端地质环境微生物这两个学科发展战略研究项目的工作组组长，负责组织了这两个战略报告的编写和出版工作，还参与了深地微生物战略研究项目及其报告编写。他是地球生物学国家创新研究群体学术带头人，组织了国内地质微生物的人才队伍，先后承担了

973 计划项目和国家重点研发计划项目。他完成的有关地质微生物的研究成果以第一作者或通讯作者身份发表在 *Nature* 和 *Science* 期刊上，获得过两项国家自然科学奖二等奖。因此，建议由谢树成教授具体负责，担任联络中心主任的工作。

建议联络中心设立学术指导委员会，成员包括焦念志、殷鸿福、金振民、陈骏、金之钧、邓子新、沈树忠、潘永信、朱永官等院士。

参考文献

丁芳芳 . 2013. 煤层中产甲烷微生物的培养及产气 . 合肥：合肥工业大学：33-39.

何飞 . 2020. 行星空间环境光学遥感 . 科学通报，65: 1305-1319.

焦念志，张传伦，李超，等 . 2013. 海洋微型生物碳泵储碳机制及气候效应 . 中国科学：地球科学，43(1):1-18.

李雄耀，林巍，肖智勇，等 . 2019. 行星地质学：地质学的"地外"模式 . 中国科学院院刊，34: 776-784.

李学恭，徐俊，肖湘 . 2013. 深海微生物高压适应与生物地球化学循环 . 微生物学通报，40(1): 59-70.

林海，李洋子，汪涵，等 . 2012. 产甲烷菌的诱变及其对煤层气产出的影响 . 煤炭学报，37(A02)：407-411.

林巍 . 2020. 临近空间生物研究及其天体生物学意义 . 科学通报，65: 1297-1304.

林巍，李一良，王高鸿，等 . 2020. 天体生物学研究进展和发展趋势 . 科学通报，65: 380-391.

邵宗泽 . 2018. 深海热液区化能自养微生物多样性、代谢特征与环境作用 . 科学通报，63(36): 3902-3910.

苏现波，徐影，吴昱，等 . 2011. 盐度、pH 对低煤阶煤层生物甲烷生成的影响 . 煤炭学报，36(8)：1302-1306.

孙伊，刘红 . 2014. 密闭舱室内微生物污染问题及其防控体系 . 载人航天，20: 543-549.

汤浩 . 2014. 煤系地下水硫酸盐还原菌和产甲烷菌的共生性研究 . 合肥：合肥工业大学：36-41.

陶明信，王万春，李中平，等 . 2014. 煤层中次生生物气的形成途径与母质综合研究 . 科学通报，11：970-978.

王爱宽，秦勇，林玉成，等 . 2010. 褐煤中天然产甲烷菌富集培养与生物气产出模拟 . 高校地质学报，16(1)：80-85.

王风平，陈云如 . 2017. 深部生物圈研究进展与展望 . 地球科学进展，32(12): 1277-1286.

吴福元，魏勇，宋玉环，等 . 2019. 从科教融合到科学引领：中国特色的行星科学建设思路 . 中国科学院院刊，7: 741-747.

徐恒 . 2017. 一种不可培养微生物原位模拟实验舱及原位研究系统 . CN206768088U，2017-05-25.

袁俊霞，印红，马玲玲，等 . 2020. 载人航天工程中的微生物科学与技术应用 . 载人航天，26: 237-243.

赵维殳，肖湘 . 2017. 多重极端环境中的生命：深海热液中的超嗜热古菌 *Thermococcales*. 中国科学：生命科学，47(5):14-25.

Abrahão J, Silva L, Silva L S, et al. 2018. Tailed giant Tupanvirus possesses the most complete translational apparatus of the known virosphere. Nature Communications, 9(1): 749.

Alcamán M E, Fernandez C, Delgado A, et al. 2015. The cyanobacterium Mastigocladus fulfills the nitrogen demand of a terrestrial hot spring microbial mat. The ISME Journal, 9: 2290-2303.

Anantharaman K, Brown C T, Hug L A, et al. 2016. Thousands of microbial genomes shed light on interconnected biogeochemical processes in an aquifer system. Nature Communications, 7: 13219.

Anantharaman K, Duhaime M B, Breier J A, et al. 2014. Sulfur oxidation genes in diverse deep-sea viruses. Science, 344(6185): 757-760.

Attistuzzi F U, Feijao A, Hedges S B. 2004. A genomic timescale of prokaryote evolution: insights into the origin of methanogenesis, phototrophy, and the colonization of land. BMC Evolutionary Biology, 4: 44.

Aubineau J, El Albani A E, Bekker A, et al. 2019. Microbially induced potassium enrichment in Paleoproterozoic shales and implications for reverse weathering on early Earth. Nature Communications, 10: 2670.

Baker-Austin C, Dopson M. 2007. Life in acid: pH homeostasis in acidophiles. Trends in Microbiology, 15: 165-171.

Baker B J, Saw J H, Lind A E, et al. 2016. Genomic inference of the metabolism of cosmopolitan subsurface archaea, Hadesarchaea. Nature Microbiology, 1: 16002.

Bar-On Y M, Phillips R, Milo R. 2018. The biomass distribution on Earth. Proceedings of the National Academy of Sciences of USA, 115(25): 6506-6511.

Battistuzzi F U, Feijao A, Hedges S B. 2004. A genomic timescale of prokaryote evolution: insights into the origin of methanogenesis, phototrophy, and the colonization of land. BMC Evolutionary Biology, 4: 44.

Beal E J, Claire M W, House C H. 2011. High rates of anaerobic methanotrophy at low sulfate concentrations with implications for past and present methane levels. Geobiology, 9: 131-139.

Beal E J, House C H, Orphan V J. 2009. Manganese-and iron-dependent marine methane oxidation. Science, 325: 184-187.

Becraft E D, Woyke T, Jarett J, et al. 2017. Rokubacteria: genomic Giants among the Uncultured Bacterial Phyla. Frontiers in Microbiology, 8: 2264.

Borgonie G, Linage-Alvarez B, Ojo A O, et al. 2015a. Eukaryotic opportunists dominate the deep-subsurface biosphere in South Africa. Nature Communications, 6: 8952.

Borgonie G, Linage-Alvarez B, Ojo A, et al. 2015b. Deep subsurface mine stalactites trap endemic fissure fluid archaea, bacteria, and nematoda possibly originating from ancient seas. Frontiers in Microbiology, 6: 833.

Bosak T, Souza-Egipsy V, Corsetti F A, et al. 2004. Micrometer-scale porosity as a biosignature in carbonate crusts. Geology, 32: 781-784.

Braun S, Mhatre S S, Jaussi M, et al. 2017. Microbial turnover times in the deep seabed studied by amino acid racemization modelling. Scientific Reports, 7(1): 5680.

Brock T D, Freeze H. 1969. *Thermus aquaticus* gen. n. and sp. n., a nonsporulating extreme thermophile. Journal of Bacteriology, 98(1): 289-297.

Brocks J J, Jarrett A J, Sirantoine E, et al. 2017. The rise of algae in Cryogenian oceans and the emergence of animals. Nature, 548: 578-581.

Brocks J J, Love G D, Summons R E, et al. 2005. Biomarker evidence for green and purple sulphur bacteria in a stratified Palaeoproterozoic sea. Nature, 437: 866-870.

Bryan N C, Christner B C, Guzik T G, et al. 2019. Abundance and survival of microbial aerosols in the troposphere and stratosphere. The International Society of Microbial Ecology Journal, 13: 2789-2799.

Cairns-Smith A G. 1982. Genetic takeover and the mineral origins of life. Cambridge: Cambridge University Press: 488.

Canfield D E, Stewart F J, Thamdrup B, et al. 2010. A cryptic sulfur cycle in oxygen-minimum-zone waters off the Chilean coast. Science, 330(6009): 1375-1378.

Castanie-Cornet M P, Penfound T A, Smith D, et al. 1999. Control of acid resistance in *Escherichia coli*. Journal of Bacteriology, 181: 3525-3535.

Chien A, Edgar D B, Trela J M. 1976. Deoxyribonucleic acid polymerase from the extreme thermophile *Thermus aquaticus*. Journal of Bacteriology, 127(3): 1550-1557.

Chivian D, Brodie E L, Alm E J, et al. 2008. Environmental genomics reveals a single-species ecosystem deep within Earth. Science, 322: 275-278.

Cockell C S, Cousins C, Wilkinson P T, et al. 2015. Are thermophilic microorganisms active in cold environments? International Journal of Astrobiology, 14: 457-463.

Colman D R, Poudel S, Stamps B W, et al. 2017. The deep, hot biosphere: twenty-five years of retrospection. Proceedings of the National Academy of Sciences of USA, 114(27): 6895-6903.

Connon S A, Lester E D, Shafaat H S, et al. 2007. Bacterial diversity in hyperarid Atacama desert soils. Journal of Geophysical Research Atmospheres, 112: G04S17.

Cortesao M, Fuchs F M, Commichau F M, et al. 2019. Bacillus subtilis spore resistance to simulated Mars surface conditions. Frontiers in Microbiology, 10: 333.

Ćirković M M. 2004. The temporal aspect of the Drake equation and SETI. Astrobiology, 4: 225-231.

Dalmasso C, Oger P, Selva G, et al. 2016. *Thermococcus piezophilus* sp. nov., a novel hyperthermophilic and piezophilic archaeon with a broad pressure range for growth, isolated from a deepest hydrothermal vent at the Mid-Cayman Rise. Systematic and Applied Microbiology, 39: 440-444.

Dartnell L R, Hunter S J, Lovell K V, et al. 2010. Low-temperature ionizing radiation resistance of *Deinococcus radiodurans* and Antarctic Dry Valley bacteria. Astrobiology, 10: 717-732.

Darwin C R. 1871. Letter to Hooker J D//Darwin F. The life and letters of Charles Darwin. New York: Basic Books: 202-203.

DasSarma P, DasSarma S. 2018. Survival of microbes in Earth's stratosphere. Current Opinion in Microbiology, 43: 24-30.

Davidson M M, Silver B J, Onstott T C, et al. 2011. Capture of planktonic microbial diversity in fractures by long-term monitoring of flowing boreholes, Evander Basin, South Africa. Geomicrobiology Journal, 28: 275-300.

Daye M, Klepac-Ceraj V, Pajusalu M, et al. 2019. Light-driven anaerobic microbial oxidation of manganese. Nature, 576: 311-314.

De Maayer P, Anderson D, Cary C, et al. 2014. Some like it cold: understanding the survival strategies of psychrophiles. EMBO Reports, 15: 508-517.

De Vera J P, Boettger U, de la Torre Noetzel R, et al. 2012. Supporting Mars exploration: BIOMEX in Low Earth Orbit and further astrobiological studies on the Moon using Raman and PanCam technology. Planetary and Space Science, 74: 103-110.

Delaney J R, Kelley D S, Lilley M D, et al. 1998. The quantum event of oceanic crustal accretion: impacts of diking at mid-ocean ridges. Science, 281: 222-230.

DeLong E F. 1992. Archaea in coastal marine environments. Proceedings of the National Academy of Sciences of USA, 89: 5685-5689.

DeLong E F. 1997. Evolutionary relationships of cultivated psychrophilic and barophilic deep-sea bacteria. Applied & Environmental Microbiology, 63: 2105-2108.

Deming J W, Baross J A. 1993. Deep-sea smokers: windows to a subsurface biosphere? Geochimica et Cosmochimica Acta, 57: 3219-3230.

Des Marais D J, Allamandola L J, Benner S A, et al. 2003. The NASA astrobiology roadmap. Astrobiology, 3: 219-235.

Des Marais D J, Nuth J A, Allamandola L J, et al. 2008. The NASA astrobiology roadmap. Astrobiology, 8: 715-730.

Desai M S, Assig K, Dattagupta S. 2013. Nitrogen fixation in distinct microbial niches within a chemoautotrophy-driven cave ecosystem.The International Society of Microbial Ecology Journal, 7: 2411-2423.

D'Hondt S, Rutherford S, Spivack A J. 2002. Metabolic activity of subsurface life in deep-sea sediments. Science, 295(5562): 2067-2070.

Djokic T, Van Kranendonk M J, Campbell K A, et al. 2017. Earliest signs of life on land preserved in ca. 3.5 Ga hot spring deposits. Nature Communications, 8: 15263.

Dodd M S, Papineau D, Grenne T, et al. 2017. Evidence for early life in Earth's oldest hydrothermal vent precipitates. Nature, 543: 60-64.

Dolfing J, Larter S R, Head I M. 2008. Thermodynamic constraints on methanogenic crude oil biodegradation. The International Society of Microbial Ecology Journal, 2: 442-452.

Dong Y, Kumar C G, Chia N, et al. 2014. Halomonas sulfidaeris-dominated microbial community inhabits a 1.8 km-deep subsurface Cambrian Sandstone reservoir. Environmental Microbiology, 16: 1695-708.

Dragosits M, Mozhayskiy V, Quinones-Soto S, et al. 2013. Evolutionary potential, cross-stress behavior, and the genetic basis of acquired stress resistance in *Escherichia coli*. Molecular Systems Biology, 9: 643.

Dubilier N, McFall-Ngai M, Zhao L P. 2015. Microbiology: create a global microbiome effort. Nature, 526(7575): 631-634.

Dundas C M, Bramson A M, Ojha L, et al. 2018. Exposed subsurface ice sheets in the Martian mid-latitudes. Science, 359: 199-201.

Edwards K J, Bond P L, Gihring T M, et al. 2000. An archaeal iron-oxidizing extreme acidophile important in acid mine drainage. Science, 287: 1796-1799.

Eigenbrode J L, Freeman K H. 2006. Late Archean rise of aerobic microbial ecosystems. Proceedings of the National Academy of Sciences of USA, 103: 15759-15764.

Eigenbrode J L, Summons R E, Steele A, et al. 2018. Organic matter preserved in 3-billion-year-old mudstones at Gale crater, Mars. Science, 360: 1096-1101.

Engelhardt T, Kallmeyer J, Cypionka H, et al. 2014. High virus-to-cell ratios indicate ongoing production of viruses in deep subsurface sediments. The ISME Journal, 8(7):1503-1509.

Ettwig K F, Zhu B, Speth D, et al. 2016. Archaea catalyze iron-dependent anaerobic oxidation of methane. Proceedings of the National Academy of Sciences of USA, 113(45):12792-12796.

Fairén A G, Parro V, Schulze-Makuch D, et al. 2017. Searching for life on Mars before it is too late. Astrobiology, 17: 962-970.

Fike D A, Grotzinger J P, Pratt L M, et al. 2006. Oxidation of the Ediacaran ocean. Nature, 444: 744-747.

Filippidou S, Junier T, Wunderlin T, et al. 2019. Adaptive strategies in a poly-extreme environment: differentiation of vegetative cells in serratia ureilytica and resistance to extreme conditions. Frontiers in Microbiology, 10: 102.

Fischer E, Martínez G M, Elliott H M, et al. 2014. Experimental evidence for the formation of liquid saline water on Mars. Geophysical Research Letters, 41: 4456-4462.

Fisk M R, Giovannoni S J, Thorseth I H. 1998. Alteration of oceanic volcanic glass: textural evidence of microbial activity. Science, 281: 978-980.

Flemming H C, Wuertz S. 2019. Bacteria and archaea on Earth and their abundance in biofilms. Nature Reviews Microbiology, 17: 247-260.

Frank Y, Banks D, Avakian M, et al. 2016. Firmicutes is an important component of microbial communities in water-injected and pristine oil reservoirs, Western Siberia, Russia. Geomicrobiology Journal, 33: 387-400.

Gibert O, Assal A, Devlin H, et al. 2019. Performance of a field-scale biological permeable reactive barrier for *in-situ* remediation of nitrate-contaminated groundwater. Science of the Total Environment, 659: 211-220.

Gilmour I, Sephton M A. 2004. An introduction to astrobiology. Cambridge: Cambridge University Press.

Glass J B, Orphan V J. 2012. Trace metal requirements for microbial enzymes involved in the production and consumption of methane and nitrous oxide. Frontiers in Microbiology, 3: 61.

Glud R N, Wenzhöfer F, Middelboe M, et al. 2013. High rates of microbial carbon turnover in sediments in the deepest oceanic trench on Earth. Nature Geoscience, 6: 284-288.

Godoy M S, Mongili B, Fino D, et al. 2017. About how to capture and exploit the CO_2 surplus that nature, perse, is not capable of fixing. Microbial Biotechnology, 10: 1216-1225.

Greaves J S, Richards A M S, Bains W, et al. 2020. Phosphine gas in the cloud decks of Venus. Nature Astronomy,5:655-664.

Grice K, Cao C Q, Love G D, et al. 2005. Photic zone euxinia during the Permian-Triassic superanoxic event. Science, 307: 706-709.

Grotzinger J P. 2014. Exploring martian habitability. Habitability, taphonomy, and the search for organic carbon on Mars. Science, 343: 386-387.

Guidi L, Chaffron S, Bittner L, et al. 2016. Plankton networks driving carbon export in the oligotrophic ocean. Nature, 532: 465-470.

Guo H, Liu R, Yu Z, et al. 2012. Pyrosequencing reveals the dominance of methylotrophic methanogenesis in a coal bed methane reservoir associated with Eastern Ordos basin in China. International Journal of Coal Geology, 93: 56-61.

Guo H, Zhang J, Han Q, et al. 2017. Important role of fungi in the production of secondary biogenic coalbed methane in China's southern Qinshui basin. Energy & Fuels, 31(7): 7197-7207.

Halevy I, Bachan A. 2017. The geologic history of seawater pH. Science, 355: 1069-1071.

Hallmann C, Schwark L, Grice K. 2008. Community dynamics of anaerobic bacteria in deep petroleum reservoirs. Nature Geoscience, 1: 588-591.

Han H J, Wen H L, Zhao L, et al. 2017. Novel coronaviruses, astroviruses, adenoviruses and circoviruses in insectivorous bats from northern China. Zoonoses and Public Health, 64: 636-646.

Haroon M F, Hu S, Shi Y, et al. 2013. Anaerobic oxidation of methane coupled to nitrate reduction in a novel archaeal lineage. Nature, 500: 567-570.

Harrison J P, Gheeraert N, Tsigelnitskiy D, et al. 2014. The limits for life under multiple extremes. Trends in Microbiology, 21: 204-212.

Hassler D M, Zeitlin C, Wimmer-Schweingruber R F, et al. 2014. Mars' Surface radiation environment measured with the Mars Science Laboratory's Curiosity rover. Science, 343: 1244797.

Hayes A, Aharonson O, Callahan P, et al. 2008. Hydrocarbon lakes on Titan: distribution and interaction with a porous regolith. Geophysical Research Letters, 35: L09204.

He H, Han Y X, Jin D C, et al. 2016. Microbial consortium in a non-production biogas coal mine of eastern China and its methane generation from lignite. Energy Sources, Part A: Recovery, Utilization, and Environmental Effects, 38(10): 1377-1384.

He Y, Li M, Perumal V, et al. 2016. Genomic and enzymatic evidence for acetogenesis among multiple lineages of the archaeal phylum bathyarchaeota widespread in marine sediments. Nature Microbiology. 1(6):16035.

Hoehler T M, Jørgensen B B. 2013. Microbial life under extreme energy limitation. Nature Reviews Microbiology, 11: 83-94.

Hoffman P F, Kaufman A J, Halverson G P, et al. 1998. A Neoproterozoic snowball earth. Science, 281: 1342-1346.

Hönisch B, Ridgwell A, Schmidt D N, et al. 2012. The geological record of ocean acidification. Science, 335: 1058-1063.

Horneck G, Walter N, Westall F, et al. 2016. AstRoMap European astrobiology roadmap. Astrobiology, 16: 201-243.

Huber J A, Johnson H P, Butterfield D A, et al. 2006. Microbial life in ridge flank crustal fluids. Environmental Microbiology, 8: 88-99.

Hurford T A, Helfenstein P, Hoppa G V, et al. 2007. Eruptions arising from tidally controlled periodic openings of rifts on Enceladus. Nature, 447: 292-294.

Idris H, Goodfellow M, Sanderson R, et al. 2017. Actinobacterial rare biospheres and dark matter revealed in habitats of the Chilean Atacama Desert. Scientific Reports, 7(1): 8373.

Irwin A J, Oliver M J. 2009. Are ocean deserts getting larger? Geophysical Research Letters, 36: L18609.

Jakosky B M, Farmer C B. 1982. The seasonal and global behavior of water vapor in the Mars atmosphere: complete global results of the viking atmospheric water detector experiment. Journal of Geophysical Research, 87: 2999-3019.

Jakubowska-Deredas M, Jurczak-Kurek A, Richert M, et al. 2012. Diversity of tailed phages in Baltic Sea sediment: large number of siphoviruses with extremely long tails. Research in Microbiology, 163: 292-296.

Jamieson A. 2015. The hadal zone: life in the deepest oceans. Cambridge: Cambridge University Press.

Jiang G Q, Kaufman A J, Christie-Blick N, et al. 2007. Carbon isotope variability across the Ediacaran Yangtze platform in South China: implications for a large surface-to-deep ocean $\delta^{13}C$ gradient. Earth and Planetary Science Letters, 261: 303-320.

Jiang Z, Li P, Wang Y H, et al. 2019. Arsenic mobilization in a high arsenic groundwater revealed by metagenomic and Geochip analyses. Scientific Reports, 9: 12972.

Joachimski M M, Lai X L, Shen S Z, et al. 2012. Climate warming in the latest Permian and the Permian-Triassic mass extinction. Geology, 40: 195-198.

Joseph R G, Dass R S, Rizzo V, et al. 2019. Evidence of life on Mars. Journal of Astrobiology and Space Science Reviews, 1: 40-81.

Jungbluth S P, Bowers R M, Lin H T, et al. 2016. Novel microbial assemblages inhabiting crustal fluids within mid-ocean ridge flank subsurface basalt. The International Society of Microbial Ecology Journal, 10: 2033-2047.

Jørgensen B B, D’Hondt S. 2006. A starving majority deep beneath the seafloor. Science, 314(5801): 932-934.

Jørgensen B B. 2000. Bacteria and marine biogeochemistry//Schulz H D, Zabel M. Marine Geochemistry. Berlin: Springer: 173-207.

Jørgensen B B. 2011. Deep subseafloor microbial cells on physiological standby. Proceedings of the National Academy of Sciences of USA, 108(45): 18193-18194.

Kallmeyer J, Pockalny R, Adhikari R R, et al. 2012. Global distribution of microbial abundance and biomass in subseafloor sediment. Proceedings of the National Academy of Sciences of USA, 109(40):16213-16216.

Kasting J F, Kopparapu R, Ramirez R M, et al. 2014. Remote life-detection criteria, habitable zone boundaries, and the frequency of Earth-like planets around M and late K stars. Proceedings of the National Academy of Sciences of USA, 111: 12641-12646.

Katoh H, Asthana R K, Ohmori M. 2004. Gene expression in the cyanobacterium Anabaena sp. PCC7120 under desiccation. Microbial Ecology, 47: 164-174.

Kieft T L. 2016. Microbiology of the deep continental biosphere//Hurst C J. Their world: A Diversity of Microbial Environments. Berlin: Springer: 225-249.

Kim H, Kaown D, Mayer B, et al. 2015. Identifying the sources of nitrate contamination of groundwater in an agricultural area (Haean Basin, Korea) using isotope and microbial community analyses. Science of the Total Environment, 533: 566-575.

Koki H. 2011. Extremophiles handbook//Kato C. Distribution of Piezophiles. Berlin: Springer: 643-655.

Krstulović N, Šolić M, Šantić D, et al. 2013. Microbial community structure in two anchialine caves on Mljet Island (Adriatic Sea). Acta Adriatica, 54: 183-198.

Krulwich T A, Sachs G, Padan E. 2011. Molecular aspects of bacterial pH sensing and homeostasis. Nature Reviews Microbiology, 9: 330-343.

Kumar S, Herrmann M, Blohm A, et al. 2018. Thiosulfate- and hydrogen-driven autotrophic denitrification by a microbial consortium enriched from groundwater of an oligotrophic limestone aquifer. FEMS Microbiology Ecology, 94 (10): fiy141.

Kumaresan D, Stephenson J, Doxey A C, et al. 2018. Aerobic proteobacterial methylotrophs in movile cave: genomic and metagenomic analyses. Microbime,6 (1)1-10.

Kuypers M M, Blokker P, Erbacher J, et al. 2001. Massive expansion of marine archaea during a mid-Cretaceous oceanic anoxic event. Science, 293: 92-95.

Lam P, Kuypers M M M. 2011. Microbial nitrogen cycling processes in oxygen minimum zones. Annual Review of Marine Science, 3: 317-345.

Lau M C Y, Cameron C, Magnabosco C, et al. 2014. Phylogeny and phylogeography of functional genes shared among seven terrestrial subsurface metagenomes reveal N-cycling and microbial evolutionary relationships. Frontiers in Microbiology, 5: 531.

Lau M C Y, Kieft T L, Kuloyo O, et al. 2016. An oligotrophic deep-subsurface community dependent on syntrophy is dominated by sulfur-driven autotrophic denitrifiers. Proceedings of the National Academy of Sciences of USA, 113: E7927–E7936.

Lazcano A, Miller S L. 1996. The origin and early evolution of life: prebiotic chemistry, the pre-RNA world, and time. Cell, 85: 793-798.

Lennon J T, Aanderud Z T, Lehmkuhl B K, et al. 2012. Mapping the niche space of soil microorganisms using taxonomy and traits. Ecology, 93: 1867-1879.

Lever M A. 2011. Acetogenesis in the energy-starved deep biosphere: a paradox? Frontiers in Microbiology, 2: 284.

Lever M A, Rouxel O, Alt J C, et al. 2013. Evidence for microbial carbon and sulfur cycling in deeply buried ridge flank basalt. Science, 339: 1305-1308.

Li C, Love G D, Lyons T W, et al. 2010. A stratified redox model for the Ediacaran ocean. Science, 328: 80-83.

Li H Z, Yang Q H, Li J, et al. 2015. The impact of temperature on microbial diversity and AOA activity in the Tengchong Geothermal Field, China. Scientific Reports, 5: 17056.

Li J T, Mara P, Schubotz F, et al. 2020. Recycling and metabolic flexibility dictate life in the lower oceanic crust. Nature, 579: 250-255.

Li P, Jiang Z, Wang Y H, et al. 2017. Analysis of the functional gene structure and metabolic potential of microbial community in high arsenic groundwater. Water Research, 123: 268-276.

Li X X, Mbadinga S M, Liu J F, et al. 2017. Microbiota and their affiliation with physiochemical characteristics of different subsurface petroleum reservoirs. International biodeterioration and Biodegradation, 120: 170-185.

Lindeboom R E F, Ilgrande C, Carvajal-Arroyo J M, et al. 2018. Nitrogen cycle microorganisms can be reactivated after Space exposure. Scientific Reports, 8: 13783.

Lipp J S, Morono Y, Inagaki F, et al. 2008. Significant contribution of archaea to extant biomass in marine subsurface sediments. Nature, 454(7207): 991-994.

Liu C H, Huang X, Xie T N, et al. 2017. Exploration of cultivable fungal communities in deep coal - bearing sediments from ~ 1.3 to 2.5 km below the ocean floor. Environmental Microbiology, 19: 803-818.

Liu H, Li P, Wang H L, et al. 2020. Arsenic mobilization affected by extracellular polymeric substances (EPS) of the dissimilatory iron reducing bacteria isolated from high arsenic groundwater. Science of the Total Environment, 735:139501.

Liu Y F, Qi Z Z, Shou L B, et al. 2019. Anaerobic hydrocarbon degradation in candidate phylum 'Atribacteria' (JS1) inferred from genomics. The International Society of Microbial Ecology Journal, 13: 2377-2390.

Lomstein B A, Langerhuus A T, D'Hondt S, et al. 2012. Endospore abundance, microbial growth and necromass turnover in deep sub-seafloor sediment. Nature, 484:101-104.

Love G D, Grosjean E, Stalvies C, et al. 2009. Fossil steroids record the appearance of Demospongiae during the Cryogenian period. Nature, 457: 718-721.

Lu P, Ma W L, Li H , et al. 2009. Seasonal dynamic changes in bacterial compositions in the Inner Mongolia desert steppe. Frontiers of Biology in China, 4: 187-191.

Luo G M, Kump L R, Wang Y B, et al. 2010. Isotopic evidence for an anomalously low oceanic sulfate concentration following end-Permian mass extinction. Earth and Planetary Science Letters, 300: 101-111.

Lyons T W, Reinhard C T, Planavsky N J. 2014. The rise of oxygen in Earth' s early ocean and atmosphere. Nature, 506: 307-315.

Macalady J, Banfield J F. 2003. Molecular geomicrobiology: genes and geochemical cycling. Earth and Planetary Science Letters, 209: 1-17.

Magnabosco C, Lin L H, Dong H, et al. 2018. The biomass and biodiversity of the continental subsurface. Nature Geoscience, 11: 707-717.

Magnabosco C, Ryan K, Lau M C Y, et al. 2016. A metagenomic window into carbon metabolism at 3 km depth in Precambrian continental crust. The International Society of Microbial Ecology Journal, 10: 730-741.

Man B Y, Wang H M, Yun Y, et al. 2018. Diversity of fungal communities in Heshang Cave of central China revealed by mycobiome-sequencing. Frontiers in Microbiology, 9: 1400.

Manga M, Wang C Y. 2007. Pressurized oceans and the eruption of liquid water on Europa and Enceladus. Geophysical Research Letters, 34: L07202(1-5).

Martin W, Baross J, Kelley D, et al. 2008. Hydrothermal vents and the origin of life. Nature Reviews Microbiology, 6: 805-814.

Martins Z, Cottin H, Kotler J M, et al. 2017. Earth as a tool for astrobiology: a european perspective. Space Science Reviews, 209: 43-81.

Martín-Torres F J, Zorzano M P, Valentín-Serrano P, et al. 2015. Transient liquid water and water activity at Gale crater on Mars. Nature Geoscience, 8: 357-361.

Mason O U, Nakagawa T, Rosner M, et al. 2010. First investigation of the microbiology of the deepest layer of ocean crust. PLoS One, 5:e15399.

Mateo M E, Galvez M S, Gil L C, et al. 2019. Pyrite-induced UV-photocatalytic abiotic nitrogen fixation: implications for early atmospheres and Life. Scientific Reports, 9: 15311.

Mayumi D, Mochimaru H, Tamaki H, et al. 2016. Methane production from coal by a single methanogen. Science, 354: 222-225.

McKay C P. 1997. The search for life on Mars//Whittet D C. Planetary and Interstellar Processes Relevant to the Origins of Life. Berlin: Springer: 263-289.

Merino N, Aronson H S, Bojanova D P, et al. 2019. Living at the extremes: extremophiles and the limits of life in a planetary context. Frontiers in Microbiology, 10: 780.

Michalski J R, Onstott T C, Mojzsis S J, et al. 2018. The Martian subsurface as a potential window into the origin of life. Nature Geoscience, 11: 21-26.

Miller S L. 1953. A production of amino acids under possible primitive earth conditions. Science, 117: 528-529.

Ming D W, Archer P D, Glavin D P, et al. 2014. Volatile and organic compositions of sedimentary rocks in Yellowknife Bay, Gale Crater, Mars. Science, 343: 1245267.

Momper L, Jungbluth S P, Lee M D, et al. 2017. Energy and carbon metabolisms in a deep terrestrial subsurface fluid microbial community. The International Society of Microbial Ecology Journal, 11: 2319-2333.

Morono Y, Terada T, Nishizawa M, et al. 2011. Carbon and nitrogen assimilation in deep subseafloor microbial cells. Proceedings of the National Academy of Sciences of USA, 108: 18295-18300.

Morrison D. 2001. The NASA astrobiology program. Astrobiology, 1: 3-13.

Nagata T, Tamburini C, Arístegui J, et al. 2010. Emerging concepts on microbial processes in the bathypelagic ocean–ecology, biogeochemistry, and genomics. Deep Sea Research Part Ⅱ : Topical Studies in Oceanography, 57: 1519-1536.

National Academies of Sciences, Engineering, and Medicine, Division on Engineering and Physical Sciences, Space Studies Board, et al. 2018. An Astrobiology Strategy for the Search for Life in the Universe. Washington (DC): National Academies Press (US).

Navarro-González R, Vargas E, de La Rosa J, et al. 2010. Reanalysis of the Viking results suggests perchlorate and organics at midlatitudes on Mars. Journal of Geophysical Research: Planets, 115: E12010(1-11).

Nigro O D, Jungbluth S P, Lin H T, et al. 2017. Viruses in the Oceanic Basement. mBio, 8: e02129-16.

Nunoura T, Hirai M, Yoshida-Takashima Y, et al. 2016. Distribution and niche separation of planktonic microbial communities in the water columns from the surface to the hadal waters of the Japan Trench under the Eutrophic Ocean. Frontiers in Microbiology, 7: 1261.

Nunoura T, Takaki Y, Hirai M, et al. 2015. Hadal biosphere: insight into the microbial ecosystem in the deepest ocean on Earth. Proceedings of the National Academy of Sciences of USA, 112: E1230-E1236.

Onstott T C, Magnabosco C, Aubrey A D, et al. 2014. Does aspartic acid racemization constrain the depth limit of the subsurface biosphere? Geobiology, 12(1):1-19.

Orcutt B N, Bach W, Becker K, et al. 2011. Colonization of subsurface microbial observatories deployed in young ocean crust. The International Society of Microbial Ecology Journal, 5: 692-703.

Orcutt B N, Wheat C G, Rouxel O, et al. 2013. Oxygen consumption rates in subseafloor basaltic crust derived from a reaction transport model. Nature Communications, 4: 2539.

Orosei R, Lauro S E, Pettinelli E, et al. 2018. Radar evidence of subglacial liquid water on Mars. Science, 361: 490-493.

Orsi W, Biddle J F, Edgcomb V. 2013. Deep sequencing of subseafloor eukaryotic rRNA reveals active fungi across marine subsurface provinces. PLoS One, 8: e56335.

Ortiz M, Legatzki A, Neilson J W, et al. 2014. Making a living while starving in the dark: metagenomic insights into the energy dynamics of a carbonate cave. The International Society of Microbial Ecology Journal, 8: 478-491.

Ortiz M, Neilson J, Nelson W, et al. 2013. Profiling bacterial diversity and taxonomic composition on speleothem surfaces in Kartchner caverns, AZ. Microbial Ecology, 65: 371-383.

Oren A. 2002. Diversity of halophilic microorganisms: environments, phylogeny, physiology, and applications. Journal of Industrial Microbiology and Biotechnology, 28(1): 56-63.

Park S Y, Liang Y N. 2016. Biogenic methane production from coal: a review on recent research and development on microbially enhanced coalbed methane (MECBM). Fuel, 166: 258-267.

Parkes R J, Cragg B A, Bale S J, et al. 1994. Deep bacterial biosphere in Pacific Ocean sediments. Nature, 371(6496): 410-413.

Parkes R J, Cragg B A, Wellsbury P. 2000. Recent studies on bacterial populations and processes in subseafloor sediments: a review. Hydrogeology Journal, 8(1): 11-28.

Pathom-Aree W, Nogi Y, Sutcliffe I C, et al. 2006. *Williamsia marianensis* sp. nov., a novel actinomycete isolated from the Mariana Trench. International Journal of Systematic and Evolutionary Microbiology, 56: 1123-1126.

Plumb J, Haddad C, Gibson J, et al. 2007. *Acidianus sulfidivorans* sp. nov., an extremely acidophilic, thermophilic archaeon isolated from a solfatara on Lihir Island, Papua New Guinea, and emendation of the genus description. International Journal of Systematic and Evolutionary Microbiology, 57(7): 1418-1423.

Pointing S B, Belnap J. 2012. Microbial colonization and controls in dryland systems. Nature Reviews Microbiology, 10: 551-562.

Polymenakou P, Mandalakis M, Dailianis T, et al. 2018. Preliminary assessment of methanogenic microbial communities in marine caves of zakynthos island. Mediterranean Marine Science, 19: 284-289.

Prangishvili D. 2013. The wonderful world of archaeal viruses. Annual Review of Microbiology, 67: 565-585.

Preston L J, Dartnell L R. 2014. Planetary habitability: lessons learned from terrestrial analogues. International Journal of Astrobiology, 13: 81-98.

Prieur D. 1997. Microbiology of deep-sea hydrothermal vents. Trends in Biotechnology, 15(7): 242-244.

Rabbow E, Parpart A, Reitz G. 2016. The planetary and space simulation facilities at DLR cologne. Microgravity Science and Technology, 28: 215-229.

Raulin F, Hand K P, McKay C P, et al. 2010. Exobiology and planetary protection of icy moons. Space Science Reviews, 153: 511-535.

Reboul G, Moreira D, Bertolino P, et al. 2019. Microbial eukaryotes in the suboxic chemosynthetic ecosystem of Movile Cave, Romania. Environment Microbiology Reports, 11: 464-473.

Ren M L, Feng X Y, Huang Y J, et al. 2019. Phylogenomics suggests oxygen availability as a driving force in Thaumarchaeota evolution. The International Society of Microbial Ecology Journal, 13: 2150-2161.

Rodriguez-Ramos L E, Rios-Velazquez C. 2018. Microbiome dataset from Clara Cave and Empalme Sinkhole waters in Puerto Rico. Data in Brief, 21: 1674-1677.

Romsdahl J, Blachowicz A, Chiang Y M, et al. 2020. Metabolomic analysis of *Aspergillus niger* isolated from the international space station reveals enhanced production levels of the antioxidant pyranonigrin A. Frontiers in Microbiology, 11: 931.

Rotella M D, Wilson C J N, Barker S J, et al. 2013. Highly vesicular pumice generated by buoyant detachment of magma in subaqueous volcanism. Nature Geoscience, 6: 129-132.

Rothman D H, Fournier G P, French K L, et al. 2014. Methanogenic burst in the end-Permian carbon cycle. Proceedings of the National Academy of Sciences of USA, 111: 5462-5467.

Rothman D H, Hayes J M, Summons R E. 2003. Dynamics of the Neoproterozoic carbon cycle. Proceedings of the National Academy of Sciences of USA, 100: 8124-8129.

Rothschild L J. 1990. Earth analogs for Martian life. Microbes in evaporites, a new model system for life on Mars. Icarus, 88: 246-260.

Rummel J D. 2001. Planetary exploration in the time of astrobiology: protecting against biological contamination. Proceedings of the National Academy of Sciences of USA, 98: 2128-2131.

Russell M J, Murray A E, Hand K P. 2017. The possible emergence of life and differentiation of a shallow biosphere on irradiated icy worlds: the example of Europa. Astrobiology, 17: 1265-1273.

Sabath N, Ferrada E, Barve A, et al. 2013. Growth temperature and genome size in bacteria are negatively correlated, suggesting genomic streamlining during thermal adaptation. Genome Biology and Evolution, 5: 966-977

Santelli C M, Orcutt B N, Banning E, et al. 2008. Abundance and diversity of microbial life in ocean crust. Nature, 453: 653-656.

Scharf C, Virgo N, Cleaves H J, et al. 2015. A strategy for origins of life research. Astrobiology, 15: 1031-1042.

Schountz T. 2014. Immunology of bats and their viruses: challenges and opportunities. Viruses, 6:

4880-4901.

Schrenk M O, Edwards K J, Goodman R M, et al. 1998. Distribution of *Thiobacillus ferrooxidans* and *Leptospirillum ferrooxidans*: implications for generation of acid mine drainage. Science, 279: 1519-1522.

Schulze-Makuch D, Airo A, Schirmack J. 2017. The adaptability of life on Earth and the diversity of planetary habitats. Frontiers in Microbiology, 8: 2011.

Schwieterman E W, Kiang N Y, Parenteau M N, et al. 2018. Exoplanet biosignatures: a review of remotely detectable signs of life. Astrobiology, 18: 663-708.

Sean M, John P. 2014. Weighing the deep continental biosphere. FEMS Microbiology Ecology, 87(1):113-120.

Setchell W A, Gardner N L. 1903. Algae of northwestern America. University of California Publications in Botany, 1: 165-418.

Shapiro B, Hoehler T M, Jin Q S. 2018. Integrating genome-scale metabolic models into the prediction of microbial kinetics in natural environments. Geochimica et Cosmochimica Acta, 242: 102-122.

Shen L, Liu Y, Wang N, et al. 2019. Genomic insights of *Dyadobacter tibetensis* Y620-1 isolated from ice core reveal genomic features for succession in glacier environment. Microorganisms, 7: 211.

Shen Y, Buick R, Canfield D E. 2001. Isotopic evidence for microbial sulphate reduction in the early Archaean era. Nature, 410: 77-81.

Siddiqui K S, Williams T J, Wilkins D, et al. 2013. Psychrophiles. Annual Review of Earth and Planetary Sciences, 41: 87-115.

Sielaff A C, Urbaniak C, Mohan G B M, et al. 2019. Characterization of the total and viable bacterial and fungal communities associated with the International Space Station surfaces. Microbiome, 7(50): 1-21.

Sim M S, Bosak T, Ono S. 2011. Large sulfur isotope fractionation does not require disproportionation. Science, 333: 74-77.

Simkus D N, Slater G F, Sherwood L B, et al. 2016. Variations in microbial carbon sources and cycling in the deep continental subsurface. Geochimica et Cosmochimica Acta, 173: 264-283.

Smith D J. 2013. Microbes in the upper atmosphere and unique opportunities for astrobiology research. Astrobiology, 13: 981-990.

Smith M D. 2002. The annual cycle of water vapor on Mars as observed by the thermal emission spectrometer. Journal of Geophysical Research: Planets, 107: H9.

Sobol M S, Hoshino T, Futagami T, et al. 2019. Draft genome sequences of *Penicillium* spp. from deeply buried oligotrophic marine sediments. Microbiology Resource Announcements, 8:

e01613-e01618.

Spang A, Saw J H, Jørgensen S L, et al. 2015. Complex archaea that bridge the gap between prokaryotes and eukaryotes. Nature, 521: 173-179.

Stegen J C , Fredrickson J K, Wilkins M J, et al. 2016. Groundwater–surface water mixing shifts ecological assembly processes and stimulates organic carbon turnover. Nature Communications, 7: 11237.

Stevens T O, McKinley J P. 1995. Lithoautotrophic microbial ecosystems in deep basalt aquifers. Science, 270: 450-454.

Strapoć D, Mastalerz M, Dawson K, et al. 2011. Biogeochemistry of microbial coal-bed methane. Annual Review of Earth and Planetary Sciences, 39: 617-656.

Stumm W, Morgan J J. 1996. Aquatic Chemistry. 3rd edn. New York: John Wiley.

Sun Y D, Joachimski M M, Wignall P B, et al. 2012. Lethally hot temperatures during the Early Triassic greenhouse. Science, 338: 366-370.

Sun Y, Shi Y L, Wang H, et al. 2018. Diversity of bacteria and the characteristics of Actinobacteria community structure in Badain Jaran Desert and Tengger Desert of China. Frontiers in Microbiology, 9:1068.

Suzuki Y, Yamashita S, Kouduka M, et al. 2020. Deep microbial proliferation at the basalt interface in 33.5–104 million-year-old oceanic crust. Communications Biology, 3: 136.

Tamburini C, Boutrif M, Garel M, et al. 2013. Prokaryotic responses to hydrostatic pressure in the ocean: a review. Environmental Microbiology, 15: 1262-1274.

Tang Y Q, Ji P, Lai G L, et al. 2012. Diverse microbial community from the coal beds of the Ordos Basin, China. International Journal of Coal Geology, 90-91: 21-33.

Tarn J, Peoples L M, Hardy K, et al. 2016. Identification of free-living and particle-associated microbial communities present in hadal regions of the Mariana Trench. Frontiers in Microbiology, 7: 665.

Tarter J C. 2004. Astrobiology and SETI. New Astronomy Reviews, 48: 1543-1549.

Timmers P H A, Widjaja-Greefkes H C A, Plugge C M, et al. 2017. Evaluation and optimization of PCR primers for selective and quantitative detection of marine ANME subclusters involved in sulfate-dependent anaerobic methane oxidation. Applied Microbiology and Biotechnology, 101: 5847-5859.

Tyson G W, Chapman J, Hugenholtz P , et al. 2004. Community structure and metabolism through reconstruction of microbial genomes from the environment. Nature, 428: 37-43.

Ueno Y, Yamada K, Yoshida N, et al. 2006. Evidence from fluid inclusions for microbial methanogenesis in the early Archaean era. Nature, 440: 516-519.

Uhran B, Conley C, Spry J A. 2019. Updating planetary protection considerations and policies for Mars sample return. Space Policy, 49: 101322.

Wang Q H, Cen Z, Zhao J J. 2015. The survival mechanisms of thermophiles at high temperatures: an angle of omics. Physiology, 30: 97-106.

Wang Y H, Li P, Dai X Y, et al. 2015. Abundance and diversity of methanogens: potential role in high arsenic groundwater in Hetao Plain of Inner Mongolia, China. Science of the Total Environment, 515-516: 153-161.

Weinstein D J, Allen S E, Lau M C Y, et al. 2019. The genome of a subterrestrial nematode reveals adaptations to heat. Nature Communications, 10: 5268.

Welch T J, Farewell A, Neidhardt F C, et al. 1993. Stress response of *Escherichia coli* to elevated hydrostatic pressure. Journal of Bacteriology, 175(22): 7170-7177.

Whitman W B, Coleman D C, Wiebe W J. 1998. Prokaryotes: the unseen majority. Proceedings of the National Academy of Sciences of USA, 95(12): 6578-6583.

Williford K H, Farley K A, Stack K M, et al. 2018. The NASA Mars 2020 Rover Mission and the search for extraterrestrial life//Cabrol N A, Grin E A. From Habitability to Life on Mars. Amsterdam: Elsevier: 275-308.

Woese C R, Kandler O, Wheelis M L. 1990. Towards a natural system of organisms: proposal for the domains archaea, bacteria, and eucarya. Proceedings of the National Academy of Sciences of USA, 87: 4576-4579.

Wright J J, Konwar K M, Hallam S J. 2012. Microbial ecology of expanding oxygen minimum zones. Nature Reviews Microbiology, 10: 381-394.

Wu Y C, Tan L C, Liu W X, et al. 2015. Profiling bacterial diversity in a limestone cave of the western Loess Plateau of China. Frontiers in Microbiology, 6: 244.

Wurch L, Giannone R J, Belisle B S, et al. 2016. Genomics-informed isolation and characterization of a symbiotic Nanoarchaeota system from a terrestrial geothermal environment. Nature Communications, 7: 12115.

Wächtershäuser G. 2006. From volcanic origins of chemoautotrophic life to bacteria, archaea and eukarya. Philosophical Transactions of the Royal Society B: Biological Sciences, 361: 1787-1806.

Xiao L, Wang J, Dang Y A, et al. 2017. A new terrestrial analogue site for Mars research: the Qaidam Basin, Tibetan Plateau (NW China). Earth-Science Reviews, 164: 84-101.

Xiao L. 2020. Mars on Earth: A Study of the Qaidam Basin. Singapore: World Scientific Publishing Company: 346.

Xiao X, Zhang Y, Wang F. 2021. Hydrostatic pressure is the universal key driver of microbial evolution in the deep ocean and beyond. Environmental Microbiology Reports, 13: 68-72.

Xiong J, Fischer W M, Inoue K, et al. 2000. Molecular evidence for the early evolution of photosynthesis. Science, 289: 1724-1730.

Xie W, Zhang C L, Wang J, et al. 2014. Distribution of ether lipids and composition of the archaeal community in terrestrial geothermal springs: impact of environmental variables. Environmental Microbiology, 17(5): 12595.

Yayanos A A, Dietz A S, van Boxtel R. 1979. Isolation of a deep-sea barophilic bacterium and some of its growth characteristics. Science, 205(4408): 808-810.

Yayanos A A. 1995. Microbiology to 10,500 meters in the deep sea. Annual Review of Microbiology, 49(1): 777-805.

Yoon H S, Hackett J D, Ciniglia C, et al. 2004. A molecular timeline for the origin of photosynthetic eukaryotes. Molecular Biology and Evolution, 21: 809-818.

Young E D, Kohl I E, Lollar B S, et al. 2017. The relative abundances of resolved $^{12}CH_2D_2$ and $^{13}CH_3D$ and mechanisms controlling isotopic bond ordering in abiotic and biotic methane gases. Geochimica et Cosmochimica Acta, 203: 235-264.

Yu T T, Wu W C, Liang W Y, et al. 2018. Growth of sedimentary Bathyarchaeota on lignin as an energy source. Proceedings of the National Academy of Sciences of USA, 115: 6022-6027.

Yun Y, Wang H M, Man B Y, et al. 2016. The relationship between pH and bacterial communities in a single karst ecosystem and its implication for soil acidification. Frontiers in Microbiology, 7: 1955.

Zahnle K, Claire M, Catling D. 2006. The loss of mass - independent fractionation in sulfur due to a Palaeoproterozoic collapse of atmospheric methane. Geobiology, 4: 271-283.

Zerkle A L, House C H, Cox R P, et al. 2006. Metal limitation of cyanobacterial N_2 fixation and implications for the Precambrian nitrogen cycle. Geobiology, 4: 285-297.

Zhang B G, Qiu R, Lu L, et al. 2018. Autotrophic vanadium(V) bioreduction in groundwater by elemental sulfur and zerovalent iron. Environmental Science & Technology, 52: 7434-7442.

Zhang C L, Pearson A, Li Y L, et al. 2006. A thermophilic temperature optimum for crenarchaeol synthesis and its implication for archaeal evolution. Applied and Environmental Microbiology, 72: 4419-4422.

Zhang X, Sigman D M, Morel F M, et al. 2014. Nitrogen isotope fractionation by alternative nitrogenases and past ocean anoxia. Proceedings of the National Academy of Sciences of USA, 111: 4782-4787.

Zhang Y, Chen W J, Riseborough D W. 2006. Temporal and spatial changes of permafrost in Canada since the end of the Little Ice Age. Journal of Geophysical Research Atmospheres, 111: D22103(1-14).

Zhao R, Wang H M, Cheng X Y, et al. 2018. Upland soil cluster γ dominates the methanotroph communities in the karst Heshang Cave. FEMS Microbial Ecology, 94(2): 1-13.

Zhao R, Wang H M, Yang H, et al. 2017. Ammonia-oxidizing archaea dominate ammonia-oxidizing communities within alkaline cave sediments. Geomicrobiology Journal, 34: 511-523.

Zhuang G C, Xu L, Liang Q Y, et al. 2019. Biogeochemistry, microbial activity, and diversity in surface and subsurface deep-sea sediments of South China Sea. Limnology and Oceanography, 64: 2252-2270.

Zobell C E, Johnson F H. 1949. The influence of hydrostatic pressure on the growth and viability of terrestrial and marine bacteria. Journal of Bacteriology, 57(2): 179-189.

关键词索引

F

G

H

J

L

N

Q

R

S

T

W

X

Y

Z

彩　　图

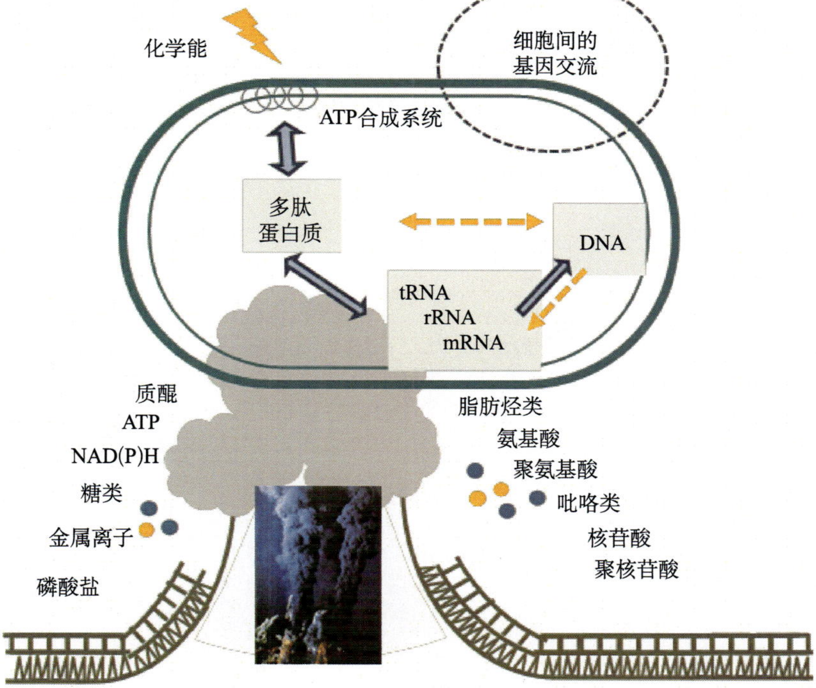

图 1-1　生命可能起源于深海热液环境

注：转移 RNA（tRNA）、核糖体 RNA（rRNA）、信使 RNA（mRNA）、三磷酸腺苷（ATP）、还原型辅酶Ⅰ（NADH）、还原型辅酶Ⅱ（NADPH）。

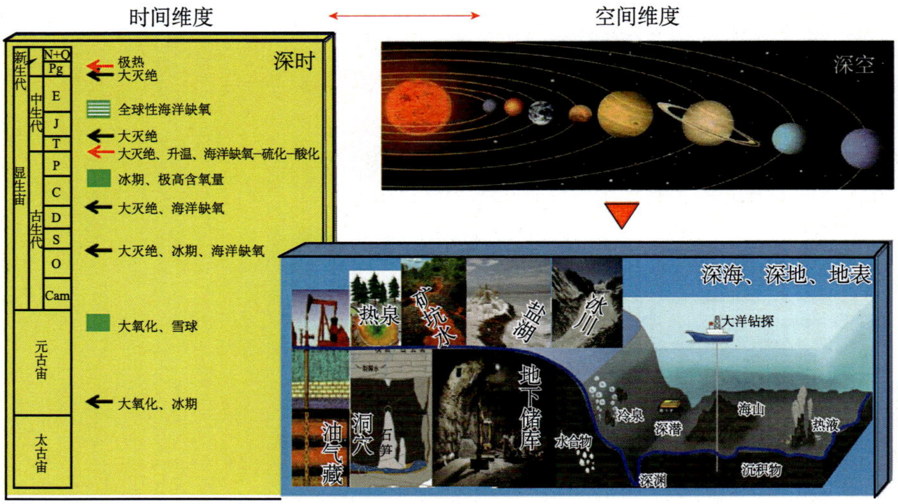

图 2-1　极端地质环境的时间和空间维度

资料来源：深空图来自 https://www.163.com/dy/article/DQSHR3B80525T0QI.html。

注：寒武纪（Cam）、奥陶纪（O）、志留纪（S）、泥盆纪（D）、石炭纪（C）、二叠纪（P）、三叠纪（T）、侏罗纪（J）、白垩纪（K）、古近纪（E）、新近纪（N）、第四纪（Q）。

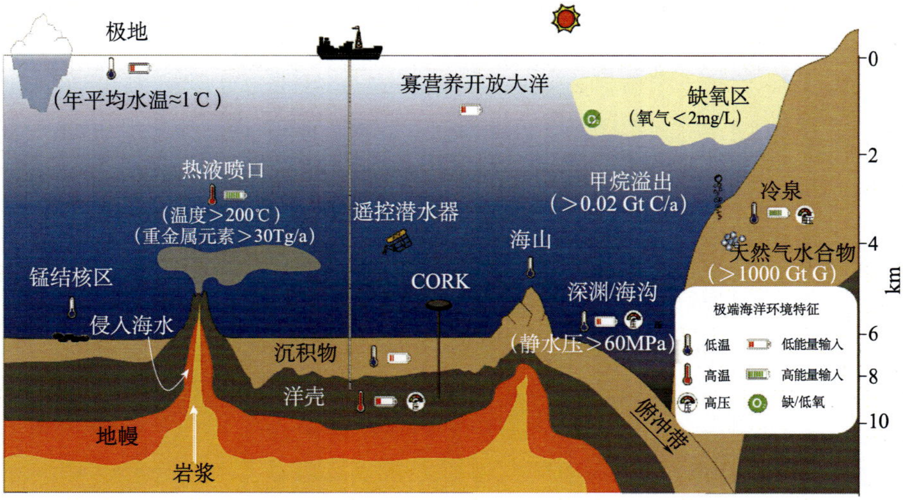

图 2-2　海洋极端环境示意图

图 2-4　常见陆地盐碱环境

注：（a）、（b）为盐碱土壤，（c）、（d）为盐湖泊。

图 2-5　各种类型的极端酸性环境

注：（a）酸性矿坑水；（b）深海热液；（c）热泉；（d）洞穴；（e）酸性硫酸盐土壤；（f）自然暴露的矿床。

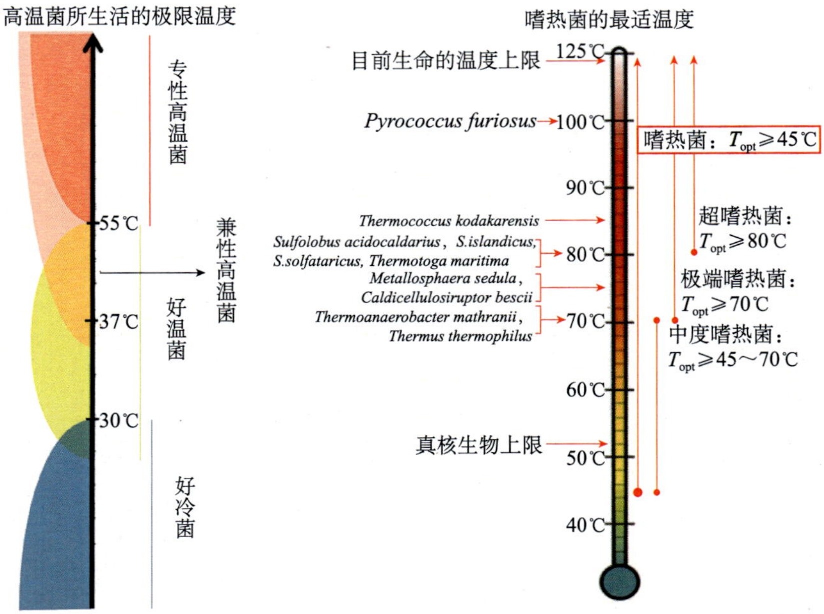

图 2-6　高温菌的生长温度

注：强烈火球菌（*Pyrococcus furiosus*）、柯达卡兰热球菌（*Thermococcus kodakarensis*）、海栖热袍菌（*Thermotoga maritima*）、热解纤维素菌（*Caldicellulosiruptor bescii*）、勤奋金属球菌（*Metallosphaera sedula*）、马氏热厌氧杆菌（*Thermoanaerobacter mahranii*）、嗜热栖热菌（*Thermus thermophilus*）。

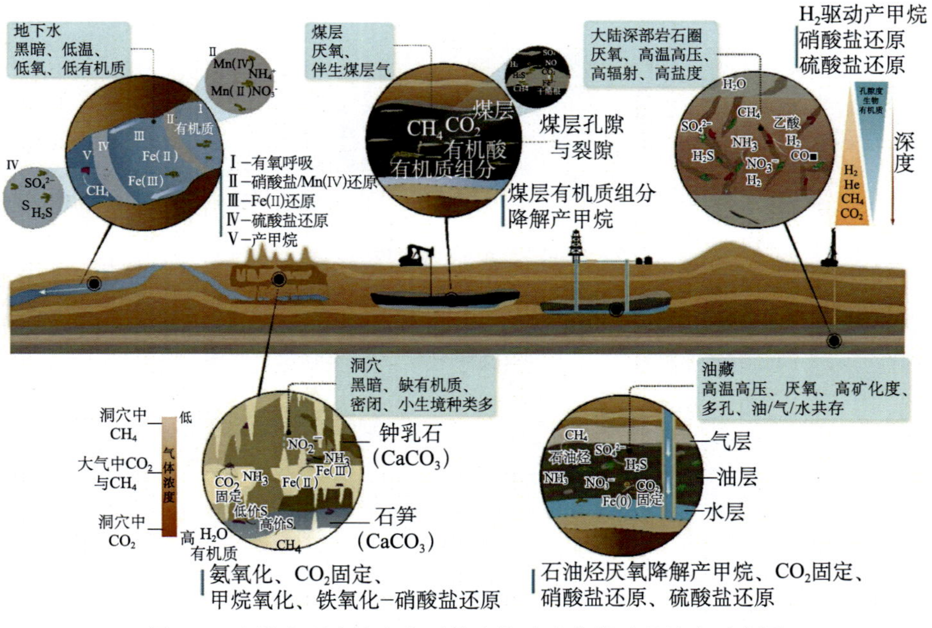

图 2-7　大陆典型生境和主要的生物地球化学过程综合示意图

注：硫酸根（SO_4^{2-}）、氨气（NH_3）、水（H_2O）、碳酸钙（$CaCO_3$）、氦（He）。

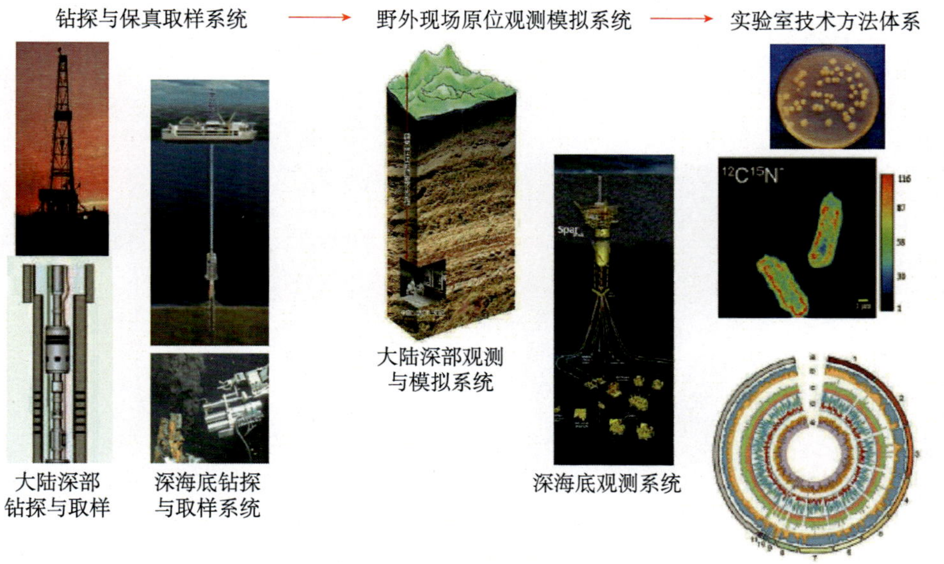

图 2-8　极端地质环境微生物研究的三大技术体系

资料来源：http://news.sina.com.cn/c/2010-12-13/070321629707.shtml；https://baike.baidu.com/item/%E7%A7%91%E9%92%BB；https://5b0988e595225.cdn.sohucs.com/images/20170628/10fd2c9458644480afa8b3c53b2fc3f7.png；https://zhidao.baidu.com/question/1927524789956804747.html；http://s12.sinaimg.cn/orignal/4be5cbafte10f301a2e8b；http://www.hnhxnh.com/upload/201905/09/20190509172539 5352.jpg；http://www.ioz.ac.cn/xwzx/kyjz/201401/W020140116345537641504.jpg。

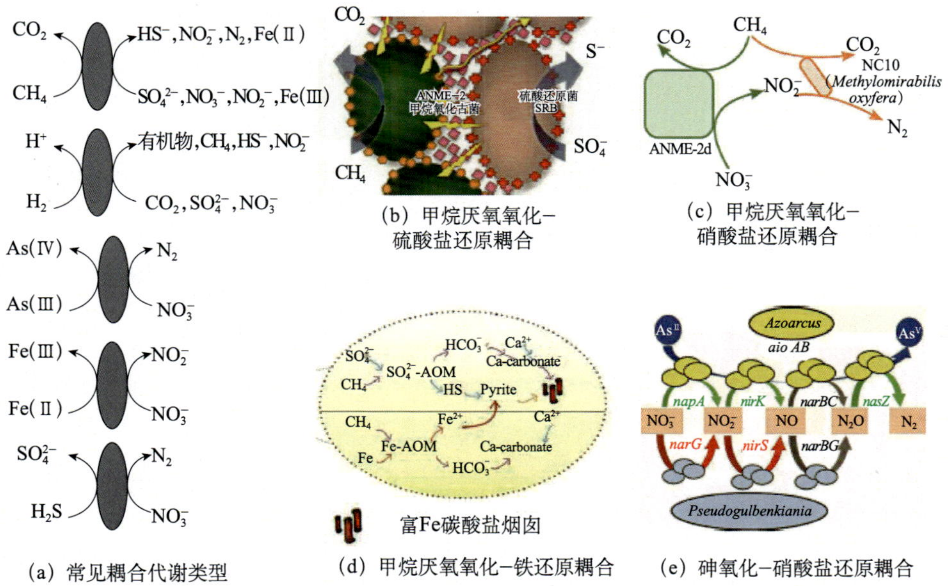

图 3-1　极端地质环境的常见耦合代谢类型及部分典型生物地球化学过程

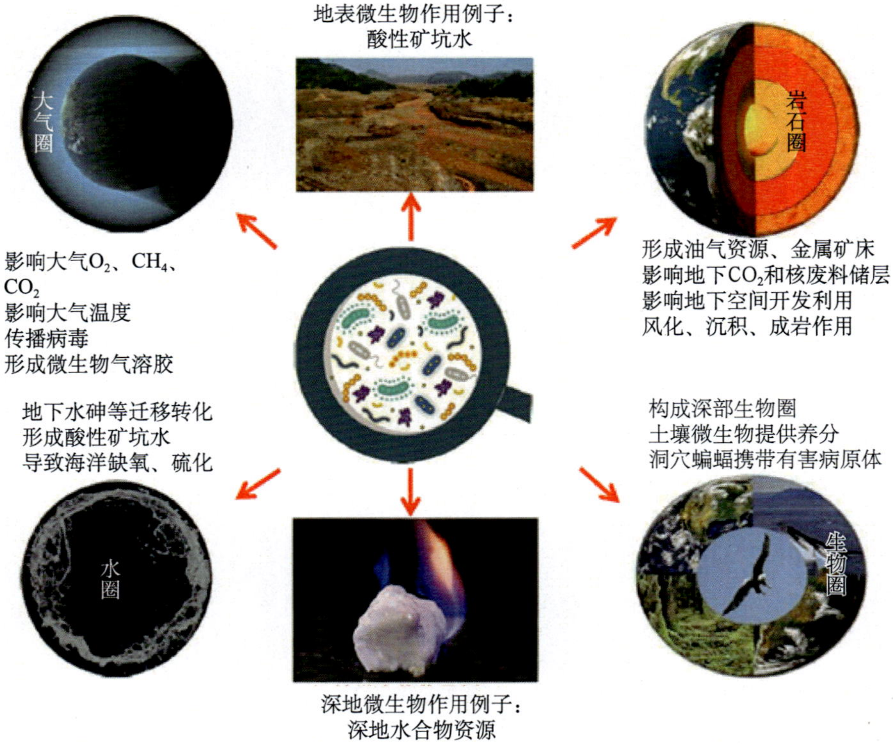

图 3-2　极端地质环境微生物对大气圈、水圈、生物圈和岩石圈的影响

资料来源：https://huaban.com/pins/1581316228；https://www.sohu.com/a/379569294_194423；http:// www.ah.xinhuanet.com/2014-12/04/c_1113512649.htm；https://bbs.le.com/thread-2107630-1.html；http://news.cnr.cn/native/gd/20170605/t20170605_523785210.shtml；https://www.21cnjy.com/H/11/93151/3912884.shtml；https://www.sohu.com/a/424813303_120012589。

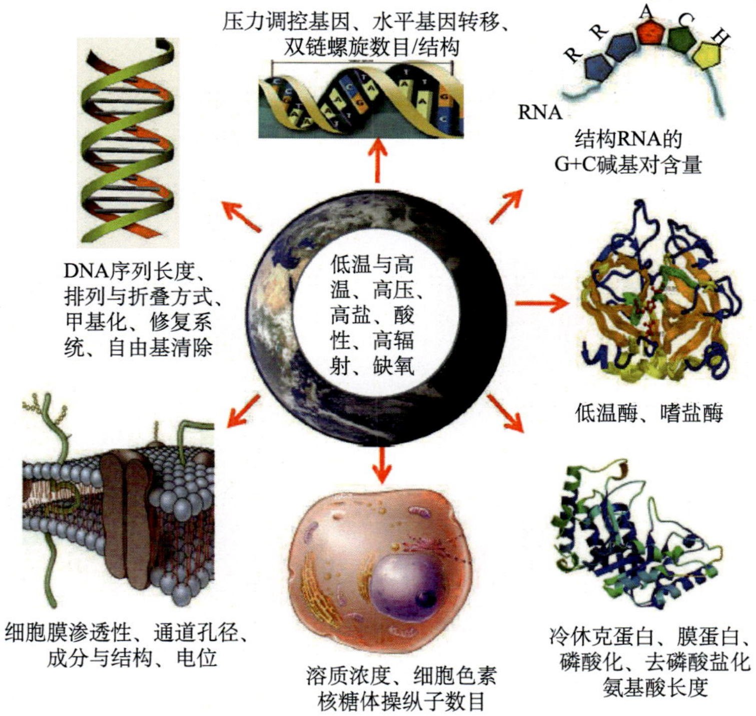

图 3-3　微生物对极端地质环境适应机制的综合图

资料来源：http://www.dashangu.com/postimg_16512606.html；https://max.book118.com/html/2016/1228/77790533.shtm；https://wenwen.sogou.com/z/q339795956.htm；http://www.dashangu.com/postimg_13440518.html；http://www.dashangu.com/postimg_13955397_12.html；https://www.zhihu.com/question/60239234/answer/174064750；https://www.biowolf.cn/m/view.php?aid=219；https://www.photophoto.cn/show/09695293.html。

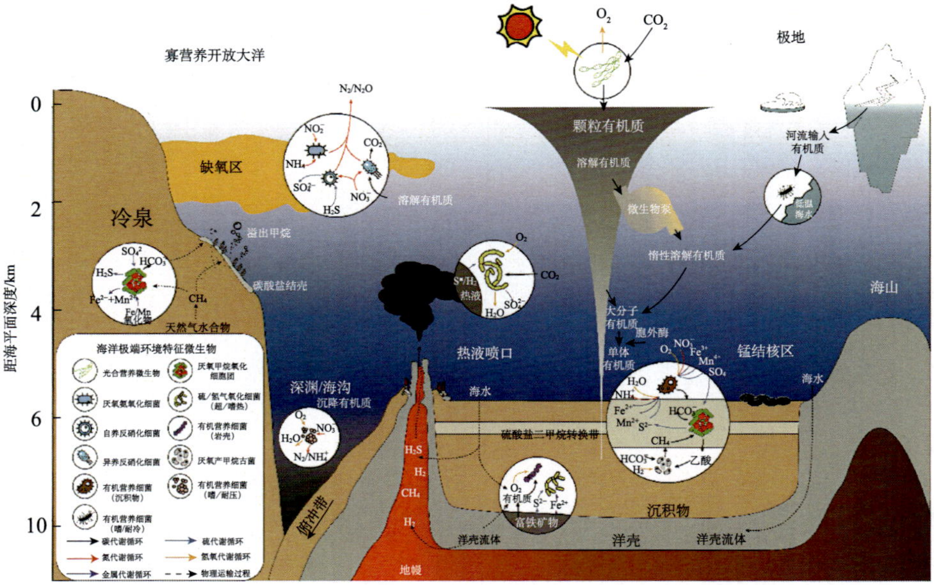

图 3-5　海洋极端环境特征微生物的主要类型与代谢途径示意图

资料来源：改自 Xiao et al.（2021）。

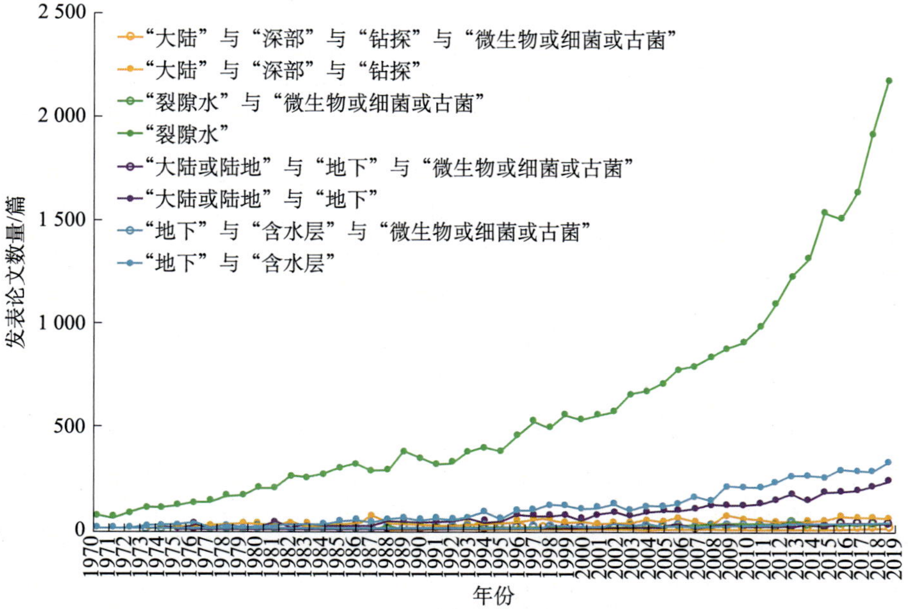

(a) 发表论文数量

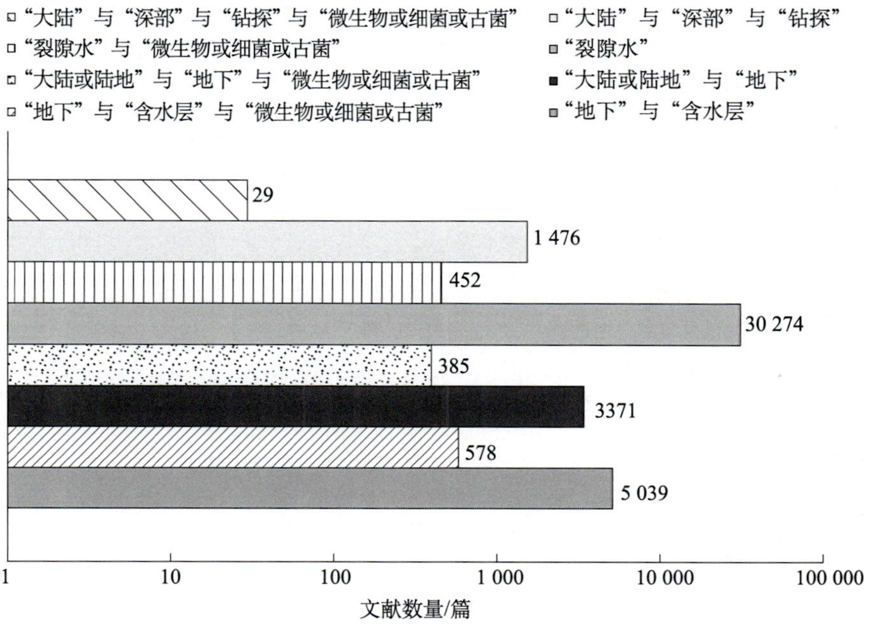

(b) 分布领域统计图

图 3-12　与岩石圈环境相关研究论文数量及分布领域统计图

资料来源：斯高帕斯（Scopus）文献数据库。

注：搜索用的词组以引号标记，而带"微生物或细菌或古菌"（microorganism or bacteria or archaea）的与生物圈有关。

图 3-16　全球用于天体生物学研究的极端环境典型区域的现场景观

资料来源：分布图改自 Preston and Dartnell（2014）。

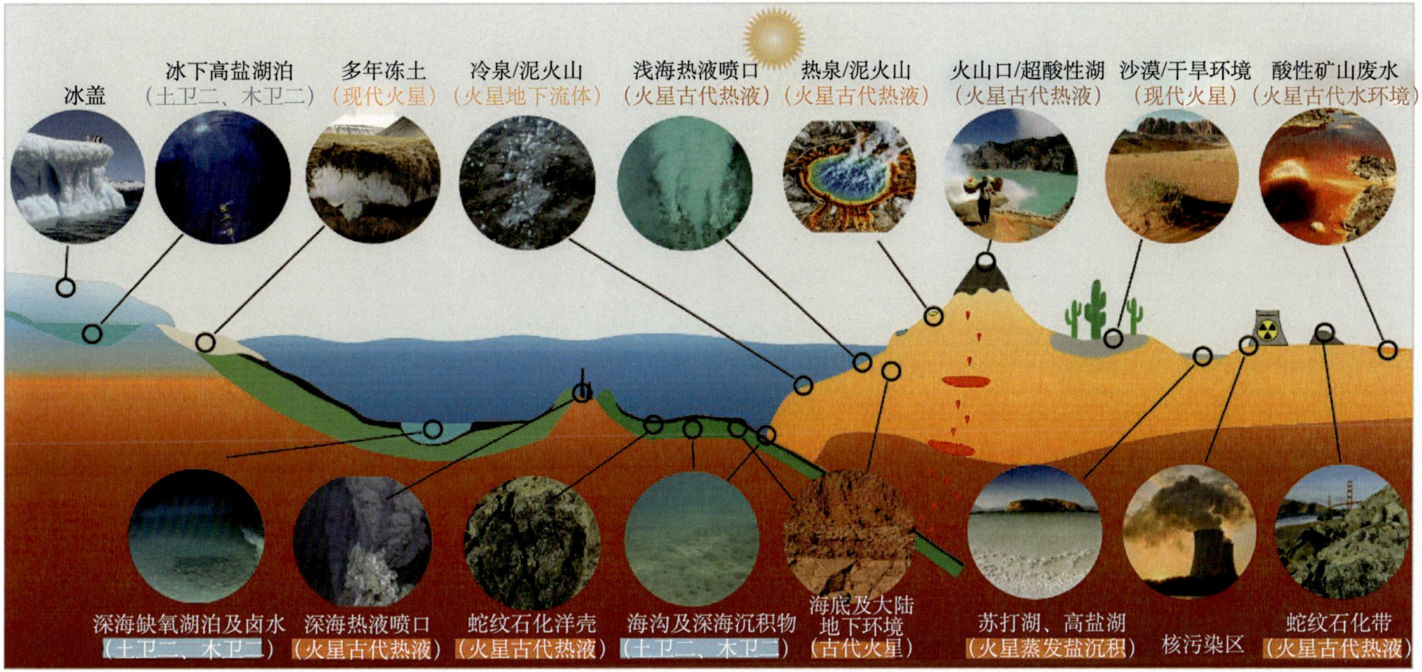

图 5-2　地球代表性极端环境

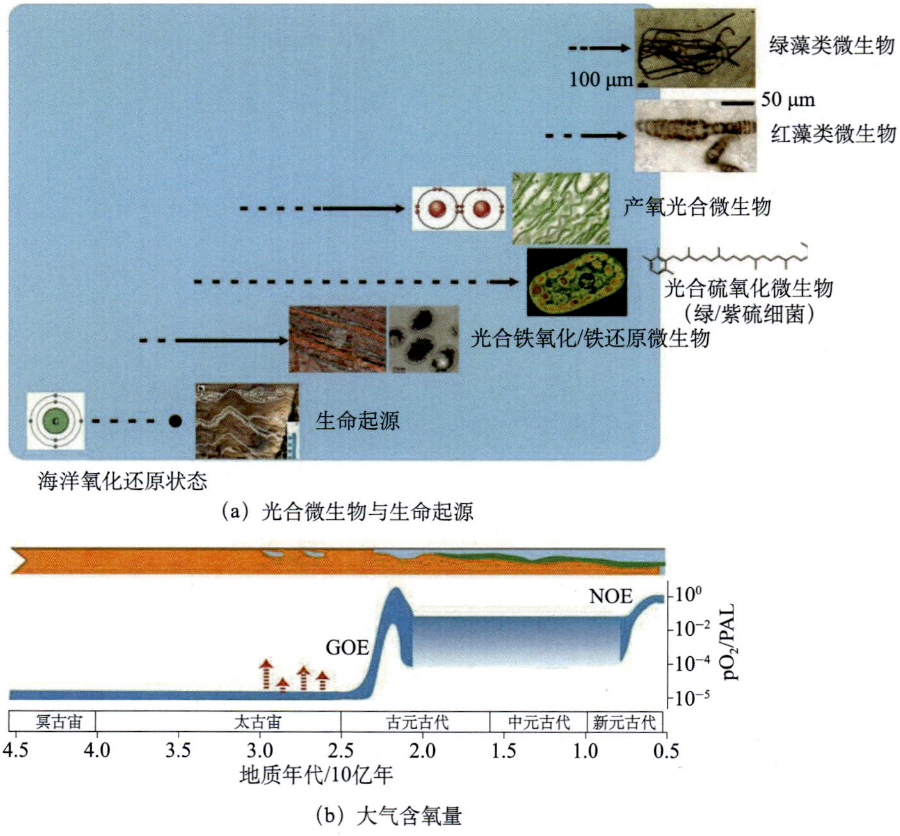

图 6-5　地质时期光合微生物、大气含氧量和海洋环境之间的演化关系图

资料来源：大气含氧量的图来自 Lyons et al.（2014）。